Troubleshooting and Repairing Consumer Electronics Without a Schematic

Troubleshooting and Repairing Consumer Electronics Without a Schematic

Second Edition

Homer L. Davidson

McGraw-Hill

New York San Francisco Washington, D.C. Auckland Bogotá
Caracas Lisbon London Madrid Mexico City Milan
Montreal New Delhi San Juan Singapore
Sydney Tokyo Toronto

Library of Congress Cataloging-in-Publication Data

Davidson, Homer L.
    Troubleshooting and repairing consumer electronics without a
schematic / Homer L. Davidson. — 2nd ed.
        p.   cm.
    Includes index.
    ISBN 0-07-015764-2 (hardcover). — ISBN 0-07-015765-0 (pbk.)
    1. Electronic apparatus and appliances—Maintenance and repair.
I. Title.
TK7870.2.D387   1997
621.389'7—dc21   96-37281
                                                        CIP

McGraw-Hill

A Division of The McGraw-Hill Companies

99105

    4 5 6 7 8 9 0   DOC/DOC   9 0 1 0 9 8

ISBN 0-07-015765-0 (PBK)
        0-07-015764-2 (HC)

The sponsoring editor for this book was Steven Chapman, the editing supervisor
was Penny Linskey, and the production supervisor was Claire Stanley. It was set in
ITC Century Light by Wanda Ditch through the services of Barry E. Brown
(Broker—Editing, Design and Production).

Printed and bound by R.R. Donnelley & Sons Company.

McGraw-Hill books are available at special quantity discounts to use as premi-
ums and sales promotions, or for use in corporate training programs. For
more information, please write to the Director of Special Sales, McGraw-Hill,
11 West 19th Street, New York, NY 10011. Or contact your local bookstore.

This book is printed on recycled, acid-free paper containing a minimum of
50% recycled, de-inked fiber.

To Robert Douglas, who helped me correct
messed-up programs that this stubborn computer and I created.

Contents

3 Repairing audio amps, large and small   65

8 Troubleshooting the TV chassis   *229*

11 Troubleshooting AM-FM-MPX circuits *335*

12 VCR mechanical and electronic problems *369*

13 Testing the remote control circuits 397

Introduction

The need to troubleshoot and repair consumer electronics without a schematic occurs every day in the life of a busy electronics technician. The technician who repairs all types of electronic equipment must make quick and practical repairs; otherwise, he or she will be out of the business within a few years. It is difficult to have every schematic of the electronic product that appears upon the service bench. Remember, these technicians turn out hundreds of electronic repairs each week, month after month, without a schematic.

Even the largest and best-equipped establishment cannot have all the schematics required to service every piece of equipment that crosses the service bench. The more experienced and better informed the electronics technician is, the more productive in troubleshooting and repairing consumer electronics he or she will be. This book will help the beginning, intermediate, and experienced electronic technician service and repair different types of consumer electronics without a schematic. Besides servicing tips and valuable information, practical case histories are found throughout the book.

The purpose of this book is to provide practical service experience and methods for servicing electronic equipment without a schematic. Of course, repairing certain types of electronic products cannot be accomplished unless a certain schematic is available. There are many repairs you can make without a circuit diagram. The tough dog and intermittent service problem are difficult to find without a schematic.

Most repair centers cannot afford every schematic on consumer equipment. Others simply do not have room for them. In addition, some schematics for import models are difficult to obtain. It might take weeks or months to get them, and the electronic product sits for days, in pieces, until the diagram arrives. Sometimes the schematic never comes; they are no longer available.

Troubleshooting and Repairing Consumer Electronics Without a Schematic begins with servicing methods. Chapter 2 shows you how to locate, test, and repair the electronic product. Repairing audio amps, large and small, are given in Chapter 3, with the list of required test equipment, symptoms, and methods of servicing these amplifiers. Chapter 4 is on servicing the auto or car radio receivers. Troubleshooting the cassette player is found in Chapter 5, with the various symptoms, tips, and actual case histories.

In Chapter 6, you will learn how to repair the black-and-white TV chassis. Chapter 7 shows you how to service the compact disc player, found in the boom box, table-top, auto, and CD changer. Troubleshooting the color TV chassis is found in Chapter 8 with the many different circuits and troubleshooting tips. Repairing power supplies is covered in Chapter 9, which covers all power sources found in the many electronic components. Chapter 10, on servicing stereo sound circuits, covers most stereo audio circuits located in the many electronic products within the consumer electronic field.

Troubleshooting AM/FM/MPX circuits is located in Chapter 11. How to service VCR mechanical and electronic problems is covered in Chapter 12, which provides various symptoms, VCR problems, and actual case histories. Chapter 13 shows you how to test the remote control and the infrared receiver circuits. The many service problems within the boom-box cassette and CD player are given in Chapter 14. Last but not least, 20 tough-dog symptoms and repairs are found in the Chapter 15.

Of course, in a book this size, it is impossible to show how to repair every type and model of consumer electronic products. However, information on how to troubleshoot and repair audio amplifiers, auto receivers, cassette players, black-and-white TVs, compact disc players, color TV chassis, stereo units, AM/FM/MPX circuits, and VCRs is found throughout the various chapters, without schematics.

Don't push that electronic unit aside and wait for the correct schematic. Apply the methods within this book to turn out more repairs and fill up that cash register. Electronic products collecting dust provide no income. Troubleshooting and repairing consumer electronics can be fun and quite rewarding, even when the schematic is not available.

Homer L. Davidson

Acknowledgments

A great deal of thanks to RCA Consumer Electronics and Radio Shack for service data. A special thanks to Nils Conrad Persson, editor of *Electronics Servicing & Technology* magazine and electronics technicians Tom Rich from Toms TV, Robert P. Saunders from Eagle Electronics, and Tom Krough from Krough Repair, who supplied troubleshooting tips and symptoms.

1
CHAPTER

Servicing methods without a schematic

Servicing consumer electronic products can be a lot of fun, if you let it be. Just like solving a great murder mystery, the smart detective must have the skill, desire, and ability to find the culprit. The electronics technician must take the symptom and isolate, locate, remove, and replace the defective part.

The anticipation of locating a dead, open, or leaky component is like finding a needle in a haystack (Fig. 1-1). Your heart beats a little faster as you zoom in on the suspected part. Then great disappointment sets in when the suspected component is not the guilty party. So you try once again.

Troubleshooting the TV chassis, VCR, CD player, and cassette player presents many rewards. Solving the intermittent or tough-dog problem is a moment of triumph. However, if you can repair and service the electronic chassis without a schematic, you receive the highest honors of all. So let's begin.

Take a good look

Take a good look at the defective chassis. Visual inspection has repaired thousands of electronic chassis. Watch out for visual scars, like a burned resistor, worn connections, arc-over parts, cracked boards, overheated components, and wear and tear (Fig. 1-2).

For example, firing or arcing inside the picture tube can indicate a cracked CRT gun assembly. Arcing between parts on PC boards results from liquid spilled inside the chassis. Smoke curling up from overheated components signals trouble ahead. When sweat forms on a carbon resistor, an overheated circuit is nearby.

Cracked or blown resistors or capacitors show signs of higher voltage or defective parts. Arcing inside the flyback transformer indicates leaky high-voltage diodes. Overheated white and dark marks on the IC body might point out a leaky part. Cracks found in ceramic capacitors and resistors indicate overheated components.

1-1 Locate the defective component with critical voltage, resistance, and diode tests with the DMM.

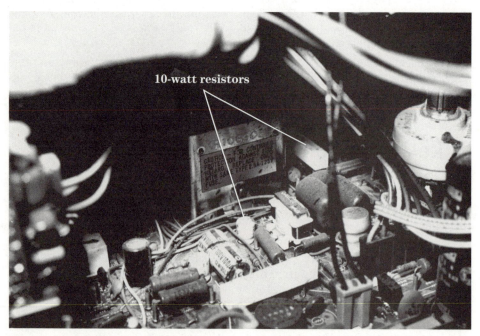

1-2 Look for large 5- and 10-W resistors with cracked areas, burned parts, and hot components upon the chassis.

The damaged or frozen voice coil of a PM speaker will not move with audio. Too-hot-to-touch transistors or ICs might be leaky or shorted.

Several burned resistors might point to a leaky transistor or IC. Taking a close-up peek through a magnifying glass might solve the intermittent problem. Dark-brown areas around the component terminal leads on PC boards can indicate a poor connection or overheated part. Just spend a little time before jumping to conclusions. Look before isolation, then take another peek at the collected symptoms.

Listen, listen

You should be able to hear the TV chassis with high voltage by listening with your ear close to the deflection yoke. Listen for arcing in the flyback or a tic-tic noise indicating horizontal problems (Fig. 1-3). Intermittent arcing noise within a defective capacitor can indicate an internal break or loose terminal wire.

Check the main filter capacitors when an extreme hum is heard in the speaker. A low hum noise can be caused by dried-up decoupling capacitors. Pickup hum might point toward a poor ground or an open base circuit. Hum with distorted audio might originate in the output transistors or IC components in stereo circuits. A mushy speaker sound might be caused by a dropped or frozen voice coil.

Some women can hear the frequency of horizontal circuits, while most people cannot hear over 10 kilocycles. The singing flyback might be caused by loose particles or loose mounting. Don't overlook a vibrating ferrite bead that might dislodge or come loose.

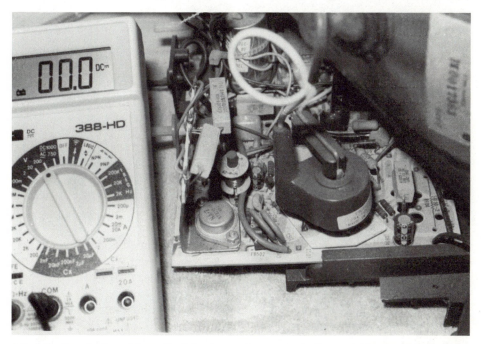

1-3 A tic-tic noise heard from the flyback transformer can indicate problems within the horizontal output circuits.

Low frying noises in large power amps might be caused by defective transistors or IC components. Especially check ceramic capacitors, transistors, and IC parts in the front end of stereo or cassette tape amplifiers. Loose grounds or soldered joints might cause microphonic noises in the input and output stages of the amplifier.

Hands have it

You can tell if the soldering iron is warm, cold, on the blink, or shut off by holding it near your arm. Do not touch it. Locate the shorted yoke by removing it and feeling the inside area. Replace the yoke assembly if hot spots are found (Fig. 1-4). Feel the outside winding of a power transformer or flyback, with the power off; heat might indicate shorted windings or overloading.

Deflection

Yoke

1-4 Leakage between the horizontal and vertical windings and hot spots in the yoke can indicate a defective yoke.

The overloaded transistor or IC component can be located by touch. The same applies to resistors, choke coils, and pincushion parts. Touch the neck of the picture tube for warmness when you cannot see any filament or heater light up. Feel large high-voltage resistors for overheating and overloading.

Feel the speaker cone, by pushing it in and out, to determine if the voice coil is dragging. Loud speaker vibration can be felt with fingers on the speaker cone area. Intermittent voice coil connection can be felt with the volume turned up and fingers pressed against the speaker cone. Vibrating components on the chassis can be located by exerting pressure on the suspected area. Touch can locate a lot of different problems in the electronic chassis.

Bright lights and Broadway

The excessively bright raster symptom without any control might result from a defective picture tube or CRT circuits, or a change in boost voltage. Usually B+ boost voltage is developed in the derived secondary circuits of the horizontal transformer. An open or burned isolation resistor or silicon diode or a dried-up filter capacitor can cause improper boost voltage. The excessive bright raster, followed by shutdown, might be caused by a defective picture tube.

High-voltage arc-over can be seen at the anode connection on the picture tube. If liquid is spilled on the CRT, arc-over between the anode and the aqueduct coating or ground spring can occur. Rubber damage on the anode button or socket can result in bright high-voltage arc-over. A pinched or cracked high-voltage anode cable results in bright arcing to parts or chassis.

Often when high voltage arcs over at the CRT anode socket, excessive high voltage is caused by a defective flyback or safety capacitor (Fig. 1-5). This capacitor, located across the damper diode or attached to the collector terminal, might open up, letting the horizontal output increase and resulting in uncontrollable high-voltage arc-over.

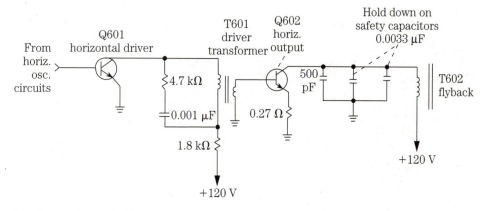

1-5 Open safety or hold down capacitors in the collector circuit of the horizontal output transistor can cause HV arcover and cracking noises.

Bright arcing with popping noises, between the tripler unit and metal chassis, may be found in the early TV chassis. The high voltage increases with a breakdown of HV diodes and capacitors within the tripler unit. Always replace the tripler unit, then try to repair it.

Sometimes arcing within the picture tube can be caused by a cracked glass or open filament. Thin bright arcing lines found in the raster might be caused by a defective focus control. White-blue arcing lines within the spark gap, found on the CRT terminals, might result from internal element leakage, excessive high voltage, or a dirty or clogged spark gap assembly. Shut down the chassis at once to prevent further damage.

Snap, crackle, and stop

Cracking or popping sounds in the speaker might be caused by a defective audio output transistor or IC. Isolate the output stage. Component replacement is the only answer. Low popping noises can be caused by defective ICs or transistors. Spray the component with coolant, and if the noise stops, replace the suspected component. Low frying or firing noises can be caused by a defective input transistor or IC.

When switched, a dirty function switch might produce scratching or cracking noises in the speaker (Fig. 1-6). You might find one or two dead circuits with dirty switch contacts. Spray cleaning fluid inside the switch area, and work the switch control on and off to help clean the contacts. Replace the switch if it cannot be cleaned or has worn contacts.

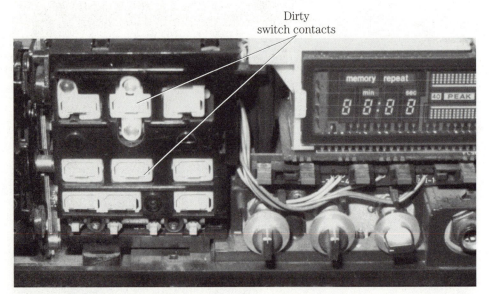

1-6 Dirty switches within the AM-FM-MPX receiver amp causes cracking and firing noises in the speaker.

High-voltage arc-over and popping noises are caused by excessive high voltage. When the TV chassis starts up or shuts down, a cracking noise can be heard. Of course, no problem exists with the yoke assembly expanding and collapsing with signal applied and turned off. This is a good cracking sound.

Smoke gets in your eyes

Where there is smoke, there is fire, so the saying goes. This also applies to troubles that occur in the electronic chassis. Suspect a shorted power transformer with smoke and odors rising from the low-voltage circuits. Often, small power transformers with leaky silicon diodes or leaky filter capacitors might open the primary winding. In large power transformers, found in high-voltage stereo amplifiers, the shorted transformer might burn or smoke.

Remove all secondary leads from the power transformer to the power circuits and let the transformer cool down. Again turn the chassis on. If the transformer begins to smoke and run excessively warm, replace it. Remember, power transformers are rather expensive and sometimes difficult to obtain.

When liquid is spilled in an area where higher-than-normal voltage is found, firing on the PC board results. Smoke will curl upwards and holes will be burned in the board if the unit is left on. Firing and arcing between components and wiring might melt down and destroy a few parts before the chassis is shut off. Arc-over between horizontal and vertical windings in the yoke assembly might be caused by liquid spilled into the yoke or breakdown within yoke assembly (Fig. 1-7).

1-7 Pepsi, Coke, or liquid spilled down into the yoke assembly can cause arcover and smoke curl upwards.

Separate PC board replacement is costly, and many times they are difficult to obtain. Repair the board if possible. Make a rough drawing or sketch of the replacement parts, PC wiring, and terminal connections. Remove defective parts. Install new parts on a terminal strip or small PC board.

Small burned spots on a board can be repaired by replacing burned or charred components. Spray on a cleaner/degreaser to wash away dirt from the parts and circuit. Scrape off burned areas with a pocket knife or flat tool. Replace burned components with new parts. Recheck for overheating, arcing, and smoke after repairs. Spray on a light coat of premium preservative fluid and let dry.

Often the horizontal-output transformer must be replaced when heavy arcing is found within the high-voltage windings or diodes. Replace the flyback with the original part number. A sweet smell of germanium, which might not include smoke, is caused by a leaky high-voltage diode in the black-and-white TV chassis.

Hit between the eyes

Most symptoms are found by sight, sound, or touch. Of course, test equipment is needed to check voltage, resistance, current, waveforms, and continuity on the various components. Isolating and pinpointing the defective part is the most difficult job in troubleshooting. It only takes a few minutes to replace a defective part after locating it.

Sometimes we try too hard to locate a problem, when the defective part is staring up at us, right under our eyes (Fig. 1-8). It's best to get away from the chassis when a defective part cannot be found in one hour. The longer you stay with it, the more dark and clouded the scene becomes.

A little common sense, electronics knowledge, the correct test equipment, and experience can help you locate the tough dog. When the mind is free and clear, early in the morning, tackle that tough repair problem.

1-8 Check the chassis over carefully, as the defective component is looking you right in the face.

Nose to the grindstone

After working on a chassis for several hours, you might want to give up. Why not try a different approach or method of troubleshooting? Each electronics technician has his or her own method of troubleshooting an electronic product and does not want to change. Have you tried asking fellow technicians if they have had the same problem? Help can be obtained from electronics distributors and service departments. Why not ask the manufacturer for help? Help is out there; just ask for it. Some manufacturers have service centers around the country where their electronic products can be serviced.

Don't give up the ship. Keep your nose to the grindstone. Sometimes you wonder if this job is all worth it. You bet it is. There is no greater satisfaction than licking that tough-dog problem.

Different model, same chassis

The electronics manufacturers use several chassis that are the same, year in and year out. The cabinet and outside dressing are changed, but the insides are essentially the same. In fact, the same chassis might be found in other brands of TVs.

Let's take the RCA XL100 chassis, which can be found in many different TVs. The CTC87, CTC97, CTC107, CTC108, and CTC109 have practically the same circuits. Although you might find some modifications, they have the same basic circuits. If you have a schematic for an RCA CTC107 chassis, why not use it for the other chassis when the correct schematic is not handy (Fig. 1-9)?

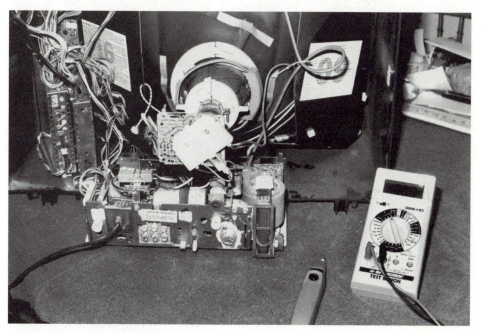

1-9 The RCA CTC107 has the same circuits as the CTC87, CTC97, CTC108, and CTC109.

Often the horizontal and vertical circuits are the same. A different tuner might be used, but the IF sections are identical. The same color IC might be used in several dozen TV chassis. So why not try to use a schematic that is similar to the chassis found on your service bench, instead of waiting for months for the exact diagram?

Same track record

Remember, the circuits found on the schematic of a manufacturer might be used in many different chassis. Some firms have their own pet circuits that are used in many different chassis, including other brands. In fact, several overseas companies make electronic units for other brands. These units might not look the same, but inside they are the same circuits.

When a breakdown occurs in one chassis, look for the same part to break down in other chassis. Sometimes the same component fails in many different chassis. You can make money on these types of repairs because they occur over and over. However, it's difficult to remember what parts break down in several hundred different chassis each year. Write it down.

Keep the service data in a notebook, in card file, or on a computer. The best place is to write it down on the service schematic. Circle the defective component, draw a line out to the side, and write down the symptom. Circle the voltages that are measured when the part is defective, with proper waveforms (Fig. 1-10).

1-10 Write the various symptoms, defective components, and voltages down on the margin of the schematic diagram.

Making a written record of your repair procedures and problems might help you and your fellow technicians troubleshoot that same symptom when it happens again. Take good clean service notes and precious service time can be saved. Besides, it's easy money the next time around.

Service notes and troubleshooting tips by the manufacturer or service center can be added to the same chassis. If you are a warranty station for several brands of electronic products, the firm passes on troubleshooting guidelines that can help in servicing the tough-dog chassis. Of course, manufacturers' service meetings and seminars are worth their weight in gold. Take a few hours away from the bench to gain valuable knowledge.

Just take a few minutes to write down the defective component on the schematic. Compare another schematic to the chassis when the exact diagram is not handy. Remember, the same part might break down in another chassis.

Highway lines

Two yellow lines down the middle of the highway divide the coming and going traffic in half, while white lines along the outside edge of the asphalt show the edge of the pavement. Unwanted lines in the TV picture must be erased for a good viewing picture. Sharp, jagged, or dotted lines across the picture can indicate outside interference or arcing in the TV chassis. Outside interference lines can be caused by motors, neon signs, and high power lines arcing over (Fig. 1-11).

Thin firing lines can be caused by improper grounding of the dag (outside area) of the picture tube. Clip a ground lead from the chassis to the block, rounded area of

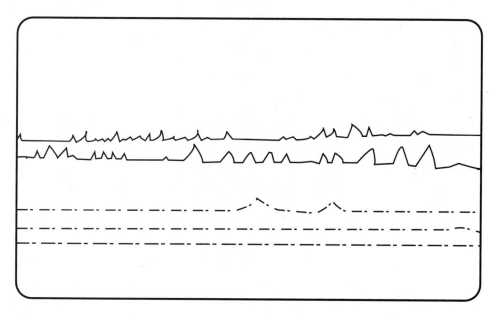

1-11 Interference running horizontally across the picture can be picked up by cable or the antenna.

the picture tube. Sometimes these small springs around the bell assembly become dirty and corroded leaving a poor ground. Check to see if the ground strap from the chassis to picture tube mounting is in place. If broken, run a ground wire from the picture tube to the metal chassis.

Arcing lines can be caused by a defective flyback transformer, tripler unit, or focus control assembly. Check for corona arcs around the outside plastic body of the horizontal output transformer. Rotate the focus control to see if the lines disappear or appear brighter. Replace the defective components that cause arcing in the picture.

A dirty spark gap around the picture tube socket can cause arc-over. Blow out the dust and dirt within the picture tube socket, or replace the socket. If one of these gaps continually arcs over, it can cause an extremely blurry or out-of-focus picture, loss of color, a raster of one color, and chassis shutdown. Check for arcing under the anode HV rubber cap assembly. Sometimes it is not hooked properly and comes loose causing HV arc-over.

Vertical lines to the left of the picture can result from Barkhausen or improper filtering in the AGC circuits (Fig. 1-12). In the early chassis, Barkhausen lines resulted from a horizontal output tube. Today, heavy-firing vertical lines can be caused by the horizontal output transistor or flyback transformer. Shunt small electrolytic capacitors in the AGC and sync circuits for vertical lines in the picture.

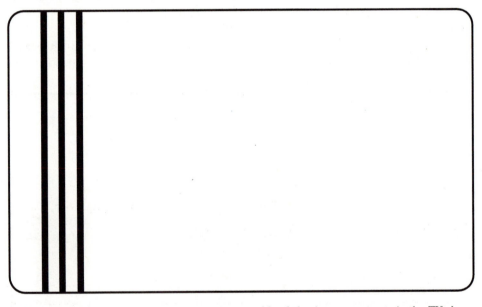

1-12 Vertical lines in the raster or picture is caused by defective components in the TV chassis.

Body part numbers

Sometimes you can locate the section that the defective component is in by looking at the part number stamped on the top. The large IC part (TA7644BP)

Large IC

1-13 The large IC component might have 40 or more pin terminals and circuits in the TV chassis.

shown in Fig. 1-13 contains the video, AGC, sync, color, brightness, contrast, horizontal, and vertical. Likewise the vertical output IC might have the part number stamped on the body area. Not only will the part number tell you what circuit it works in, but also the correct replacement.

Simply look up the part number in a semiconductor manual for the correct universal replacement. You can look at another chassis of the same brand, locate the same IC, and compare the voltage measurements with the other schematic.

For example, if no horizontal waveform is found at the driver transistor, look up the correct pin number on the TA7644BP IC in the semiconductor manual. Check the horizontal waveform at pins 34 and 24. Measure the horizontal supply voltage on pin 33 (Fig. 1-14). The supply voltage (VCC) is pin 3 (12 V). Not only have you located the suspected part, correct waveforms and voltage measurements are shown on the replacement linear ICs. Also, correct terminals and part outlines are given in the replacement manual.

Part numbers and letters found on ICs and transistors will help you locate the correct circuit or component, and sometimes the operating voltages. RCA, GE, Sylvania, Workman, and NTE semiconductor replacement manuals cover American, European, and Japanese solid-state devices. These manuals can be obtained from manufacturers, electronics stores, dealers, and mail-order firms.

If the exact schematic is not available, look on another schematic from the same manufacturer using the same part number. Compare the voltage measurements to

IC TA76448P, SK7676, or SLY 1547 replacement

Color cont. — 1	42 — C. cont. out
E. cont. out — 2	41 — Bright cont.
V_{CC} 12 V — 3	40 — Clamp in
Delay in — 4	39 — H. sync sep.
Contrast cont. — 5	38 — V. sync sep.
Video inv. in — 6	37 — Sync out
Video inv. out — 7	36 — H. osc. discharge
Hue control — 8	35 — AFC
Chroma in — 9	34 — H. osc. timing
A_{CC} filter — 10	33 — H. V_{CC} 8 V
Gnd. — 11	32 — Gnd.
Chroma out — 12	31 — V. sync in
Killer filter — 13	30 — Timing
Demo in — 14	29 — Height cont.
APC filter — 15	28 — Ram capacitor
Xtal drive — 16	27 — NFB in
Xtal in — 17	26 — V. drive out
1¼ in — 18	25 — Phase comp.
B-Y out — 19	24 — H. drive out
G-Y out — 20	23 — X ray protector
R-Y out — 21	22 — Y out

Top view

1-14 Cross-reference the part number of the IC using the universal replacement manual to locate the correct part, circuit, and voltage measurement.

those of the defective chassis. Sometimes the same part in other brands will have about the same voltage measurements.

Pills and more pills

For the headache, you can take an Aspirin or Tylenol to relieve the pain. When servicing a tough-dog problem in the TV, audio, or VCR chassis nothing relieves the pressure or pain until the component culprit is located (Fig. 1-15). The tough-dog symptom refers to a very difficult service problem that involves a lot of valuable service time. A tough dog might end up with more than one problem or component in the TV chassis.

1-15 The tough dog repair might take many hours and several days before repairs are made.

Tough dogs are very difficult to locate without a schematic. In many cases, the tough dog is right under our nose. Isolate the symptom to a given section of the chassis and take critical voltage and waveforms. Often, the difficult chassis might have an intermittent component that will only act up once a week or twice a month. These sets should stay in the home until the intermittent state acts up several times a day. Just wait, they will begin to act up more often.

The intermittent chassis can result from poorly soldered connections, intermittent components, cracked boards, and poor terminal connections. Twist and push around upon the board until you have located the intermittent section. Check all component soldered connections in that area. Sometimes a wholesale soldering up of the entire section might uncover the intermittent connection (Fig. 1-16).

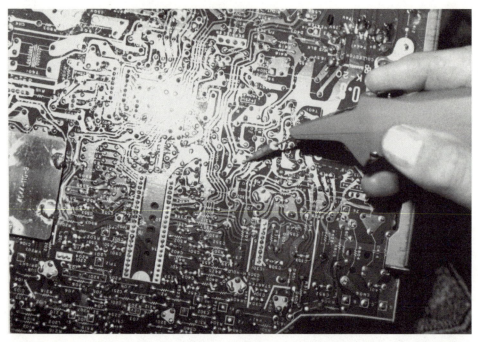

1-16 You might need to solder up every terminal lead in a certain area before you can find the intermittent connection.

A large blob of solder might contain a poorly tinned component lead. Often, a component lead that is not properly tinned or that solder will not cling to results in a poor contact. Remove the large blob of solder with a solder wick or sucking tool. Remove the terminal lead. Clean it up and tin the connections. Remount the connection, and solder up.

Intermittent operation in the TV, cassette player, VCR, and audio equipment can be caused with bad sockets or socket connections (Fig. 1-17). Poor tuner contacts were caused by a crimped socket, making a poor connection, producing intermittent TV reception in the TV chassis. Check where the socket is mounted upon the PC board for poorly soldered connections. Foil wiring breakage around board stand-offs can cause intermittent operation in the VCR. Inspect the PC wiring around where heavy components—such as tuners, heat sinks, and transformers—are mounted. If the unit is dropped in shipment or on deliveries, these heavy components can cause cracked PC wiring.

Apply heat and cold applications upon components and PC wiring that are suspected for intermittent operation. Intermittent transistors, ICs, and processors might act up with several coats of cold spray. First apply coolant and spray directly upon the body of the component or parts that require an hour or so of heat-up time. Then apply hot air from an air gun or hair dryer and see if the part returns to the intermittent state.

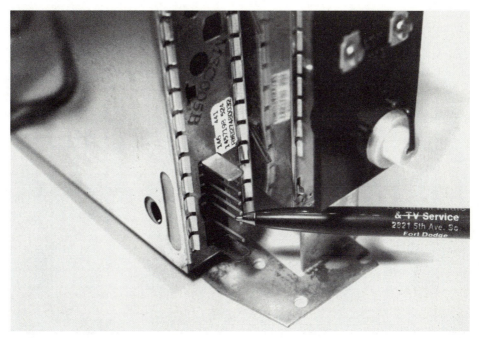

1-17 A poor socket connection on a tuner module caused intermittent tuning and reception.

Double-trouble

Here, we are not talking about a chewing gum when two or more components are defective within the TV, VCR, or cassette player. Often the electronic chassis has only one trouble symptom and one defective component. When you repair one problem and another acts up, this is double-trouble. Sometimes you might find more than two defective components in one section of the TV chassis.

For example, a leaky flyback transformer can destroy the horizontal output transistor, isolation resistor, line and B+ fuse, isolation line resistor, silicon diodes, and scan-derived voltage sources in the secondary winding of the horizontal output transformer (Fig. 1-18). After repairing the horizontal circuits, the sound begins to cut up and down. So you go to the sound circuits and find a defective electrolytic speaker coupling capacitor. If an intermittent component is found in the double-trouble problem, several hours are needed to repair the defective chassis. Leave that chassis run for several days, making sure that every repair has been completed.

Down under

Look for possible overheated terminals, brown board areas, and poorly soldered connections on the PC board wiring side. These areas will show overheated parts and intermittent connections. Check the terminals on the component side to locate a possible defective component (Fig. 1-19).

1-18 The IHVT flyback transformer has separate secondary windings and provides several dc voltage sources.

1-19 You can locate poor terminal connections by taking voltage and resistance measurements.

Burned or stripped PC board wiring might indicate defective components, power surge, or lightning damage. Several different blown-off areas of PC board wiring might indicate lightning damage. Often this means the chassis is totaled and complete set replacement is required.

Cracked or damaged boards might be visible by looking closely with a magnifying glass. A strong light above the board, showing light through the board, can help locate fine cracks in PC board wiring. The magnifying circle light is ideal for PC board wiring inspection. Going from terminal to terminal with the low-ohm scale of the digital multimeter (DMM) might help locate open PC board wiring.

Outer space

Check for defective components outside the regular chassis when all other tests are normal. For example, a horizontal white line or no vertical sweep might be caused by an open winding or poor connection of the yoke assembly. Red missing from the raster might be caused by a defective color transistor or R-Y circuit located on the CRT socket board (Fig. 1-20).

1-20 The red color was missing from the raster in a Sharp 195B60R portable.

Don't overlook possible problems in separate boards mounted horizontally above the regular PC board chassis. Poor vertical or horizontal linearity might be caused by defective pincushion parts beyond the regular circuits. Look outside the regular circuits when all other tests fail.

Don't put it off to tomorrow

The difficult or intermittent chassis may be set aside for several weeks before you attempt to repair it. Callbacks or electronic products returned for the second time can be put off for tomorrow. Tackle these kind of problems at once. Often, the repair and time involved might take less than one hour. Don't put off it today and do it tomorrow. It might never get done.

When you have repaired a boom-box cassette player for distorted sound and the unit is returned with another problem such as the cassette player is now eating tape, make another charge. Of course, the customer is angry as the repair did not last only for a few days. Any electronic technician knows that the sound distortion problem is a long ways away from the mechanical tape eating symptom.

However, did you clean up the tape heads, capstan, and pinch roller while the player was in for repair? (See Figure 1-21.) Sometimes we forget to do preventive maintenance. A good clean-up and a thorough preventive maintenance check-off can save a lot of headaches. Besides a good clean-up, one should check the type of cassettes and also the condition of the suspected cassette, which might be defective. The customer is never wrong, but he or she might not know how the cassette operates or might mishandle the cassette.

1-21 The cassette and boom box player might take longer to repair for some TV technicians.

Time, precious time

Time is one of the most important factors for the electronics technician. Time marches quickly when you are stuck with a difficult problem and no schematic. It might take months to receive the correct diagram, and sometimes you might never receive a schematic for some older products. Wasted time is wasted dollars.

Servicing a radio or stereo chassis might require more time to repair for a TV technician (Fig. 1-22). Some were not trained for audio-type service. Time can be wasted and never recovered when working on a tough-dog problem. In the following pages, I hope to reduce the time that it takes you to troubleshoot and repair consumer electronic products without a schematic.

1-22 A dirty tape roller can cause a call-back when preventive maintenance is not rendered while making other repairs.

Test points

The manufacturer has placed into the schematic several test points for easy servicing. These test points are for taking waveforms, signal tests, and voltage measurements. The test point might consist of a lug, spot, or raised area that any test instrument can be connected upon the chassis.

The test points within the CD player provide quick attachment of the scope for troubleshooting and making critical adjustments. Test points are found throughout the camcorder circuits for adjustments, waveforms, and testing procedures. The test points in the TV chassis might be located in the video, RF/SIF, auto kinne bias, and CPU circuits (Fig. 1-23). Clip the scope to the video test point to determine if the front-end circuits are functioning. Measure the RF AGC voltage at the test point fed to the tuner and IF AGC test point to determine if the AGC and IF circuits are working.

If the TV chassis has an MPX/stereo sound system, you can find test points throughout the stereo IC circuits. The test point at the VCO pin terminal (32) is to

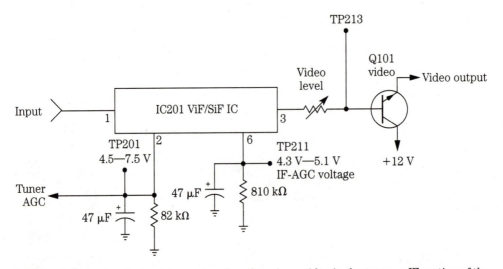

1-23 Look for various test points or tabs when there is no video in the tuner or IF section of the TV.

set the VCO adjustment control (Fig. 1-24). A SAP BPT out pin terminal 6 (TP606) can scope the SAP circuits. The multiplex input pin terminal (1) might have a test point (TP603) to determine if audio is found at the input terminal. A SAP level control has a test point to adjust the voltage at snap level detector. Many test points are placed throughout the electronic circuits to make signal checks, voltage test, and critical adjustments.

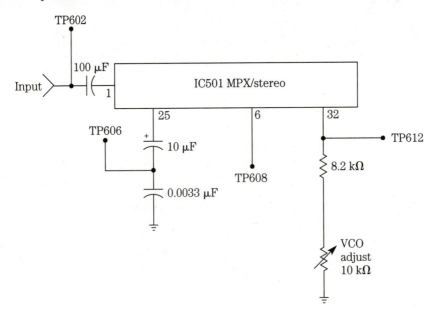

1-24 The different test points or tabs are used to make critical adjustments within the MPX/stereo section.

Say it isn't so

Have you ever checked the electronic chassis, ordered out a new part, waited for the new replacement, installed it, turned on the switch, and the results are the same? This just cannot happen after installing a new IC component. Look at all the time lost in removing the old IC and installing the IC replacement. Except this can happen to the best electronic technician when all components tied to the terminal of the suspected IC are not thoroughly checked (Fig. 1-25).

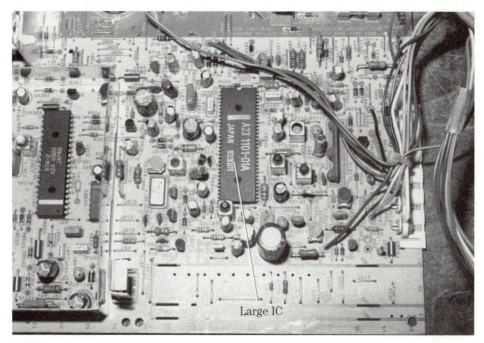

Large IC

1-25 Always check all components and circuits tied to each pin terminal of the suspected IC before replacement.

What if the voltages measured upon each pin terminal of an IC were quite normal with a fairly normal input signal and no output waveform? Even the supply voltage (VCC) was quite normal. The IC just has to be defective. You install a new IC after removing the old one and cleaning up each PC terminal connection. The results were the same: no output signal.

So, you again make voltage, resistance, and waveform tests. Everything appears normal. Maybe a piece of foil was cracked or broken. All terminal pins are soldered up again. Resistance tests were made from pin terminal to a common point on the same PC foil. All tests were good. What went wrong?

No doubt, the new IC replacement was defective. However, this never happens, as you test every component before installing for this same reason. Of course, it's difficult to test the IC with so many terminals. So you just install it and take a chance. Nine times out of ten, you never have to replace a replacement. These operations just happen in the life of an electronic technician. Always test each component before installation.

Ten points of servicing

Where do you start when servicing a TV, stereo, or VCR chassis? Start with the symptoms provided by the electronic product. In the TV chassis, use the symptoms found on the TV screen and speaker. An insufficient vertical raster indicates trouble in the vertical section. Pulled width might be caused by the low-voltage power supply, horizontal circuits, or pincushion circuits. Intermittent sound might be caused by a defective audio IC.

Start with the symptom, and isolate the defective component. Break the circuits down, and list the probable circuits the defective part might be in. Then make voltage, resistance, and solid-state tests in those circuits to locate the defective part.

To locate the defective section, you must know where they are located on the chassis without the help of a schematic. For example, if a tape player has weak and distorted sound, with normal AM and FM stereo reception, the audio frequency (AF) and audio output circuits are normal (Fig. 1-26). The defective component must be between the tape head and the stereo function switch. The most obvious things to check are a dirty tape head, electrolytic coupling capacitors, pre-amp transistors or ICs, and the supply voltage to the tape head circuits.

Locate all large components on the chassis to find the defective circuits. Around large filter capacitors are the low-voltage power supply components. Look at the large transistors or IC audio output components for defective sound circuits. Vertical circuits are nearby when locating the vertical output transistors. The horizontal output transistors and regulators might be found on a large separate or chassis heat sink. Try to locate large parts in each section that point to the correct symptom.

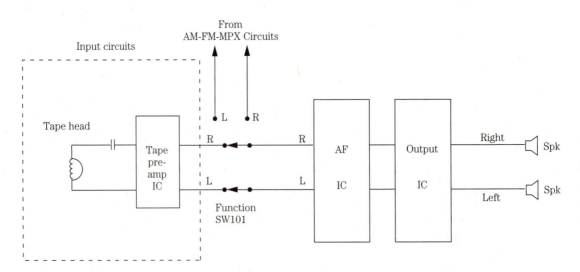

1-26 Isolating the input circuits of the tape cassette player can solve the weak and distorted audio problem.

After locating the defective circuits, take voltage and resistance measurements. Some technicians check solid-state parts first. The most accurate instrument for taking voltage and resistance measurements within the solid-state chassis is the DMM (Fig. 1-27). Voltage and resistance measurements less than 1 V or 1 Ω are required within the solid-state chassis. Accurate base bias voltages (0.3 to 0.6 V) between base and emitter terminals not only indicate if the transistor is good, but if it is a germanium or silicon type.

1-27 This Fluke DMM is ideal for taking critical voltage, resistance, and current measurements within the electronic chassis.

Next, scope waveforms are needed to signal trace the various circuits. When the horizontal output circuits are functioning, the horizontal waveform can be traced from the vertical oscillator, amp, and output to the vertical yoke winding. Low distortion can be found in the stereo circuits with a sine or square wave at the input. Scope the audio stages for clipping and distortion.

After several tests, the defective part must be removed from the circuit. Desoldering equipment and solid mesh or wick material can be used to remove excess solder. Be careful when removing transistor pins so as not to pull or pop off thin PC board wiring connections. Too much heat might dislodge or destroy small PC board wiring. The controlled electronic soldering system or station is ideal for removing and replacing PC-board-mounted components (Fig. 1-28).

Check the suspected part after it is removed from the chassis. Sometimes the intermittent transistor might be normal after heat is applied and it is removed from the

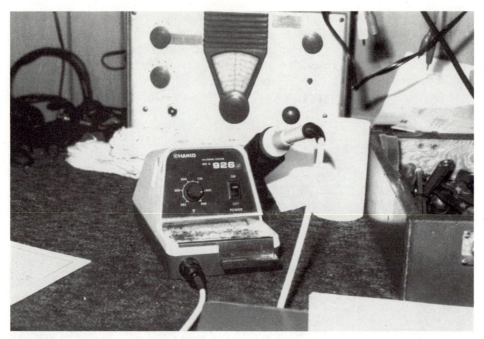

1-28 The electronic controlled soldering iron is ideal when removing and replacing components in the electronic chassis.

circuit. Replace it. Transistor, continuity, resistance, and special equipment tests might indicate a defective part after removal.

Test the new component before installing. Check transistors, capacitors, resistors, choke coils, diodes, and zener diodes before installing them. Make sure they are good before soldering. By checking the part before installation, you might save a lot of valuable service time and confusion.

Install the new part with careful soldering of the terminals. Too much solder can overlap between terminals and short out the component when the chassis is turned on. If too much heat is applied, it might damage a semiconductor. Clean out all solder and rosin flux between terminals.

Double-check all terminals. Are they all placed on the correct PC board wiring? Remember, the transistor or IC component can be soldered with reversed terminals. Check the wiring connections. Make sure no cold joints or interconnected connections are made. Check and double-check before turning on the chassis.

The following steps must be taken to locate, isolate, repair, and replace a defective component:

1. Check the symptoms.
2. Isolate the circuits.
3. Locate the part.
4. Take voltage and resistance tests.
5. Perform scope tests.
6. Remove the defective part.

7. Check the defective part.
8. Test the new part.
9. Install the new part.
10. Clean up and double-check.

Yesterday's hero

You have just serviced a tough dog that two other electronic shops could not fix, the customer is happy, you are elated, and in comes another one a week later. Tough dogs seem to come in bunches. One week you are loaded down with difficult repair jobs, and the next month you get hardly any. So you go to work, roll up your sleeves, and begin again.

Of course, this case is a little different. You have never seen a TV chassis like it before, you have no schematic, and you just know parts are not available. The TV set might act up once a week and then play normal for several months. In fact, the symptom is intermittent operation and chassis shutdown. You let it sit there for several days. This morning, you feel great, and the tough dog is in for a rough time.

Just about any component in the TV chassis can cause shutdown (Fig. 1-29). The back cover is removed, and the chassis operates normal with a very nice color picture. The only thing to do is to let it run until it quits. You come back from lunch, and it's still running. Every once in awhile, you look at the picture while servicing other electronic units. After nine days, the screen goes black.

1-29 Check for a voltage regulator IC and horizontal output transistor upon the TV chassis for chassis shutdown problems.

Check the horizontal and low-voltage power supplies because most shutdown cases occur in those circuits. Carefully look over the chassis for the horizontal output transistor and line-power supply regulator. Take a critical voltage test upon each component. No voltage was found at the horizontal output collector terminal located on a heat sink, and no voltage out of the regulator IC was found upon a separate power board. A large flanged heat sink upon a separate board held an IC that looked like a regulator and power transistor (Fig. 1-30).

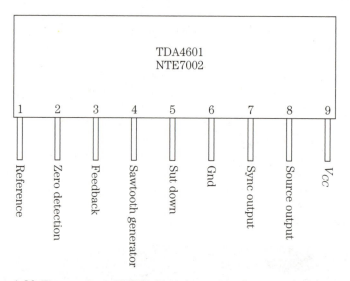

1-30 The regulator (SMPS) IC upon the heat sink can be determined by the number upon the part and checking in the replacement manual.

The power IC number TDA4601 was checked in the NTE universal semiconductor replacement manual as an NTE7002 regulator. The manual listed the IC as a switched-mode power supply control IC, working at 15-V source, which meant a power transformer and power transistor must be close by. The switched-mode power supply (SMPS) transformer was found on the same board. Voltage measurements upon the transformer winding indicated 167 V, which no doubt fed to the SMPS driver or power transistor. The voltage upon the power transistor bolted to the same heat sink had 167 V to the collector terminal. No voltage was found on the base or emitter terminals.

When the voltage probe touched the base terminal, the TV chassis came alive. At first, the power driver transistor was suspected and tested normal. Now a negative voltage (-1 V) was found at the base of the power transistor. The chassis was shut down, and the PC wiring at the base terminal went to pin 7 of the IC regulator. All voltages were now tested at each pin of the SMPS regulator (Fig. 1-31). A scope waveform at output pin 7 represented a square wave around 2 V.

The scope probe was attached to the base terminal of the power driver transistor to monitor the SMPS waveform with the DMM testing the supply voltage source

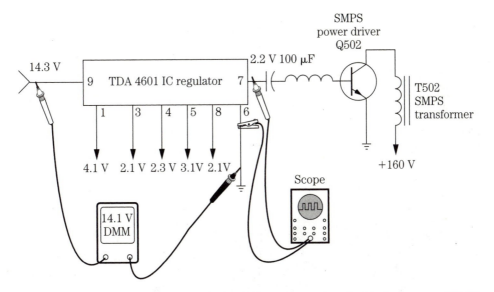

1-31 The switch-mode power supply (SMPS) IC can be monitored with the scope and DMM.

at pin 9. When the chassis shut down again, all voltages upon the regulator IC were very low, except pins 4 and 5, which had high voltage. All components led to the pin terminals of the regulator IC were tested and appeared normal. Perhaps the TDA4601 regulator was breaking down under load.

Because there was no collector voltage at the horizontal output transistor (measured 121 V), no waveform upon the base of power drive transistor, and high voltage (167 V) on the collector of power SMPS transistor, the IC voltage regulator was suspected of breaking down or going open. After the TDA4601 regulator IC was replaced, the TV chassis has been operating ever since. With previous experience from the RCA and Sylvania SMPS power supplies, no 120-V source applied to the horizontal output transistor, helped solve the foreign TV chassis, without a schematic.

Loss of life

Excessive lightning damage or power outage can destroy a TV set, VCR, receiver, and cassette player. Extensive damage to the low-voltage circuits and tuner in the TV can total out the entire chassis (Fig. 1-32). When the 220 neutral wire breaks before entering the home, it can place extra ac voltage upon any electronic chassis. High-line voltage wires touching one another can result in high-power outage.

A TV set exploding and starting a house fire can be very dangerous and fatal to the homeowner. If the fire spreads to the window curtains and spreads to other rooms, someone can be trapped in the raging house fire. Houses can be rebuilt, but a loved one is lost forever.

The electronic technician should always be careful when servicing the electronic chassis. Replace all safety marked components with exact replacements. Never re-

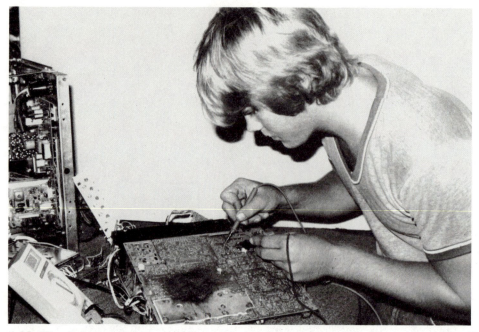

1-32 When lightning damages the TV set, the chassis might have to be totaled out instead of repaired.

place a flame proof resistor with a carbon resistor. Replace all line-power supply and horizontal circuits with correct components. Make sure that cables and wires will not lay upon hot components in the chassis. Inspect and replace the defective power cord.

Do not drop metal screws and small metal parts down into the maze of wiring and components, then forget to remove them before buttoning up the chassis. Make good, clean soldered connections. Blotched bare wiring and lousy soldered connections might give the fire department and state commission an excuse to blame the electronic technician. In case of a death, you might lose everything you have worked for. Be careful out there.

2
CHAPTER

How to locate, test, and repair

You must locate what section the defective component is in before you can remove and replace it. With several different symptoms, you can isolate the trouble to a given section of the TV, radio, VCR, or CD player. After determining where the trouble is, locate the suspected part.

Remember, you don't have a schematic diagram to locate the suspected component with accurate voltage measurements. Take voltage, resistance, scope, continuity, and semiconductor tests to determine if a part is defective (Fig. 2-1).

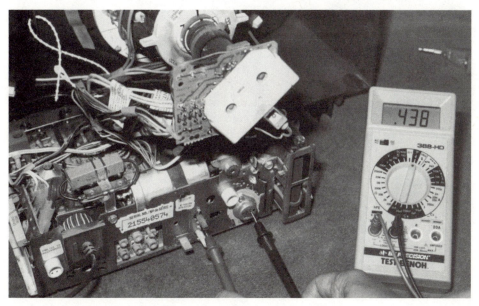

2-1 Taking voltage and resistance measurements in the RCA TV chassis.

Correct test instruments

Today the most useful test instrument on the service bench is the digital multimeter (DMM). This meter will accurately check the small voltage, resistance, continuity, and current required in solid-state circuits (Fig. 2-2). Some DMMs will also test capacitors, transistors, and inductance. The DMM comes as pocket, portable, or expensive desktop test instruments. Purchase a DMM, even if you own a volt-ohmmeter (VOM).

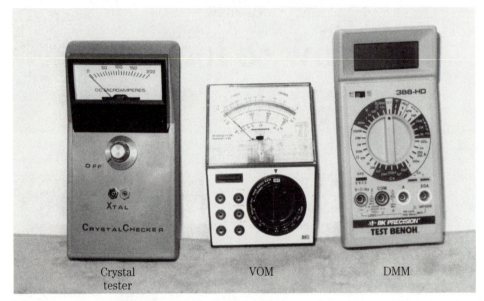

| Crystal tester | VOM | DMM |

2-2 The pocket DMM can take critical voltage, resistance, and current measurements.

The oscilloscope is the next most used instrument found on the service bench. You can easily locate a defective component with critical waveforms in many different consumer electronic products. A dual-trace oscilloscope is needed with at least a 35- or 40-MHz frequency bandwidth (higher in some cases). Knowing how to use the scope will save a lot of time on the service bench. Don't let the scope sit idle; use it. Besides regular hand tools, special and additional test equipment is listed in each chapter.

The following test equipment is needed:

- DMM
- Oscilloscope
- Transistor-semiconductor tester
- Audio signal generator
- Sweep generator
- RF signal generator
- The frequency counter
- Capacitance meter
- Color pattern generator
- Isolation transformer (Fig. 2-3)

2-3 Today, the isolation transformer should be used when you service all electronic products.

What's the symptom?

You must know how the unit is acting up before troubleshooting the chassis. For example, a dead TV might be caused by an open fuse, the low-voltage power supply, or the horizontal output circuits. A white line on the raster of a portable TV indicates a problem of insufficient sweep in the vertical section. Sides pulled in on the raster might indicate a defective low-voltage power supply, horizontal sweep, or pincushion circuits (Fig. 2-4).

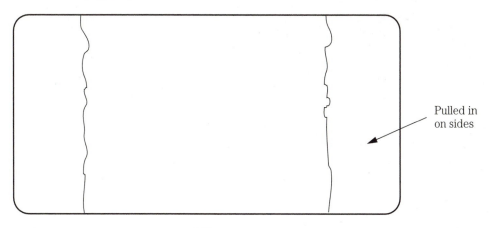

Pulled in
on sides

TV raster

2-4 The pulled-in raster of a TV can be caused by defective components in the low-voltage source.

A dead symptom in stereo radio circuits might indicate a defective low-voltage power supply to the audio output transistor, IC, or speaker. Excessive hum within a large stereo amplifier might be caused by defective filter capacitors. Pickup hum might point to worn input cables or circuits. No rotation of the disc in a CD player might be caused by a bad spindle motor or drive circuits. Use all the symptoms to locate the defective section within the electronic product.

Case histories

Case histories should be looked at after determining what section the symptoms point to. Keep all cases in a card file, on computer, or right on the schematic. Remember, the case history of one unit might help to solve problems in other chassis. Don't forget to write down and record all tough-dog case histories (Fig. 2-5).

2-5 A TV case history can help service other TV chassis.

Isolation

Defective parts can be isolated as burned or damaged components (Fig. 2-6). Overheated parts can be isolated by touch—the too hot or extra warm components. Touching the chassis and parts can indicate the intermittent component.

After isolating the possible defective section, isolate the defective part with transistor, IC, resistance, scope, and continuity tests. Test the transistors in or out of the circuit with a transistor-diode test using a DMM or a beta transistor tester. Check all suspected ICs with signal in and out, voltage, and resistance measurements. Be-

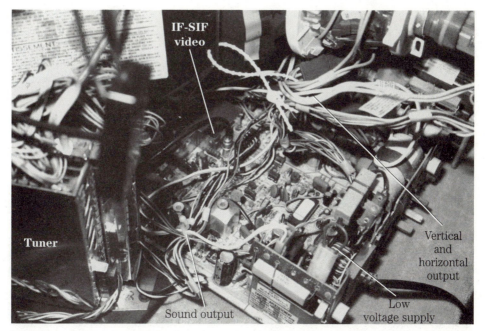

2-6 Isolate the circuits, and check for burned or damaged components.

fore you can pinpoint the defective part you must locate it on the chassis without a schematic.

Locating components on the PC board

After isolating the various symptoms, find the correct section that the trouble might occur in. Look for large components, such as large filter capacitors for the low-voltage power source, transistors mounted on separate heat sinks inside the chassis for vertical output transistors or ICs, and the horizontal output transistors on separate heat sinks or metal chassis for the horizontal output circuits (Fig. 2-7).

Another method for determining if you are in the correct section is to take the numbers located on the transistors and IC components and look them up in a semiconductor replacement manual. The manual tells you what pin goes where and what circuit it ties into on the chassis.

Look at the outline drawing for the correct pin connections. For example, the part number AN5435 lists as a color deflection signal processor in the RCA series replacement manual. The universal replacement is an SK9299 IC with pin 15 as the supply voltage pin (V_{CC}). The horizontal drive output is pin 6, and the vertical output is pin 9. Also, the SK9299 has a total of 18 pins (Fig. 2-8). Double-check with the IC mounted on the chassis for the correct number of pins, the right section, and correct universal replacement.

You might want to double-check the PC board wiring from each pin to where the parts tie into the troublesome circuit. Simply trace the wiring on the connected PC

2-7 Isolate the flyback and horizontal output transistor upon the TV chassis.

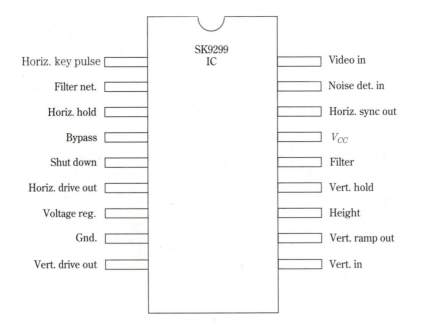

2-8 The outline and pin drawings of an RCA SK9299 deflection signal replacement IC (AN5435).

board wiring. Although this might take extra time, you might save tying up a TV chassis or cassette player for a long time until the schematic is available.

Let's say you have a Panasonic portable TV come in without a model number. The model number has been torn off, but it looks like a fairly new set, one that you have never worked on before. The symptom is a tic-tic sound with B+ present at the horizontal output transistor and no high voltage.

By supplying a horizontal pulse at the horizontal driver, you make the set run, indicating that the horizontal output stages, high voltage, and flyback circuits are normal. You think the AN5435 IC is the horizontal/vertical countdown IC. However, what pin is the voltage source (V_{CC})? You look it up in the semiconductor replacement manual.

The AN5435 IC is a color TV deflection signal processor and can be replaced with an RCA SK9299 IC. Now you have located the correct IC and section that the trouble is in. Before replacing the IC, take voltage measurements on all the IC pins. Usually the highest voltage is the supply pin (V_{CC}). Supply voltage to the deflection IC is usually from 10 to 25 V. In this case, all pins are low in voltage. So you check the highest voltage pin and trace it back to the low-voltage source. If the low-voltage source is supplied through the flyback derived secondary sources, you are out of luck. Look up the supply voltage pin on the outline drawing of the replacement SK9299 IC. Pin 6 ties to the supply voltage source.

Inject an external 10 or 12 V at pin 6 with the scope probe on the horizontal driver (pin 15). If the horizontal waveform is normal, the IC is good. Double-check the vertical output drive waveform at pin 9. If there is no waveform at either pin, suspect a defective IC or leaky components tied to the IC pins and common ground. Replace the IC with a universal replacement if the original part number is not available. Sometimes the IC's voltages and resistance measurements are normal but there are still no waveforms. Replace the IC at once.

Check the voltage at pin 6. Low or no voltage can indicate a leaky IC, improper supply voltage, or no voltage from the flyback secondary source. Trace the PC board wiring back to the supply source to locate a defective isolation resistor, silicon diode, or small electrolytic capacitor. By simply checking the number on the IC or transistor, you can determine what section the trouble is in, the correct pin, and the total number of pins.

Defective parts off the PC board

Don't overlook defective components that are not on the main chassis. Look for defective transformers, power transistors, voltage regulators, ICs, capacitors, and separate metal heat sinks or chassis. Check for problems in circuits on separate PC boards that are soldered to the main PC board (Fig. 2-9). Suspect poor soldered connections from the horizontal PC board to the main chassis.

Burned or damaged components

Burned or damaged parts are easily seen on the PC board. You might have to blow off dust to see the burned components. Sometimes you can smell burned or charred parts. The PC board might have to be repaired if it has been burned and damaged.

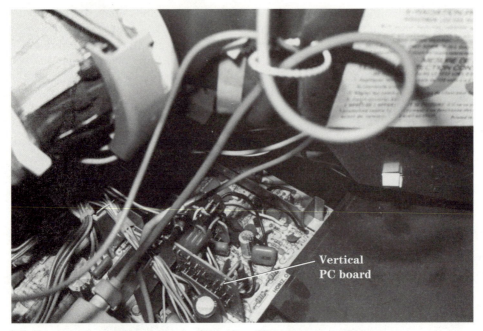

2-9 Check the vertical board for possible broken PC wiring connections.

Large parts—such as transformers, electrolytic capacitors, and power output transistors—can easily be seen. Check for a part number on the body. It is difficult to determine the total resistance in charred resistors. If the color code cannot be seen, try to locate a schematic similar to the chassis on the bench.

In many cases, the resistance might be close in value for several different known chassis in the same circuit. For example, two charred boost-dropping voltage resistors were found in a Sony portable TV chassis. Because the correct schematic was not available, another Sony portable (KV-1747R) circuit was used that was quite similar. There was no doubt that R720 and R715 were damaged when the picture tube arced over with excessively high voltage (Fig. 2-10). Make sure the circuits are similar with the correct operating voltage.

Hot ground

A common point known as a *hot ground* is a live circuit or above the common ground. The common ground in any electronic circuit can be referred to as a *cold ground*. All circuits are common to regular ground except the hot ground. When voltage or resistance measurements are made from a transistor or IC with a hot ground circuit to regular ground, these measurements are incorrect and misleading.

Many circuits of the switched mode power supply (SMPS) have a common hot ground. Look for a hot ground in switching power supplies. The ac input circuits are above ground or contain a hot ground. You will find hot grounds in the RCA, Sylvania, Goldstar, and several other TV chassis (Fig. 2-11).

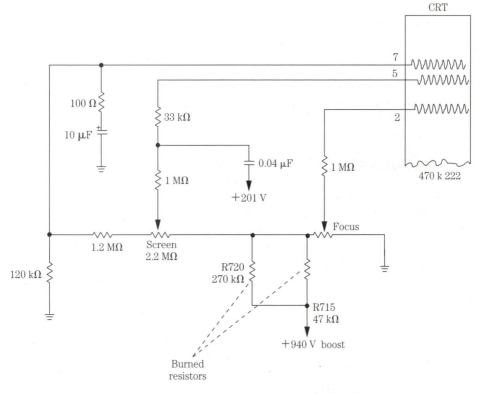

2-10 Burned resistors can be located by checking a similar schematic of the same brand.

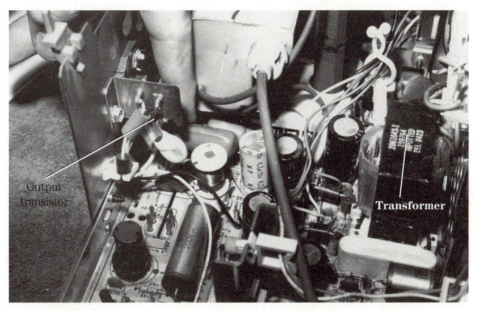

2-11 The finger points at a hot ground transistor and transformer in an RCA TV chassis.

Look for a switching or SMPS transformer found in the power supply circuit, which might indicate a circuit with a hot ground. Most schematics will show a hot ground with a triangle instead of the ground symbol. When servicing a chassis without a schematic, check for a separate transformer and switching or VIPUR output transistor upon a separate heat sink.

The primary winding of the output transformer will have a hot ground and the secondary at ground potential. Take a resistance measurement from the large filter capacitor's ground or emitter terminal of the output transistor to common ground. If there is a resistance measurement between these two points, these components are above ground or have a hot ground. When the resistance is zero, a common ground to all common components is referred to as *chassis ground*.

The hot ground in the low-voltage power supply circuits isolates the power line input voltage from the secondary voltages of the switching transformer. The VIPUR or switching voltages of the power transformer supplies a higher voltage to the horizontal output circuits with several lower voltage sources to other TV circuits (Fig. 2-12). Just remember, when taking voltage or resistance measurements on the power line voltage source, make sure the negative voltmeter terminal is on the hot ground for accurate readings.

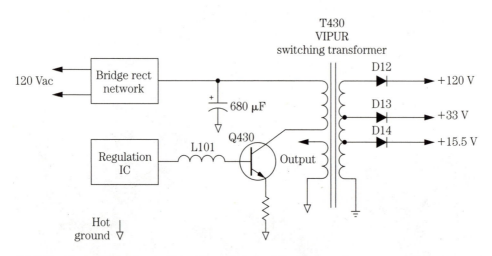

2-12 Notice the input circuits have a hot ground, while the secondary circuits of T430 have a regular common ground.

Running hot

Smoking resistors indicate a leaky transistor, IC, or capacitor. Power transformers running red hot might be caused by leaky silicon diodes or electrolytic capacitors. Red-hot power output transistors and IC components might be leaky with burned bias resistors. Replace overheated IC components that have a chipped area or white and brown spots on the body.

Check the PC board where overheated parts are tied into the PC board wiring. Often, brown spots on the top, around heated terminals, indicate a poor soldered joint. Clean the area and resolder. If the terminal lead will not take solder, cut off the bad terminal end, insert the lead through the PC board hole and solder.

Too hot to touch

A dead high-powered amplifier with a loud hum can be caused by a red-hot output transistor. Red-hot output transistors in the car radio might have burned open the fuse and charred the hot wire to the receiver. When the horizontal output transistor in the TV chassis runs red hot, suspect insufficient drive voltage or a leaky output transistor. Suspect too hot to touch transistors that are leaky or have improper bias voltage.

Check the transistor in the circuit for leaky condition. You might find one transistor leaky and the other push-pull transistor open. Remove both transistors, and test out of the circuit. Always replace both transistors when one is found leaky or shorted (Fig. 2-13). Inspect the bias resistors for burned or charred areas. Check each bias resistor and diode while the transistors are out of the circuit.

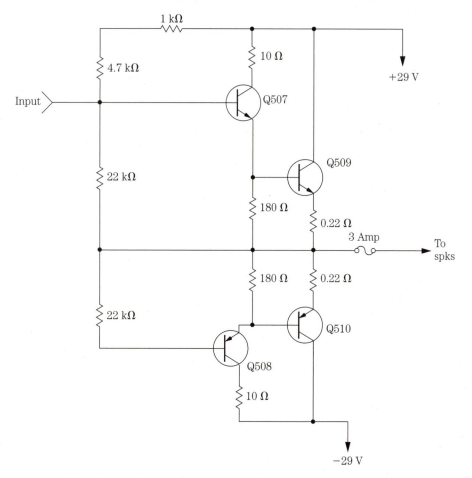

2-13 Check and test all directly driven transistors for open or leaky conditions with a hot output transistor or IC.

Red-hot audio output ICs indicate a leaky IC or excessive supply voltage. Most output ICs and transistors will run warm, but not red hot. Check for a change in bias resistors and high-power supply voltage. Often, when a red-hot IC or transistor is found, the bias resistors are burned or changed in value. Check for leaky directly driven transistors when the output transistor or IC runs red hot.

Another chassis

If you have a new TV chassis that has quit and have no schematic, compare it with another chassis. Sometimes, even with a schematic, certain voltages are not listed, and with another chassis, you can compare voltage and resistance readings. Of course, the chassis must be the same make and have the same chassis setup. Comparison tests can solve a difficult problem when a schematic is not handy.

Often the circuits within TVs, stereos, and radios are the same for several years. If you have a schematic for an older unit, use it. You can also compare the chassis with another brand. If the defective chassis has marked part numbers, use a schematic that is similar (Fig. 2-14).

Take voltage measurements of a transistor from each terminal to common ground. With an NPN transistor, the positive lead will go to each terminal with the negative probe grounded. Just reverse the test leads for voltage measurements of a PNP transistor. The positive probe is at ground potential with a PNP-type transistor. Use a DMM for accurate low-voltage tests.

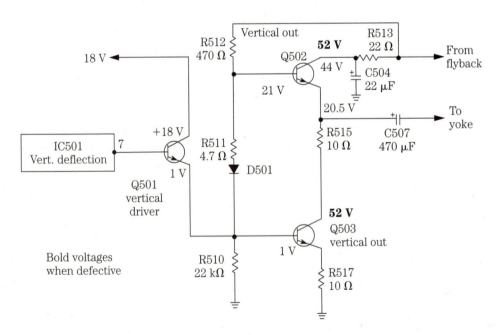

2-14 A schematic of a vertical output circuit of a case history with actual voltage measurements.

Most voltage tests on a suspected IC should be made with the positive lead at the IC terminals and the negative probe clipped to ground. Write down the voltage measurements on each terminal pin. Then determine if these voltages are correct. Always check the voltage supply pin first. If the voltage is very low, suspect a leaky IC.

Remove the voltage-supply terminal from the PC board wiring to see if the voltage increases. Simply unsolder the pin with solder wick and a hot iron. Flick the pin to make sure it is loose and not touching the PC board wiring. Now take another voltage measurement. If the voltage has increased to normal or higher, replace the leaky IC.

In most linear transistor voltage tests, the NPN transistor collector voltage will be the highest reading, with the base terminal voltage next, and the emitter voltage is the lowest (Fig. 2-15). Likewise in a PNP transistor, the emitter terminal has the highest voltage, the base terminal is next, and the collector voltage is the lowest. Zero voltage on the emitter terminal can indicate an open transistor or emitter resistor.

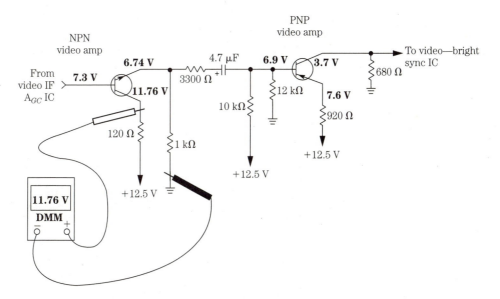

2-15 The collector terminal of an NPN transistor has the highest positive voltage.

When making high-voltage tests, such as on the anode terminal of the CRT, use either a high-voltage probe or a high-voltage probe attached to the VTVM. High-voltage tests on microwave ovens should be made with a high-voltage probe or Magnameter. Make sure the ground clip is attached to the chassis, or you will receive a shock from the high-voltage probe.

Resistance measurements

Critical resistance measurements on the terminals of a transistor or IC to common ground might locate the leaky part. Resistance measurements in transistor and diode tests with the DMM might determine if a suspected part is open or leaky. Crit-

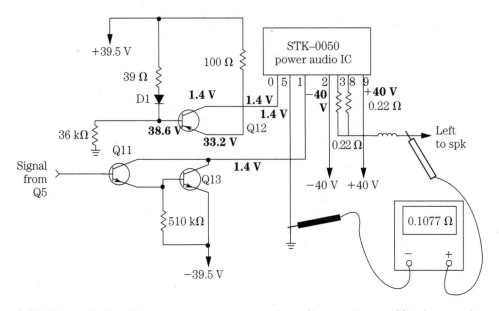

2-16 Make critical resistance measurements upon the audio output power IC to locate a defective AF or driver transistor.

ical resistance measurements on the emitter or bias resistors might indicate a leaky transistor or IC or a change in resistance (Fig. 2-16).

Within the audio amplifier, accurate resistance measurements on the speaker output terminals, with the speaker removed, can indicate a defective transistor or IC or a change in resistance. Safe resistance measurements can be made with the power off when an output IC runs hot. Comparable resistance measurements in the stereo channels might help locate the defective part. Of course, a resistance or continuity check of each component might turn up a defective part. Very few schematics show the resistance measurements to ground, except Sams TV Photo-Facts (Fig. 2-17).

Continuity checks

A continuity or resistance check of a component is a quick way to determine if the part is open or leaky. Check the speaker terminals to see if the voice coil is open or grounded to the speaker magnet or framework (Fig. 2-18). When taking speaker measurements with a DMM, the resistance should be quite close to the speaker impedance. With a VOM the resistance might be off 1 or 2 Ω.

Continuity tests on the tape head can determine if the winding is open. Place extra pressure on the tape head terminals if the winding is suspected of intermittent playback. The continuity test between the tape head terminal and the outside shield might help clear up the problem of weak or distorted tape player sound (Fig. 2-19).

Item	E	B	C
Q201	8190 Ω	68 kΩ	5030 Ω
Q202	960 Ω	2650 Ω	470 Ω
Q203	351 Ω	640 Ω	360 Ω
Q204	0	1 kΩ	57 kΩ

2-17 In most Sams Photofacts, resistance measurements are found from various terminals of transistors and ICs to ground.

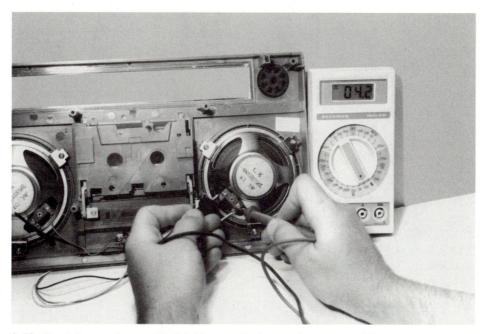

2-18 Check the speaker terminals with a continuity test for open conditions.

Continuity checks of any component can determine if the part is open. For example, the resistance of a tape, VCR, or CD motor might not be known, but a continuity test across the terminals can indicate if the motor winding is open or if there is a broken connection. Likewise, check switches, coils, and transformer connections with a continuity test.

After installing a flyback transformer or IC component that solders directly to the PC board, you should make continuity tests of the PC board wiring. Check from the actual terminal pin to the next part tied to the same wiring to determine if the PC board wiring is intact. Sometimes the PC board wiring breaks at eyelets or terminal pins. Bridge the wiring and resolder.

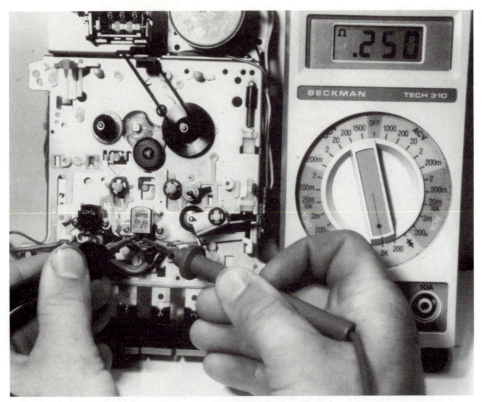

2-19 A resistance tape head measurement can determine if winding is open, shorted, or grounded.

Waveform tests

Critical waveforms can determine if a signal is missing or present within the TV, stereo amplifier, or computer. Waveforms taken on the deflection IC or processor can find an open IC without any horizontal or vertical output pulse. Check for a defective vertical oscillator transistor or IC when the sawtooth waveform is not found on the input terminal (Fig. 2-20).

The horizontal waveform can be checked from the countdown (deflection) IC or transistor oscillator circuit to the horizontal driver transistor. Check the horizontal pulse at the driver collector terminal and base terminal of the horizontal output transistor. Of course, a quick waveform check of the flyback will indicate if the horizontal circuits are functioning (Fig. 2-21). Knowing where and how to take critical waveforms can save a lot of time. Check the video waveform upon the base terminal of the first video amp with on the air broadcast signal (Fig. 2-22).

Time to vaccinate

Injecting an audio or RF signal into a radio or audio amplifier can isolate the dead or weak stage. Signal RF injection in the radio receiver front-end section can

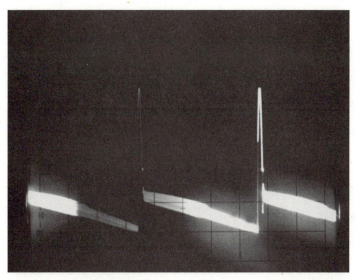

2-20 The vertical output waveform will indicate if the output IC or transistor is defective.

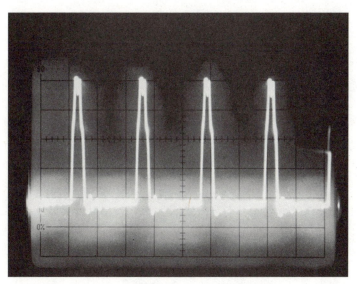

2-21 Place the scope probe near the flyback to determine if the horizontal circuits are functioning.

quickly indicate a defective circuit (Fig. 2-23). A 1-kHz audio signal injected throughout the audio circuits of a dead or weak amplifier can quickly locate the defective stage. The radio or amplifier speakers can be the indicator.

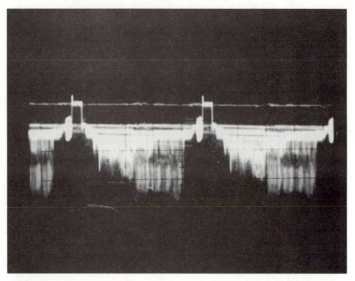

2-22 The on-air broadcast waveform at the base of the first video amp in the TV circuits.

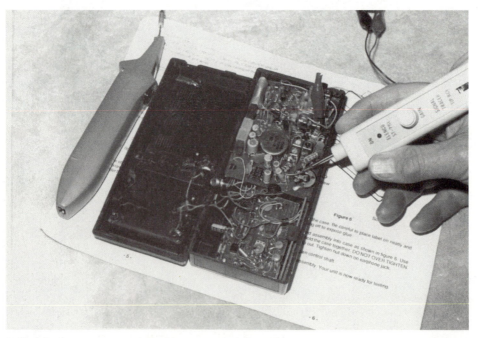

2-23 Signal injection of a radio, cassette, or amplifier will quickly determine what stage is defective.

When injecting an audio signal into the amplifier, a separate audio amplifier can be used to hear the weak or distorted sound. Signal injection from the audio signal generator into the input of each audio stage can quickly locate the defective circuit. Audio comparison tests can be made within the stereo audio circuits by injecting a 1-kHz or 3-kHz signal into the input of each left and right channel, then comparing the audio signal at several points in the circuit and comparing the signals.

When isolating a defective audio circuit within the TV, cassette player, and amplifier, start at the volume control by injecting a 1-kHz signal at the center top of control, with no schematic. If the sound is normal, suspect the input audio signal circuit. Double-check the audio signal at both center top terminals of the left and right stereo volume controls. Proceed towards the speakers with weak or no signal. Go from base to base of each audio transistor. Remember, as you proceed to the last set of transistors, the audio signal becomes higher. When the weak, distorted, or dead stage is found, take critical voltage and resistance measurements upon transistors or IC components.

External amp

The audio external amplifier can consist of a single IC amp or two or more transistors with self-enclosed pin speaker (Fig. 2-24). The external audio amplifier can be used to quickly locate a defective stage or circuit in the audio stages, by injecting a 1-kHz audio signal at the input of the amplifier circuits from a generator and using the external amp to trace the audio signal through the audio circuits. Go from base

2-24 Use the external amp for troubleshooting audio circuits.

to collector terminals to note the increase in audio signal. Keep the gain of the audio signal generator as low as possible for an audible tone. This audio signal can be signal traced right up to the speaker terminals.

When checking for a loss of audio signal in the cassette player, insert a 1-kHz or 3-kHz cassette and check the audio signal at the tape head ungrounded connection with the audio amp. Check the amplifier signal at the volume control. Test both channels for audio signal at the volume control and compare the signals. If one channel is lower than the other, suspect a defective pre-amp transistor or IC (Fig. 2-25). Go from base to base of the pre-amp transistors or IC to locate the defective component. A pre-amp IC between the tape head and volume control is easy to locate upon the PC board.

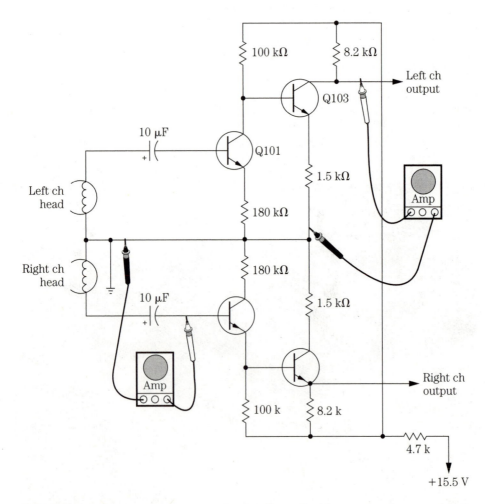

2-25 The cassette pre-amp circuit can be signal-traced with an external amp and cassette.

How to test transistors

Check each transistor in or out of the circuit with the transistor test or transistor-diode test of the DMM. When checking transistors in the circuit, watch for choke coils, low-ohm resistors, and diodes within the circuit. Low or improper measurements might occur. Don't overlook a defective transistor in the circuit that might be open or leaky but that, when removed, tests normal. Sometimes heat on the transistor terminals can restore the transistor. Replace it anyway.

Always test a transistor after removing it from the PC board. Spray coolant or heat on the transistor to make it act up. Don't forget to check the replacement before soldering it onto the board. Check the bias resistors before replacing a transistor.

The suspected transistor can be checked in the circuit with voltage tests. Low collector voltage can indicate a leaky transistor. All close voltages on the three terminals of the transistor can indicate a leaky transistor. Higher-than-normal voltage on the collector terminal can indicate an open transistor. No emitter voltage can indicate the transistor is open.

Take a voltage measurement between the base and emitter terminals for correct bias voltage. An NPN transistor should have a 0.6-V measurement, and a PNP transistor should have a 0.3-V reading. If an improper measurement is found, the transistor is defective.

Accurate transistor tests can be made with the diode test of the DMM. Most of the latest DMMs have this diode test for checking the junction resistance between terminals. A very low measurement between the collector and emitter terminals indicates a leaky transistor (Fig. 2-26). Most leaky conditions found in transistors are between the emitter and collector elements.

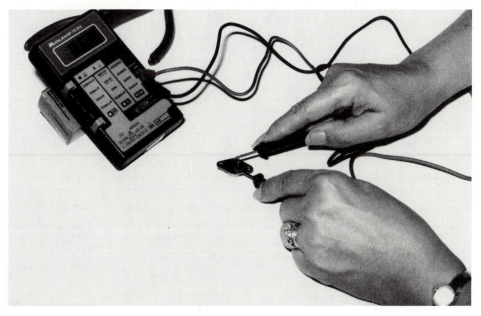

2-26 Check the suspected transistor with DMM diode-transistor test.

When checking an NPN transistor, place the red probe on the base terminal and place the black probe on the collector terminal. If you get a low resistance reading, under 100 Ω, suspect leakage between the base and collector elements. The normal measurement is above 500 Ω.

Let's say the measurement is 877 Ω (Fig. 2-27). Place the black probe on the emitter terminal and leave the red probe at the base terminal. The transistor is normal if a comparable measurement is noted (845 Ω).

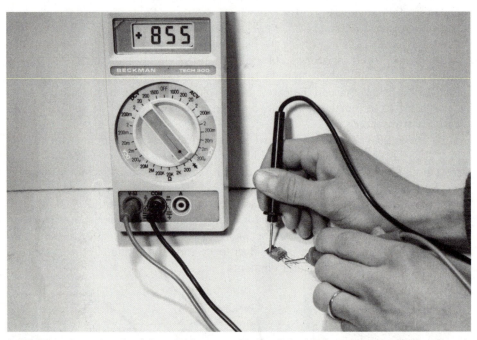

2-27 Place the red probe (+) to the base terminal and the black probe (-) to either the collector or emitter for a comparison junction-diode test.

When the reading is low between any two elements, the transistor is leaky between these elements. You might find that a transistor leaks between all three elements. Often a leaky transistor occurs between the collector and emitter terminals. When a junction test is high on one set of elements and not the other, it is likely that the transistor has a high and poor junction; replace it.

When making transistor tests on a PNP transistor, the black probe stays on the base terminal while taking measurements with the red probe on the collector and emitter terminals. Remember, low resistance measurements can be caused by small coils, chokes, diodes, and low-ohm resistors in the base circuits. In this case, remove the base terminal from the PC board wiring and take another test.

NPN or PNP

Although you will find that most transistors in later chassis are NPN types, be prepared for a few PNP types. When no schematic is handy and you don't know which terminal is the base or collector, try the diode test. Determine if the transistor has a part number on the body or marked terminals. Look up the part number in the replacement manual to tell if it is an NPN or PNP type, what its working voltage is, and what circuits it is found in.

If there are no part or terminal numbers on the transistor, try to locate the base terminal with the diode test. In a normal transistor, the base is tested with the positive probe of the DMM, and the resistance is common to both the collector and emitter terminals. Usually a higher resistance measurement is from base terminal to emitter. For example, in testing an SK3710 horizontal deflection output transistor, the resistance, with the red probe on the base terminal and the black on the collector, is 609 Ω. The resistance measured from the base to the emitter terminal is higher at 667 Ω.

Also, this measurement indicates that the transistor is normal and an NPN type. When the measurement is common with the red probe at the base terminal, it is an NPN transistor (Fig. 2-28). Try checking any transistor out of the circuit with this method.

Another method to tell which terminal is base or collector is to take accurate voltage measurements. The highest voltage is found on the collector terminal of an NPN transistor and common ground. Check (with the power off) for the lowest resistance to ground, which indicates the emitter terminal. Now use the remaining terminal (base), and test the transistor with a beta transistor in-circuit tester.

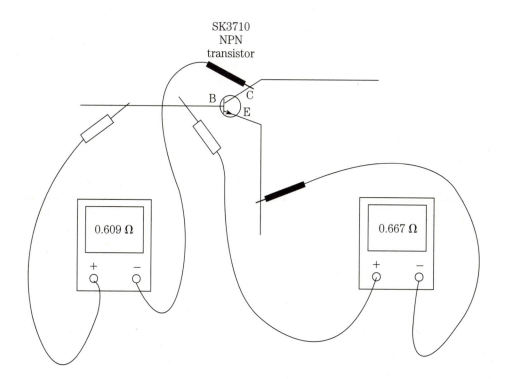

2-28 Normal diode transistor DMM test of an RCA SK3710 output transistor.

Checking diodes

Fixed diodes can be checked with the ohmmeter or diode test of the DMM. The DMM is quite accurate in determining if a diode is leaky, open, or of the correct polarity. Connect the red probe of the DMM to the anode (–) terminal of the diode and the black probe to the cathode (+) terminal. You should have a measurement of about 500 Ω or higher. Measurements below 100 Ω indicate leakage. Now reverse the test leads; no reading means the diode is normal (Fig. 2-29). A low measurement in both directions indicates the diode is leaky.

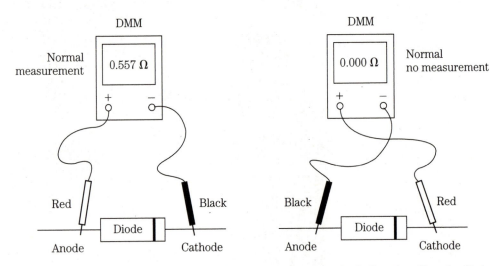

2-29 Suspect a leaky diode with a low resistance measurement in both direction. Here the diode tests good.

Zener, detection, and damper diodes must be checked in the same manner. The resistance of a normal RF diode (1N34 or 1N60) can have a higher resistance in one direction above 1 kΩ. All diodes checked on the resistance scale of a VOM will have a low resistance reading with the red probe on the cathode terminal and the black probe on the anode. Notice this reading is just the reverse of a DMM diode measurement.

High-voltage diodes found in black-and-white TVs and microwave ovens will not test with the DMM diode test. High-voltage diodes on a 2000-MΩ resistance scale will show a measurement above 10 MΩ with the black probe at the cathode terminal, and infinity with the test leads reversed.

Checking ICs

Integrated components can be checked with signal in and out tests. Likewise, scope waveforms taken at the input terminal of a vertical output IC can be seen and amplified at the output pin. An audio signal found on the output IC can be heard on the output terminal if the IC is normal.

Critical voltage measurements on each IC terminal will indicate if the part is open, leaky, or shorted. Check the supply voltage pin (V_{CC}) for higher voltage. Often, low supply voltage at the voltage source terminal indicates a leaky IC or improper applied voltage. Take critical resistance measurements from each pin to the chassis ground to locate leaky parts tied to the IC circuits (Fig. 2-30). Visually trace the pin wiring when a low measurement is found.

2-30 A critical leak measurement of IC pins to ground can identify a leaky component.

In the stereo audio chassis, compare the voltage and resistance measurements to the normal IC or transistor. If the voltage and resistance readings are way off, suspect a leaky IC or directly coupled transistor. Leaky or open transistors within the pre-amp and driver directly coupled circuits can change the voltage measurements on the output IC.

Defective boards

Cracked boards can cause intermittent operation. Broken boards are difficult to replace. Sometimes placing extra pressure on sections of the board can cause the intermittent to act up. Place a strong light over the board, and use a magnifying glass to help search for cracked wiring. Fine breaks can occur around heavy parts or chassis standoffs.

Bad soldered connections at terminal pins of components and ICs can produce intermittent sound in a TV chassis (Fig. 2-31). Large blobs of solder on part termi-

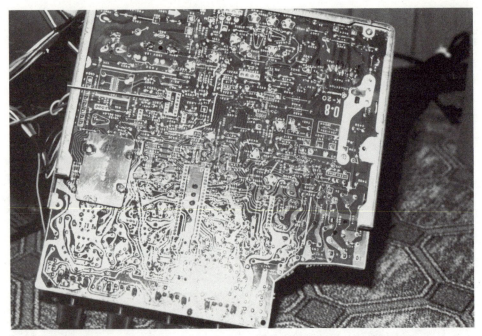

2-31 Check for a poor board connection for intermittent operation in a TV chassis.

nals can cause a dead or intermittent chassis. Brown areas around terminals can mean poor soldered connections or overheated parts. Sometimes resoldering the entire board section can solve this problem. Be careful not to overlap solder to another terminal or PC board wiring.

Intermittent problems

The intermittent chassis is the most difficult problem to repair and can be very time-consuming. Try to locate or isolate the intermittent with all the symptoms that are produced. Monitor the suspected section with voltage and scope tests. Within the audio system, use the speaker as an indicator. An oscilloscope can be used in areas where waveforms are found.

Coolant and heat sprayed on the suspected capacitor, transistor, or IC might make it act up or quit. Several applications might be needed before you get any results. Most intermittent components are found by monitoring the chassis with a voltmeter and scope until the unit acts up.

Intermittent parts in the audio stereo chassis can be located by comparing the signal in the normal channel. Critical voltage measurements on each channel at a given point can uncover the intermittent part. Sometimes just moving the board or component can cause the sound to come and go. Check for intermittent coupling capacitors, transistors, and IC components in the stereo channels.

Overheated transistors or IC components in the TV chassis can produce intermittent symptoms. Intermittent horizontal circuits are very difficult to find because derived secondary voltages are required to make the circuits operate. Look for small things, such as poor connections, bad soldered joints, poorly soldered transistor pins, defective or corroded sockets, and bad board connections.

Part replacement

After locating the defective part, locating a replacement can be most difficult. Sometimes the correct part is hard to find. Substituting another component might be the only answer. Electrolytic and bypass capacitors, resistors, and large-wattage resistors are easy to find. Special transistors and IC parts can be replaced with universal replacements found in the semiconductor manual. Transformers, coils, motors, and original replacement parts might be difficult to locate.

Always try to replace defective parts with the original component. Critical parts found in special safety areas should be original part numbers. It's difficult to substitute for the special tape motors found in VCRs or CD players (Fig. 2-32). Tuners found in TV chassis or VCRs can be replaced with new units, or they can be sent in for repair at a manufacturer's service center.

Besides local wholesale electronics parts stores, manufacturer's service centers, and parts distributors, try obtaining certain parts from mail-order firms. It might take a few days, but they might have the part to fix that chassis that is collecting dust. Don't overlook the fellow technician down the street. Sometimes certain parts might be found just next door.

2-32 Replace all motors and the VCR or camcorder head cylinder with the original part number.

Way down under

Surface-mounted components (SMD) are mounted under the chassis in the TV set. These components are found mounted and soldered directly upon the PC wiring, while standard parts are found upon the top side of the chassis board (Fig. 2-33). Extreme care must be exercised not to flex the PC board or drag the chassis across the work bench to prevent damaging the SMD components.

2-33 The pen points at a small SMD component.

The surface-mounted devices are soldered at each end to the foil of PC wiring. An SMD resistor or capacitor can be flat or round. It can be found mounted anywhere under the TV chassis. Notice that several flat SMD parts are even mounted between the large IC that is mounted on top with the pin terminals soldered to the PC wiring (Fig. 2-34).

SMD parts

The surface-mounted device might consist of a fixed resistor, capacitor, diode, transistors, ICs, and microprocessors. You might find more than one resistor or diode inside one SMD component. Of course, transistors have three separate terminals, and IC components have many pin terminals (Fig. 2-35). ICs, CPUs, and microprocessors might have gull-type mounting wings. Do not be alarmed when a resistance measurement upon a diode- or resistor-looking device turns out to be nothing more than a solid feedthrough. These feedthrough devices attach two circuits together.

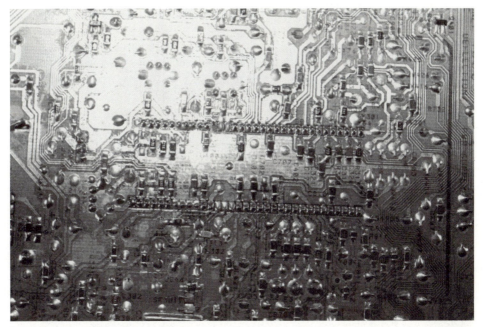

2-34 Here the SMD parts are found mounted between an IC pin terminal upon the PC wiring side.

SMD components are tested in the very same way as those mounted upon the top side of the chassis. To take resistance, voltage, and transistor measurements, sharpen the points of the test probes. If a schematic or parts layout is not available, it's very difficult to tell the difference between parts. Of course, a resistor will have resistance while a capacitor should not. A feedthrough SMD will have a dead short or zero resistance.

The digital transistor might have the base and bias resistors mounted inside the surface-mounted component, while the standard resistor has leads directly to the internal elements. Note that the PNP digital transistor has the base and bias resistors tied to the emitter terminal (Fig. 2-36). Likewise, the NPN transistor is the same. When making transistor tests on digital transistors, allow for the resistor in series with the base terminal and a resistance leakage between the base and emitter terminals.

Removing surface-mounted components

To remove a defective resistor or capacitor, remove the solder at one end. A solder wick and the soldering iron can be used at each end to remove excess solder. Hold the part with a pair of tweezers, and twist as the component is removed. Throw away all surface-mounted parts removed from the chassis. They should not be used, once removed.

To remove a transistor or diode SMD, melt the solder at one end and lift the leads upward with a pair of tweezers. Do this to each terminal until all are removed. Some larger components might be glued underneath. Cementing the new SMD is not

Semiconductor lead layouts

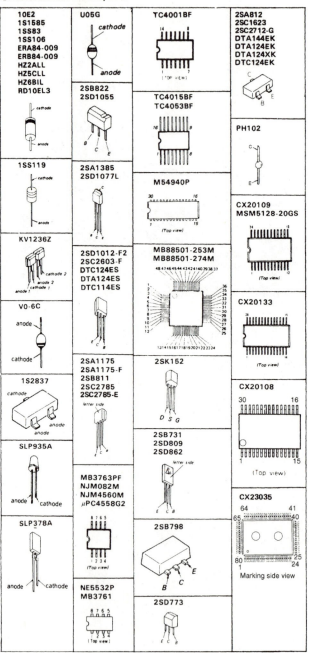

2-35 The various SMD ICs and transistors found in a consumer electronic product.

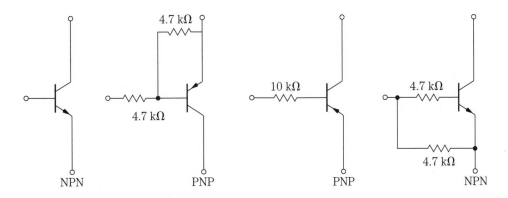

2-36 Digital transistors have base and emitter resistors inside the same body.

necessary. Remove jagged or overlapped solder from the PC wiring. Clean off lumps of solder from the wiring with the soldering iron and solder wick.

In removing flat pack ICs, apply the soldering iron to each pin terminal, and lift solder with solder mesh material. Another method is to unsolder one terminal with the soldering iron and pry up the pin terminal with a sharp tool. Some electronic technicians use a butane-fueled soldering pen, which is found at most electronic distributors. The pen is held about ½" from the flat-pin terminal and pried up with a small screwdriver blade or sharp-pointed tool.

With the gull-wing IC or microprocessor, many terminals can be removed by cutting each pin terminal near the body of the IC. Place a long resistor terminal alongside the body and then cut each pin towards the IC body. When all pins are cut, lift the leadless body off the board. Remove each cut terminal with the soldering iron and tweezers. Be very careful not to cut into the foil pattern. Also, do not apply too much heat on one area to lift the foil from the PC board.

SMD replacement

Universal SMD fixed capacitors and resistors can be used to replace a defective component. These universal components can be purchased with a card of different values (Fig. 2-37). Always use direct factory replacements with dual resistors and diodes, transistors, ICs, and microprocessors.

Remove any solder left upon the PC wiring with a desoldering braid or solder wick. Clean the entire area with cleaning fluid, acetone, and a stiff brush. Make sure the mounting area is clean and level.

When installing new SMD ICs and microprocessors, connect an electrostatic discharge strap to your arm and clip to the ground area of chassis. Position the replacement SMD IC exactly in the correct place. Observe terminal number 1. Place solder flux over each SMD pin. Tack down each corner terminal of the SMD part with solder. Carefully solder up each terminal pin, making certain two separate terminals are not soldered together. Check the connections out with the low resistance of the DMM.

2-37 A card of SMD universal resistors.

EST FCC ID numbers

Every VCR, cordless telephone, personal computer, and microwave oven must carry an FCC ID number. The first three characters of that ID identifies the manufacturer of the product. When a unit comes in and the make is not known, look it up in Fig. 2-38. Often the standard life of TVs, VCRs, stereos, or other consumer electronic products is between 7 and 10 years. Of course, there are many electronic components out there that are quite a few years older. You might find that parts and service literature are not available over the 10-year period. Check the correct prefix of the manufacturer and contact them for special components.

Clean up

Blow out dust and dirt from on top and inside the electronic chassis. Clean up the radio dial numbers and frequency letters before placing the product upon the finished bench. Wipe off the front of the TV screen, and wash the dial assembly and dirty knobs with window spray. Clean off the plastic or wood cabinet that might have color marks, dirt, and fingerprint smudges. Replace all chassis and back cover screws. Make that electronic component shine like new before it is returned.

FCC ID numbers	
Code Prefix	**Manufacturer**
A3D	NEC
A3L	Samsung
A7R	Orion
AAL	Phone Mate
AAO	Radio Shack
AAY	Midland International Corporation
ABL	Hitachi
ABW	JC Penney
ABY	Motorola
ACA	Yorx Electronics
ACB	Phonotronics
ACJ	Matsushita
ADF	Carterfone
ADT	Funai
AES	Uniden
AEZ	Sanyo
AFA	Fisher
AFL	Sharp
AFR	Curtis Mathes
AGI	Toshiba
AGV	Montgomery Ward
AHA	RCA
AIH	Litton Microwave Cooking Products
AIX	Sylvania
AJU	GE
AK8	Sony
AKC	Superscope Inc.
AKE	Marantz Co Inc.
ALA	Wells Gardner Electronics Corporation
ALI	Kenwood USA Corporation
ANV	Capetronic Int'l Corporation
API	Harman Kardon Inc.
ARR	AOC Int'l of America Inc.
ASH	Akai
ASI	Victor Company of Japan
ATA	Sharp
ATO	Zenith Electronics Corporation
ATP	Advent Corporation
BEJ	Goldstar
BGB	Mitsubishi
BOU	Philips
E0Z	Shintom

2-38 The FCC ID numbers and letters found on every VCR, cordless telephone, personal computer, and microwave oven. (Courtesy *Electronic Servicing & Technology* magazine)

3
CHAPTER

Repairing audio amps, large and small

Servicing the audio circuits within consumer electronics products is quite simple compared to TV horizontal circuits. Repairing the audio amp stages might take a few minutes longer without a schematic. However, with a few test instruments, simplified instructions, and the test procedures found throughout this chapter, troubleshooting the audio stages can be easy.

Required test equipment

The following equipment is needed to test audio amps:
- DMM
- Transistor tester
- Capacitor tester
- Signal injector
- External audio amp
- Frequency counter
- Oscilloscope
- Audio signal generator

A good VTVM, VOM, or DMM can quickly locate a defective component with continuity, resistance, voltage, and current measurements (Fig. 3-1). You need a DMM that takes accurate low-voltage and resistance measurements. Besides voltage and resistance measurements, the DMM might have transistor, capacitor, diode, and frequency ranges. A low audible continuity buzzer is found in some models.

Another important test instrument is a small audio amp. This tester is used to signal trace weak audio signals and distortion found within the audio stages. You can use a mono or stereo amp for signal tracing. In fact, you can build your own audio tester with only a few electronic components (Fig. 3-2). With these two test instruments, you can locate and repair most audio problems.

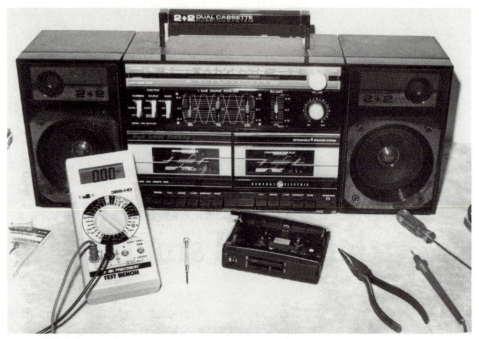

3-1 Choose a DMM for accurate resistance and voltage measurements.

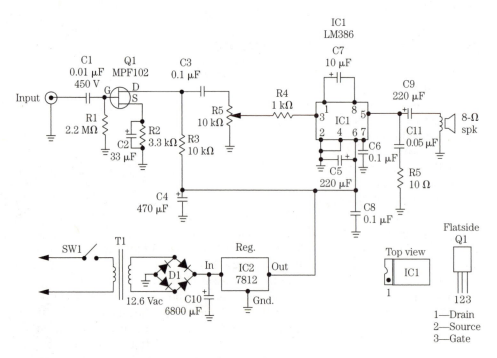

3-2 Build a signal-tracing amp with only a few electronic components.

Test records, cassettes, and discs come in handy when signal tracing audio through the various circuits. A simple test speaker, cables, and test clips provide quick audio connections. High-wattage, low-ohm resistors switched in parallel or series can provide speaker loading at the output terminals.

Optional test equipment includes a frequency counter, scope, capacity tester, audio signal generator, audio analyzer, audio wattmeter, wow and flutter meters, and dual stereo signal indicator. Additional audio test instruments are only required with certain audio problems.

The various sound problems

Sound problems include a dead channel and no, weak, intermittent, and noisy sound. The dead-channel or no-sound symptom is easy to locate in the audio circuits. The no-sound symptom can be caused by an open transistor, IC, transformer, or coupling capacitor. Don't overlook improper or no low voltage from the power supply.

Distorted sound can occur in one channel or both. Very low distortion is difficult to locate. Use a sine- and square-wave generator and oscilloscope to locate a small amount of distortion. Usually distortion occurs in the audio output stages. Distortion can be caused by leaky transistors, ICs, coupling capacitors, a change in resistance, or burned bias resistors.

Weak, intermittent, and noisy sound is more difficult to locate. Weak audio can be caused by open capacitors, a change of resistance, leaky transistors, ICs, or bias resistors. Locate the weak stage with an external audio signal and voltage and resistance tests.

The intermittent or noisy sound symptoms must be located by eliminating each audio stage. A low frying noise can be caused by a leaky transistor or IC. Ceramic bypass capacitors in the front end of the audio circuit can produce a low hissing noise. Signal trace each stage until the intermittent component is located. Just about any component within the audio circuits, including poor PC board connections, can cause intermittent sound.

Ears for a day

When servicing audio circuits, your ears are king of the day. You can hear distorted, weak, or no audio within the speakers. You can hear a popping or frying noise caused by a defective transistor, IC, or ceramic bypass capacitor. Often, the frying noise is a constant low-level noise that can be heard all the time in either or both speakers. A popping noise can result from a defective output transistor or IC component.

A weak stage within the audio circuit can cause unbalanced channels. Often, the volume must be turned up on the other channel to compensate for a weaker channel. If a dual-volume control is used in the audio circuits, the weak stage must be located and repaired. Check for small defective electrolytic coupling capacitors and a change in bias resistors within the weak circuit (Fig. 3-3). The unbalanced channel can be checked with a scope or external amp and test (3 kHz) cassette, within the cassette player.

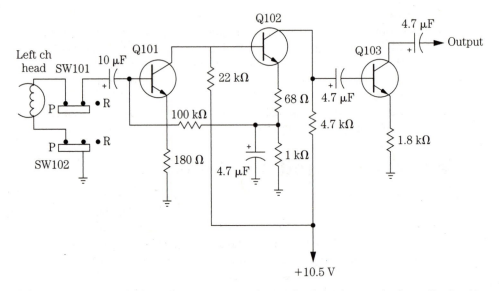

3-3 Use your ears and a few test instruments to locate the defective part in the audio circuits.

Servicing the small phono amp

The audio circuits within small inexpensive amps consist of three or four transistors or a couple of IC components (Fig. 3-4). In the latest phono amps, one large IC might be found. Digital and larger phono turntables are switched to the same stereo amplifier as the cassette deck and the AM-FM-MPX receiver.

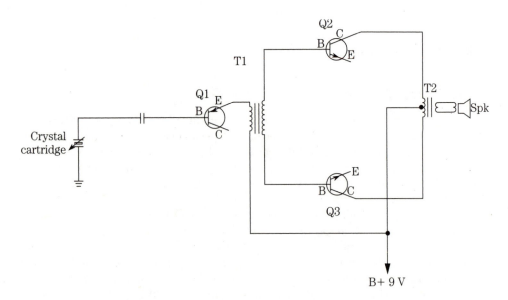

3-4 Only three transistors and coupling transformers are found in the early phono amp.

The small phono amp can operate from batteries or a separate power adapter. Weak batteries produce weak and distorted sound. Likewise, improper voltage applied to the audio circuits can produce the same results. Don't overlook a cracked or old crystal cartridge when weak sound is a problem.

Dead, no sound

Schematics for low-price phono amps are very difficult to locate. If the phono amp is battery operated, check the batteries. Measure the dc voltage applied to the audio output transistors. Usually, the two output transistors are mounted near the driver and audio output transformers in a transistorized circuit. Clip a speaker across the suspected speaker to test it.

If there is still no audio, rotate the volume control wide open and touch either lead removed from the crystal cartridge. You should hear a loud hum. If a hum occurs, suspect a defective cartridge.

Because you don't know which terminal you are taking voltage on, write down each voltage measurement. The highest positive voltage should be the collector terminal on an NPN transistor. Likewise, the highest negative voltage should be the collector terminal of a PNP transistor. The two low-voltage terminals should be the emitter and base terminals. A low resistance measurement from common ground to the transistor terminal indicates the emitter lead.

Measure the voltage between the two unmarked terminals of the lowest voltage. If the measurement is 0.3 V, the transistor is a germanium type with a high negative collector voltage. An NPN transistor will have a 0.6 V between the base and emitter terminals with a higher positive voltage at the collector terminal (Fig. 3-5). Usually the transistor is good if it has these voltage measurements.

Distorted sound

Go directly to the audio output transistors or ICs with distorted sound. You might find only one large IC in the latest phono amplifiers. Place the metal tip of a screwdriver blade on the top of the volume control or crystal cartridge. You should hear a loud hum with distortion. If there is only a hum, suspect a defective crystal cartridge. Substitute another speaker to test the suspected speaker.

Take critical voltage measurements on all IC terminals (Fig. 3-6). Notice if the supply voltage is very low. This voltage should be close to the battery or adapter voltage. Suspect a leaky IC component. The supply terminal should have the highest voltage (3 to 12 V).

Apply an audio signal from the signal generator on all terminals, and notice if distortion can be heard on any of them. If you measure a low voltage at the supply terminal and the output signal is distorted, the IC usually should be replaced.

Weak sound

Inject an audio signal at the volume control, and determine if the sound is normal. Check the audio output coupling capacitors and ICs for weak conditions. Insert an external audio generator signal on both sides of the capacitor. Go from the base to the collector of each transistor with injected signal. Apply external audio at the input and output terminals of the IC for a loss of audio.

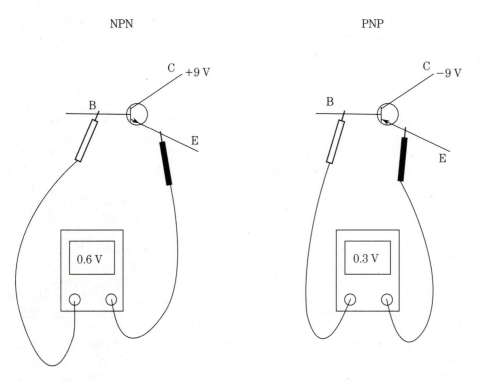

3-5 The PNP transistor has a 0.3-V bias between the emitter and base, while the NPN transistor has 0.6 V.

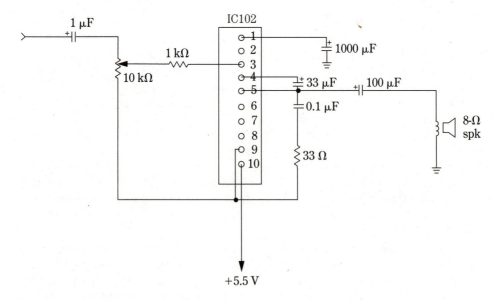

3-6 Take critical voltages upon each terminal of amp IC to locate a defective IC.

Repairing the audio amp

Today's electric clock radio and small portable radio amplifier circuits consist of transistors or ICs. IC components are found in radios with AM-FM-MPX stereo and cassette players. Transistors were found in inexpensive early portable and table model radios.

The typical transistor audio circuits consist of three transistors, with two in push-pull operation and the other as a driver or AF transistor (Fig. 3-7). Often the driver is an NPN type, while the output transistors consist of NPN and PNP transistors. The sound signal can be directly or capacity coupled in the output circuits.

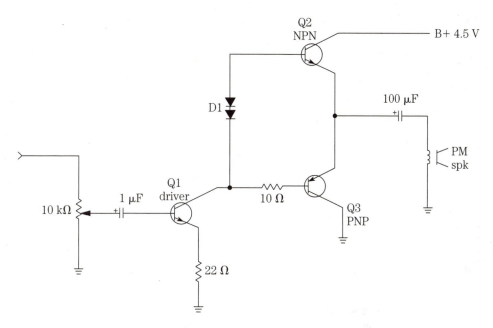

3-7 A driver and two transistors in push-pull operation in the radio amplifier.

Distorted sound

Check the two audio output transistors for leaky conditions within the circuit. Remember, when diodes or low-ohm resistors are found in the base circuits, the measurements might not be accurate. Remove the base terminal for accurate transistor tests. Very low voltage measurements on all transistors can indicate a leaky transistor.

Often both output transistors are found close together and have separate heat sinks. If in doubt, locate the large coupling capacitor to the speaker, then trace the positive end of the capacitor back to the output transistors. Another method is to eliminate the driver transistor by tracing the small coupling capacitor back to the base terminal.

Mushy sound

Suspect the small speaker for mushy or distorted sound. If a headphone jack is available, insert the phones and check for mushy audio. Replace the speaker if the headphones sound normal. Remove one speaker wire, and clip a good speaker across the terminals.

Dead, no sound

Inject an audio signal at the center terminal of the volume control to determine if the audio output circuits are normal. If there is weak or no sound, proceed to the input terminal of the IC. When a schematic is not available, try several different terminals. Trace the PC board wiring from the volume control through the coupling capacitor to the input terminal (Fig. 3-8).

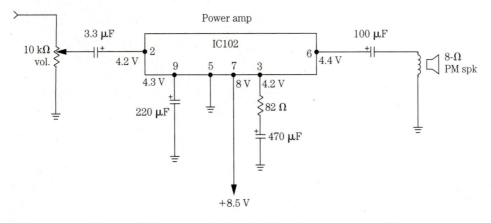

3-8 Suspect a defective IC when there is signal in (2) and none at output (6).

If there is no signal at the output terminal, suspect a defective IC, surrounding components, or improper supply voltage. Take voltage measurements on all terminal pins, and write them down. The highest voltage is usually the supply voltage. Compare this voltage with the total battery or power supply. Measure the dc voltage across the largest filter capacitor and compare.

No audio can be caused by an open speaker. Clip a good speaker across the suspected one. Don't overlook large speaker coupling capacitors for open or intermittent conditions.

Cassette amp repairs

Small cassette players might have three or four transistors, one large IC, or a combination of both. Cassette deck audio circuits might consist of several IC components or a combination of transistors and ICs. For example, the equalizer or preamp tape head stages might consist of two transistors or one IC in a stereo audio circuit.

Surface
mounted
processor or IC

3-9 There might be only one large IC in the small cassette player.

Look for a transistor within the recording bias oscillator circuits. Transistors can be used as AF or driver stages, while the latest stereo cassette decks might have several IC components. The audio output stage can be one large IC for both stereo channels or separate ICs (Fig. 3-9).

Distorted left channel

If the right channel is normal and the left channel is distorted, go directly to the audio output IC. Try to locate the audio stages by looking for power ICs on heat sinks. You might find more than one heat sink, indicating separate audio output circuits. Sometimes both output ICs are mounted on one large metal heat sink or chassis (Fig. 3-10).

You can compare voltage and resistance measurements within the stereo circuits against the defective channel. Often the left channel components are on the left side when looking from the front of the cassette deck. At other times, they are not in line but are placed in a group of ICs, transistors, and decoupling electrolytic capacitors.

After locating the audio components on the PC board, take critical voltage measurements or inject audio signal into the audio circuits. Start at the center terminal of the volume control. Apply external audio signal at each IC pin. When on the good channel, you will hear a loud tone in the speaker. The other IC output component is the distorted one. If you don't have an audio or sweep generator and don't plan to purchase one, build a 1-kHz tone generator with parts found at Radio Shack (Fig. 3-11).

Distortion can be caused by a leaky IC, a change of resistance, or open or leaky bypass capacitors. Check each resistor tied to the power IC terminals and compare

3-10 A large IC might contain both audio amp channels.

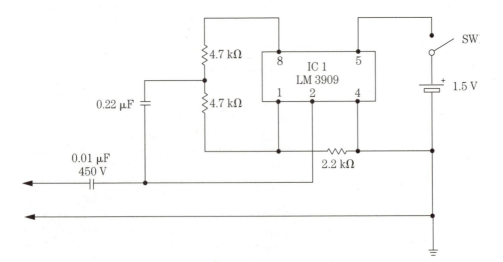

3-11 Build a 1-kHz audio signal generator with one IC and six parts.

them with the normal channel. Shunt each bypass or electrolytic capacitor or IC terminal to check for distortion. Check for high leakage across each capacitor terminal.

Noisy sound

Notice if the sound is present when the volume control is turned down. If not, the noise is created in the front end of the audio circuits. Transistors and IC components can cause a low frying noise in the sound. Locate the noisy components by inserting a signal from a signal tracer or amp to the base terminal and then the collector of each audio transistor. When the noise is heard on the collector terminal and not on the base terminal of the same transistor, you have located the noisy transistor.

Likewise, signal trace the noise at the input and output terminals of ICs to locate a noisy IC. Sometimes applying coolant to the body of an IC can cause the noise to increase or go away. Heat applied from a hair dryer or heat blower might increase the noise. Replacement of the IC is the only answer.

Warbling or squealing noises in playback mode can be caused by a dirty or worn function switch (Fig. 3-12). Spray cleaning fluid into the switch area. Work the function switch back and forth to clean the contacts.

Dual IC amp

In the small audio amp, you might find one large dual IC or two separate ICs upon a heat sink (Fig. 3-13). Today, one IC might be found in the audio stereo out-

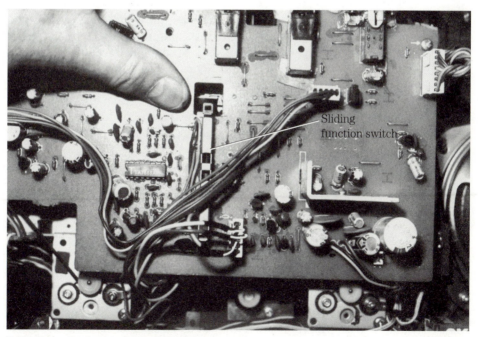

Sliding
function switch

3-12 Clean the function switch by spraying cleaning fluid down inside the switch area for intermittent, noisy, and warbling sound.

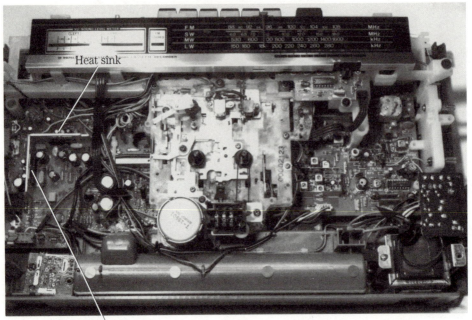

Heat sink

Defective left
channel IC

3-13 A dual IC or separate ICs might amplify the audio output sound.

put circuits. The dual IC contains both input and output circuits. A separate pre-amp IC driver stage might amplify the audio for the output circuits, while, in other dual IC circuits, the AF and audio output circuits are contained in one IC.

Often, the dual volume control rotates the input audio to terminal pins 3 and 13 (Fig. 3-14). A dirty or worn volume control can produce intermittent or a scratching sound in the speaker. Check the small 1-μF electrolytic capacitors for a weak or unbalanced stage. Check all electrolytic capacitors under 10 μF. The audio output pins of IC106 are 7 and 9. Two 1000-μF electrolytic capacitors couple the audio from each channel output to the speaker jacks. Pin 8 supplies 15.5 V to IC106.

A defective dual-power IC can produce weak, distorted, or dead audio or weak audio with distortion in both channels. You might find only one channel distorted or weak while the other channel is normal with a dual-power output IC. Suspect the power output IC when both audio stages are defective. Also, check pin 8 for the voltage supply source. Improper supply voltage can result in weak audio in both stereo channels.

Signal trace with test cassette

The 1- or 3-kHz test cassette can be used to signal trace a defective audio stage in the cassette player. Simply insert the test cassette, and use the scope or external amp to signal trace the weak, distorted, or intermittent audio channel. Start at the volume control, and work either way or at the tape head. Usually, the audio circuits are laid out with the audio transistor or IC pre-amps in line with the correct output circuits.

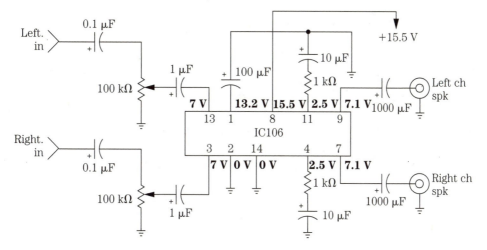

3-14 Dual IC106 contains the left and right stereo audio output channels.

By going from base to base of each audio AF or driver transistor, the volume should increase as you proceed through the audio output circuits (Fig. 3-15). Check on each side of a coupling capacitor for weak audio reception. When the stage becomes weak or distorted, check the proceeding circuits. Locate the base and collector pin terminals and test for audio. When the signal is weak at the collector compared to the base terminal, suspect that transistor or components within the circuit.

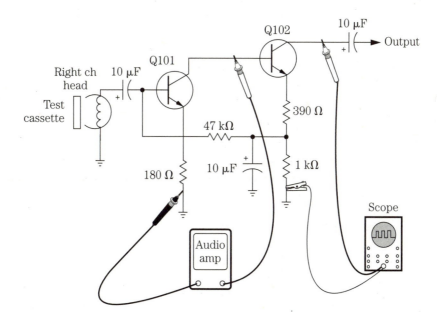

3-15 Signal trace the cassette audio circuits with a test cassette, scope, and external amp.

Very weak sound

The sound was very weak in the left channel of a J.C. Penney cassette player. After cleaning the tape head, the left channel was still weak, while the right channel was normal. Signal injected at the volume control of each channel was normal, indicating the weak stage was in the tape head pre-amp IC.

IC5 was located with signal injected at the right tape head and on terminal 1 of the IC. The sound was weak at coupling capacitor C79, coupling the tape head sound to the AF circuits. Check the audio tone on each pin of the IC of the good channel. Accurate comparison tests of the two channels will indicate the defective stage (Fig. 3-16).

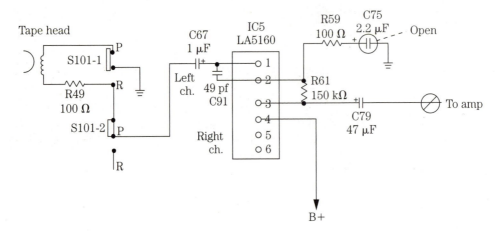

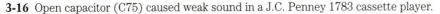

3-16 Open capacitor (C75) caused weak sound in a J.C. Penney 1783 cassette player.

IC5 was suspected of open conditions. Voltage tests on the right and left channels of IC5 were fairly normal. Although very weak audio was found at pin 3, all resistors and capacitors that were tied to the IC were checked. Capacitor C75 was found to be open. It is wise to check all the components tied to the pins of the suspected IC before replacing the IC.

Blown speaker fuse—JC Penney 3223

The speaker fuse might blow or open when dc voltage is found upon the speaker connection, when the transistor output is unbalanced, and when too much volume is applied to the speakers. A speaker fuse is inserted to protect the speakers from voice-coil damage. High volume placed upon the speaker can damage the voice coil and blow it loose from the speaker assembly (Fig. 3-17).

A leaky transistor can upset the balanced output and apply dc voltage directly upon the speaker terminals. The speaker voice coil becomes warm and is frozen against the magnet pole piece. This cannot occur in audio circuits with large electrolytic coupling capacitors. The dc voltage is isolated from the speakers. Some di-

Spk fuse

3-17 The speaker fuse can open with extreme volume, unbalanced amp circuit, and dc voltage applied to the voice coil.

rectly connected amps do not have a speaker fuse or blocking capacitor, resulting in damaged speakers.

Before installing a new speaker, always check the speaker terminals for a low dc voltage. Do not install another speaker, or it can be damaged at once. Turn the volume way down, or insert a 7.5-Ω 10-watt resistor across the speaker terminals to act as a speaker load (Fig. 3-18). Now measure the voltage across the resistor. Zero voltage should be measured across the speaker terminals. If voltage is found, repair the audio output circuits before damaging another speaker.

Loud hum: blown fuse

One of the main fuses was blown in a Pioneer SX-950 amplifier. FU2 (1 amp) fuse was black inside the glass area of the fuse. After replacing the fuse, a loud hum was heard and the chassis was shut down. Most line fuses are caused by leaky audio output transistors and ICs, and one transistor was found leaky.

A quick test of the bridge rectifier circuits indicated normal positive 50.7 V and a high negative −64.5 V (Fig. 3-19). Two voltage regulator transistors were found in the negative power source. Both transistors were tested in the circuit, and a 2SA720 transistor appeared open. The transistor was removed and tested again. After replacing Q1, the negative power source returned to −50.7 V.

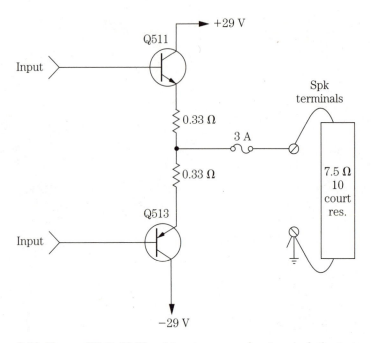

3-18 Place a 7.5-Ω, 10-W resistor across speaker terminals for test purposes.

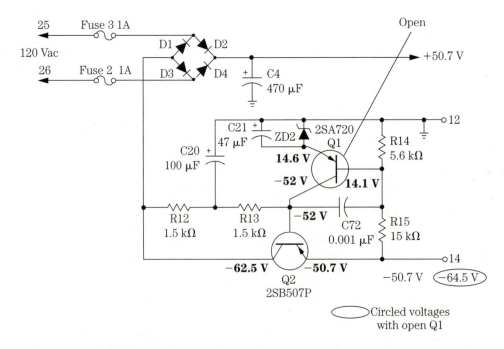

3-19 An open (2SA720) regulator transistor in the negative voltage source increased voltage to –64.5 V.

Car radio problems

Fifteen years ago, transistors and IC components were found in deluxe AM-FM-MPX stereo car radio audio circuits. You might find one large IC as pre-amp for the cassette tape head circuits, pre-amp IC audio, and power output transistors (Fig. 3- 20).

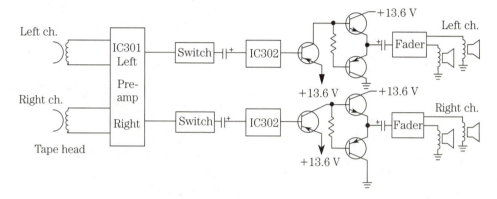

3-20 One large pre-amp IC contains both stereo tape channels.

Today, IC components are found in most audio circuits of the common car radio. Besides IC components, processors and surface-mounted parts are found in car receivers. In fact, several layers of PC boards must be removed before getting to the audio board (Fig. 3-21).

Dead right channel

Locate the dead channel with either an external audio amp or an audio generator. Start at the volume control and work both ways through the circuits. Besides defective transistors, ICs, and capacitors, check for defective speakers and wiring in the dead channel. Often, a low hum can be heard in the speaker with no sound, indicating the amplifier circuits are defective. Don't overlook a dirty function switch, burned voltage-dropping resistors, or leaky coupling capacitors.

Place a cassette in the tape deck and signal trace the audio through the various circuits. Compare each signal with the normal channel. Shorted output IC components will show signs of operating warm and might have burned or stripped PC board wiring going to the supply terminal. An overheated power IC can indicate a leaky IC or burned bias resistors.

Both tape head channels distorted

Suspect a dual pre-amp, leaky IC, tape heads, switching, or common power source when both channels are distorted. If only one large IC is found as the power output in lower-priced amps, take critical voltage measurements on each IC terminal.

Often the IC power components are mounted on a heat sink or metal radio chassis. Sometimes these power ICs have heat sinks bonded right to the component.

3-21 Several layers of PC boards might have to be removed to get at the audio board.

The dual IC amp is located on the PC board. Look for this IC with other audio components. If in doubt, look up the IC part number (on top of the IC) and cross-reference the part number with the semiconductor replacement guide (Fig. 3-22). The guide will indicate the exact replacement and what stage it works in.

Intermittent right channel

Make sure the intermittent right channel is in the speakers or radio. Clip an outside speaker to the intermittent channel. Suspect a poor PC board connection or a cracked board. Sometimes parts will break loose or vibrate from the PC board. Try to isolate the intermittent section within the audio section with signal tracing.

With intermittent sound problems, more time is required to locate the defective component. Sometimes coolant sprayed on the parts within the intermittent channel can make parts act up or quit. Lightly pushing on small components with a pencil or pen can cause the intermittent to occur. Twisting and moving the PC board can isolate the intermittent to a certain section on the board. Don't overlook worn volume controls, poor soldered connections, and cables when intermittent sound is a problem.

Popping and cracking sounds

Suspect the output power IC, module, or transistor when popping or cracking noises occur. Sometimes these power ICs will pop and crack while music is playing and then become intermittent. Locate and signal trace the popping noise with an external speaker amp. Spray coolant on the suspected component and see if it quits.

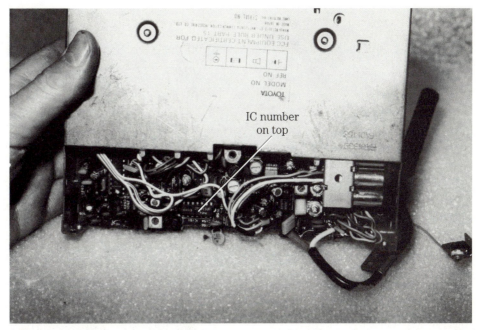

IC number
on top

3-22 Check the part number on top of the IC, and cross-reference in the universal semiconductor guide for the correct replacement.

Very loud whistling and howling from the speaker can be caused by a defective filter capacitor. Shunt a similar capacitor across the suspected one and notice if the noise disappears. Tack the capacitor in with the power off. Howling can result from a red-hot leaky output IC. An annoying hum can be caused by leaky ICs or transistors. A leaky or dried up filter or decoupling electrolytic capacitor can cause a hum in one or both channels.

Unusual dead left channel

The left channel was dead with only a low hum in a Sanyo FTV92 car radio. With a cassette playing, sound was signal traced at the dual volume control (VR1-E). No sound was found at the volume control of the left channel (1) and normal channel (2) (Fig. 3-23). This means that the suspected component is ahead of the output circuits.

Although some signal was heard ahead of R44 (3), why was there no signal at the top or center of the volume control and loud music before R44? A resistance measurement between the center terminal and ground indicated a direct short. VRE-1 was shorted internally and replaced with the exact part number.

Red-hot transistors

Audio output transistors that are too hot to touch can be leaky or shorted. In the car radio, audio output transistors are mounted upon a large heat sink or chassis. Remove the suspected output transistor, and test out of the circuit. Suspect directly coupled transistors of being leaky or open. Test all transistors in the directly coupled

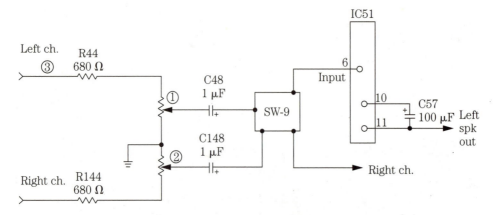

3-23 The unusual dead left channel in a Sanyo FV92 car radio.

circuits. While the transistors are out of the circuit, inspect and measure the resistance of the bias resistor. Often, bias resistors are burned or overheated with leaky output transistors.

Likewise, when an auto receiver with a loud hum and red-hot IC is found, suspect a leaky output IC (Fig. 3-24). Sometimes the PC foil and "A" lead might be overheated and burned with leaky output transistors or ICs. The supply voltage source can be low with leaky output components. Check all resistors and capacitors tied to each IC pin for leakage or a change in resistance.

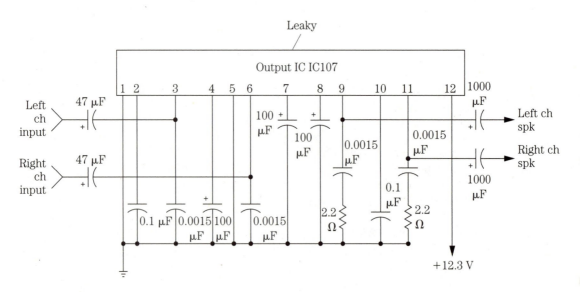

3-24 IC107 operates too warm and was found leaky.

Stereo amp repairs

Usually stereo amps are easy to repair, because one of the channels might be normal and can be used as a reference. The channels are systematically laid out for location of components when servicing without a schematic. Voltage, resistance, and signal tests in the normal channel can be compared to the defective channel (Fig. 3-25).

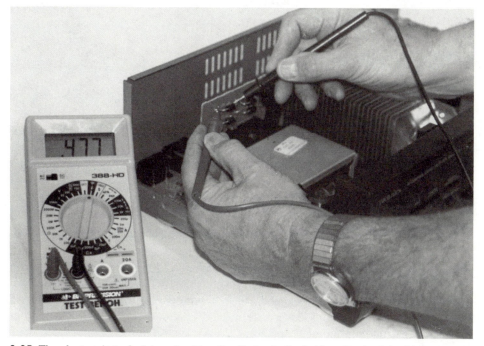

3-25 The electronic technician checking the diodes in the bridge circuit of the high-powered amp.

Higher-wattage amplifiers might have transistors or ICs in the output circuits. These components are located on a large heat sink or metal chassis. Often the bias resistors and decoupling capacitors are found close by (Fig. 3-26). The AF, driver, and pre-amp stages are found on a separate board or are mounted ahead of the power output components.

No left channel

Signal trace the audio at the volume control. If the signal is normal on both channels, proceed to the output transistors. Inject an external audio signal into the base terminal of each output transistor or the input terminal of the IC, and listen for a tone in the speaker. The audio signal on the output transistor will be low when normal. The audio should be high at the input terminal of the power IC when normal.

Proceed toward the driver and pre-amp stages with the signal injection methods. Likewise, audio signal can be checked with a cassette tape or with the unit

3-26 A large IC mounted upon a heavy heat sink combines both audio channels.

turned to radio in the AM-FM-MPX receiver. Use an external test amp for the latter method of signal tracing.

Check the left channel speaker for open connections. An open driver or AF transistors and IC pre-amps can cause a dead channel. Don't overlook open or dried-up electrolytic coupling capacitors. Leaky decoupling capacitors can open voltage supply resistors with no or improper voltage at the audio circuits. Often, leaky or shorted output transistors and IC components will have weak and distorted sound.

Both channels distorted

When both channels are distorted in a stereo amp, look for something in common with both channels. A dual stereo power input or output IC can become leaky and cause distortion in both channels. Improper voltage from the power supply circuits can cause distortion. Either negative or positive voltage sources can cause distortion in both if one voltage is missing. Dirty or corroded function switches can apply incorrect voltage to the audio section, causing distortion.

Distorted right channel

Go directly to the audio output transistors or ICs for extreme distortion in the audio (Fig. 3-27). Locate the distortion in the right channel by checking each output transistor with a signal tracer. Keep the signal tracer amp volume as low as possible. Likewise check the input terminals of each power IC for distortion.

Power transistors

Heat sink

3-27 Check for leaky and open output transistors for weak and distorted sound.

Distortion in the audio output components is often caused by leaky transistors or ICs, burned bias resistors, leaky bias diodes, and improper voltages. Look for a change in the bias resistors or blown electrolytic or coupling capacitors. Small, corroded screwdriver-adjusted bias resistors can cause distortion.

A leaky power output IC can cause distortion in a given channel. Burned bias resistors or leaky driver and Darlington transistors directly coupled to the output IC can cause distortion. Signal trace the audio through each stage until the distortion clears up, then take voltage and resistance measurements. Always check bias resistors when a leaky IC or transistor has been removed from the circuit.

Keeps blowing fuses

Blown speaker fuses can be caused by too much power applied to a set of speakers. A defective or grounded voice coil can open the fuse. The speaker fuse might keep blowing if a leaky component in a directly coupled amplifier is applying raw dc voltage to the speaker fuse (Fig. 3-28). Always check for dc voltage at the speaker terminals before connecting another speaker; otherwise, the speaker and fuse might be damaged.

Keeps damaging the power output IC

After a new power IC (STK-0050) was installed in a large Pioneer amplifier, the IC became warm and extreme distortion was noted before the replacement was destroyed (Fig. 3-29). DC voltage was found on the right speaker terminals. R263, a

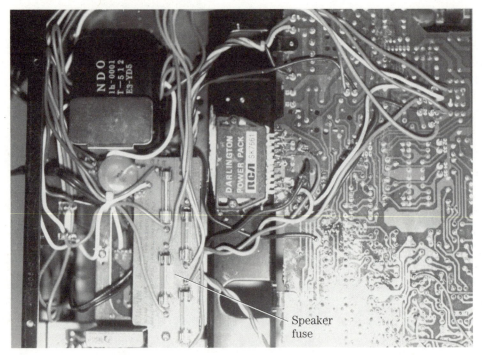

3-28 Check for dc voltage upon speaker terminals after locating a defective voice coil.

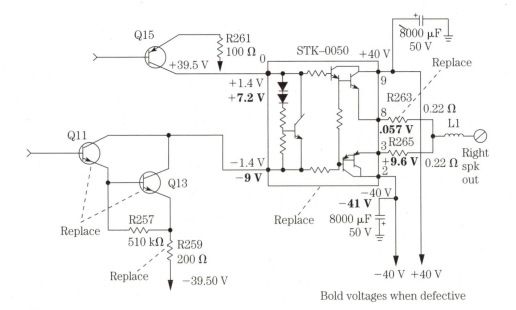

3-29 Leaky driver transistors keep destroying the large output IC in a Pioneer amplifier.

0.22- Ω resistor, was found burned and was replaced. When the right channel IC was defective, the circled voltages (see Fig. 3-29) were found on the terminal pins.

All driver transistors connected to pins 0 and 1 were tested in the circuit. Q11 and Q13 indicated leakage. Both transistors were removed and tested. Because Q13 was directly shorted, both transistors were replaced. At the same time bias resistors R257 (510 kΩ) and R259 (200 Ω) were checked. R259 had changed resistance and was replaced. Zero voltage was found on the speaker load resistor (10 Ω, 10 W), indicating normal sound, after replacing the power IC (STK-0050) with an original replacement, Q11, and Q13 and resistors R259 and R263.

Troubleshooting CD player audio circuits

Signal trace the audio from the output of the digital-to-analog (D/A) converter IC through the sample hold IC and audio amp to the line output jack with an external audio amp. The stereo audio circuits begin at the output terminals of the D/A converter stage. Usually the audio amp in both channels is found in one IC component. Determine if the trouble occurs in the CD player or the sound amp connected to the line output jacks (Fig. 3-30).

If one channel is normal and the other is dead or distorted, use the good channel for voltage and reference measurements. Check the sound at points 1 through 4 (see Fig. 3-30) to locate the defective component. Check the input and output signal of each IC. When the signal becomes weak or distorted in the external amp, you have located the defective stage.

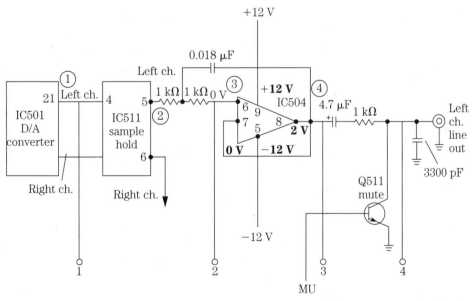

Bold voltages when defective

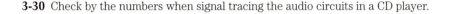

3-30 Check by the numbers when signal tracing the audio circuits in a CD player.

Distortion can be caused by one positive or negative voltage missing from each IC. Start at the audio line output jack, and locate the audio amp IC. Backtrack signal tests until you hear sound in the external speaker. Don't overlook a defective mute switch transistor. Usually the emitter terminal of the mute transistor is at ground potential. Remove the emitter terminal, and notice if the audio is now present at the line output terminal (Fig. 3-31). Take voltage measurements on the stereo amp, and compare both channels.

3-31 Suspect mute transistors with no sound on the line output jacks.

Distorted right channel: headphones

In a Realistic 42-5029 compact disc player, the right channel was distorted with weak sound. The headphones were replaced with another pair, and indications were the same. The right headphone circuit was traced from the headphone jack back to a small 10-Ω resistor and a 220-µF coupling capacitor (Fig. 3-32). The electrolytic coupling capacitor was found connected to pin 10 of a dual output IC. The voltage supply pin measured 1.7 V.

New batteries were installed, and the results were the same. Audio was signal traced into pin 3 of IC1, going to the volume control with distorted output at pin 10. The volume was up and normal in the left channel. IC1 was replaced, restoring normal CD headphone reception.

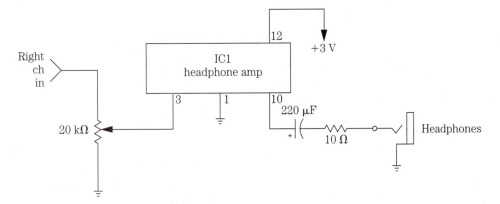

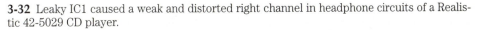

3-32 Leaky IC1 caused a weak and distorted right channel in headphone circuits of a Realistic 42-5029 CD player.

Servicing the TV audio amp

Sound symptoms in the audio output stages can include no, intermittent, weak, muffled, distorted, and noisy sound. Try to locate the sound discriminator coil and audio output IC. Often the IC is located close to the shielded discriminator coil assembly (Fig. 3-33). Several audio amp ICs in the TV chassis have their own heat sink on the PC board. Another method is to signal trace the speaker wires back to the audio output circuits.

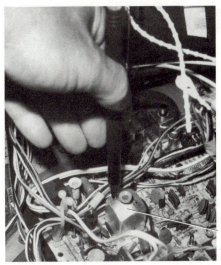

3-33
Locate the shielded coil of the discriminator, and adjust for clean audio.

Discriminator
shielded audio
coil

Distorted sound

After locating the discriminator coil, lightly touch up the core until the sound is clear. Sometimes weather changes or extremely wet weather can cause the coil frequency to change. If the sound needs to be adjusted again, and the coil is off-frequency, suspect a small ceramic bypass capacitor within the coil circuit.

Extreme distortion can be caused by a leaky audio output IC or components. Take critical voltage measurements on each IC terminal. Check each component that ties to ground. Look for open electrolytic capacitors or resistors in the IC circuits. Check for leaky audio output transistors and bias resistors that could cause distortion.

Hum in the sound

Check for poor ground connections within the pre-amp audio stages. A low hum can be caused by dried-up decoupling capacitors and a loud hum by open main filter capacitors. Check for a defective IC or audio output transistor when a low hum occurs.

In a General Electric AC-B chassis, a low hum was heard with some distortion. The hum noise acted like a pickup hum in a microphone. Finally R175 was located as the cause of the low-hum symptom off of pin 15 of the sound IC (Fig. 3-34).

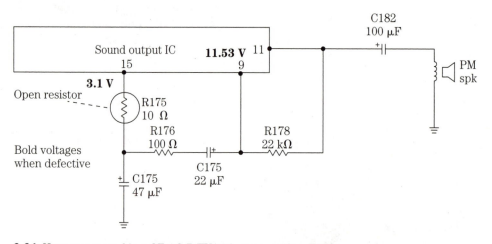

3-34 Hum was caused in a GE AC-B TV with open resistor R175.

Intermittent and weak sound

Weak and intermittent sound can be caused by defective transistors, ICs, and open electrolytic coupling capacitors. Signal trace the in and out of each component. Replace the suspected IC or output transistor if the sound coming in is normal but weak sound is found at the output terminal. Don't overlook a defective speaker-coupling capacitor for weak and intermittent sound (Fig. 3-35). Clip in another speaker to determine if the speaker is normal.

Popping noises

Look for defective audio output ICs or transistors when popping and snapping noises occur. Sometimes the popping noise occurs after the TV set has warmed up.

Discriminator
disc coil

Audio
IC

Electrolytic
coupling
capacitor

3-35 Intermittent, weak, and no sound can be caused by a defective electrolytic coupling speaker capacitor.

Apply coolant on the suspected component to see if the noise disappears. Signal trace the noise at the input IC terminal or the base of the transistor with the external amp connected. If the noise is traced to the output component, replace it.

No audio

Because the Goldstar CMT2612 set had stereo sound, several different stages of audio must be found upon either the main chassis or separate PC boards. The left channel was dead with a normal right audio channel. Because most sound problems occur in the audio output circuits, the speaker leads were traced back to P903 on the TV chassis. Nearby was a large IC mounted upon a large heat sink. At the other end, another power transistor was mounted.

The audio was signal traced with the scope, on input pin 2 and output pin 12 of the right channel. Pin 7 of the left channel was connected to a 470-µF electrolytic capacitor. Audio appeared normal at pin 7, which no doubt was the left output pin terminal. A scope test on both sides of the electrolytic capacitor was normal and found no sound at the speaker. The speaker wires were traced to P903 and to an external-internal switch. When the ohmmeter lead was touched upon the switch, the left audio channel was normal. A poorly soldered connection upon the switch terminal caused the left audio channel to go dead (Fig. 3-36).

Audio distorted

The sound was distorted and weak in a Sharp portable TV. The speaker wires were traced back to a speaker-matching transformer from two small audio transis-

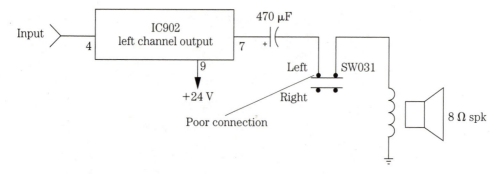

3-36 A poor contact on SW031 resulted in no left channel audio in a Goldstar CMT-2612.

tors in a push-pull circuit. Next, four small transistors were located toward the transformer. Besides an output transformer, high voltage was found upon the audio output transistors.

The base emitter and collector voltages were high in both transistors, indicating a leaky directly coupled transistor. Both transistors were removed, and the NPN transistor indicated leakage (Fig. 3-37). While out of the circuit, a driver transistor tested normal. Although the PNP transistor tested good, it was replaced with an ECG32 universal replacement, and the NPN was replaced with an ECG31. Remember, these two transistors operate in a higher-than-normal voltage output audio circuit.

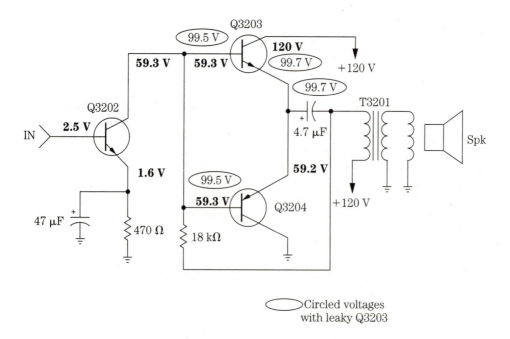

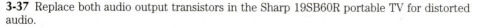
Circled voltages
with leaky Q3203

3-37 Replace both audio output transistors in the Sharp 19SB60R portable TV for distorted audio.

Servicing high-power amps

The very high-power amps (100 to 1000 W) contain flat gold contact areas for maximum current handling conditions. Total amperage with individual amps can exceed 1750 A. Each stereo section is lined up with gold-plated, solid-copper bar strips down the center. The power output ICs and transistors might have extra heavy heat sinks or outside chassis heat sinks. Each stereo or mono amp might have its own power supply, removing channel-to-channel interaction and crosstalk.

High-speed switching FETs supply input power to the power toroids, keeping RF from entering or exiting the power amp supply. Fast-switching diodes might be found in some units. High-powered Darlington and triple Darlington output devices are found in some output stages.

Usually high-powered output stages are in line and easy to locate. Audio signal tracing methods can locate most defective components. Check for burned boards, flat bar connections, and burned toroids or resistors that might indicate leaky components. Compare the good channel with the defective one. Separate high-powered mono amps can be checked against a normal one.

Transistor and diode tests within the circuit can locate an open or leaky component. Actual voltage tests on power ICs can determine if an IC is leaky. Compare the voltage and resistance tests of a suspected component against the good channel. Substitution of high-powered ICs and Darlington transistors might be necessary after signal tracing the audio to the defective component.

Always replace high-powered amp transistors, ICs, Darlington transistors, FETs, and toroids with the exact part number. Remember, these are special parts.

Power output resistance measurements

After testing diodes and transistors in the circuit, take critical resistance measurements on the output ICs and transistors to common ground. This test can help locate defective transistors, FETs, and IC components that break down under load. Connect a speaker bank-load of resistors across the defective speaker terminals, especially in directly coupled circuits. The output voltage at the speaker terminals should be zero in directly driven circuits.

When the directly driven circuits have dc voltage at the speaker terminals, suspect a leaky or shorted component in the audio output stages. With a power load connected to the speaker terminals, the resistors might run warm but will still give the technician time to make some tests. This will prevent voice-coil speaker damage. Just connect the high-wattage resistance load to the speaker terminals.

Resistance measurements within the audio output circuits can uncover the defective component, especially when the unit keeps knocking out fuses or destroying output ICs or transistors. Make resistance tests of each transistor or IC terminal to common ground. Compare each measurement with the good channel.

The VTVM or VOM works best in resistance tests because electrolytic capacitors will charge and discharge. It takes time for the numbers to count down or add up with a DMM. Of course, the DMM is far more accurate in these tests. Write down each resistance measurement for comparison.

When you find a big difference in resistance measurements, the defect is in that circuit. Often the power ICs or transistors have been replaced and the trouble still exists. Sometimes voltage comparison tests might not show the defective component, but accurate resistance measurements within the circuit can indicate problems.

Audio amp troubleshooting chart

Check the power amp troubleshooting chart shown in Table 3-1.

Table 3-1. Power amp troubleshooting chart.

Trouble	Remedy
Location—Output	Locate the power output transistors and ICs on the large heat sinks.
Location—Input	Check for the pre-amp transistors or IC connected to the tape heads or phono input. Check for a driver IC or transistor connected to the volume control.
Signal trace	Start at the volume control and work toward the front-end or speaker.
Signal trace with test cassette	Insert cassette—signal trace at the volume control and output and input stages.
Signal trace with signal generator	Trace the 1 kHz signal from the input to the output speaker terminals. Go from base to base of each transistor and input to output terminal of the IC.
Locate the defective stage	When the signal stops, suspect the preceding stage.
Transistor tests	Check each transistor and IC where the signal stops. Test each transistor in and out of circuit.
IC tests	Signal trace the signal at input. If there is no output, take critical voltage measurements upon each terminal. Take critical resistance measurements on each terminal.
Weak audio	Suspect defective coupling capacitors and bias resistors. Check for leaky or open transistors.
Distorted audio	Go directly to the audio output circuits. Check the transistors for open or leaky conditions. Check the IC voltages and resistance measurements. Suspect leaky coupling capacitors. Do not overlook an improper voltage source.
Weak and distorted sound	Check the audio output transistors and IC. Suspect bias resistors and bypass capacitors.
Intermittent sound	Suspect transistors, ICs, poor terminal connections, poor board connections, and cracked boards. Do not overlook the electrolytic coupling capacitors.
Too hot to touch	Suspect transistors and leaky output ICs. Test the transistors for leaky conditions. Look for burned components. Measure all bias resistors for a change in value.

4
CHAPTER

Servicing auto receivers

Drastic changes have been made in the car radio since tubes and vibrators were king. Today the car radio has many new features, including digital varactor and quartz tuning, AM and FM preset stations, Dolby B and C, auto-reverse cassette player, CD changer, and high-powered output. The radio might have a CD changer control with the changer located in the trunk.

The car radio used to have tubes, then transistors and a few IC components (Fig. 4-1). Now they contain ICs and processors in the front-end section. These radios might have boards with miniature parts and surface-mounted components. Sometimes several boards must be removed to get at the audio and power circuits.

Required test instruments

The following equipment is needed to test car radio circuits:
- DMM or voltmeter
- Transistor tester
- RF-IF signal generator
- Sine- and square-wave generator
- Oscilloscope
- External amplifier
- Frequency meter (optional)
- Audio analyzer (optional)

Front and rear section

The audio radio can be divided into two separate sections for troubleshooting. The rear section consists of tape player, audio, and power circuits from the volume control to the speaker. Start at the top of the control, and move toward the antenna plug for the front section. Today's car receiver might have a separate amplifier, crossover networks, and several connected stereo speakers (Fig. 4-2).

4-1 The latest cassette-radio receivers have transistors and IC components.

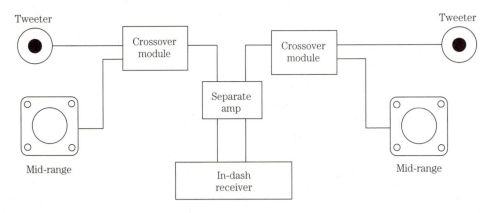

4-2 A block diagram with a high-powered radio amp and speakers.

Start at the volume control center terminal with the external amp or audio signal generator to signal trace the audio circuits. The sine- or square-wave generator can determine if distortion is found in either stereo audio channel. Connect the signal generator to the RF or IF section to signal trace the front end. Injecting signal from a white-noise generator can trace signals in both the RF and audio circuits (Fig. 4-3).

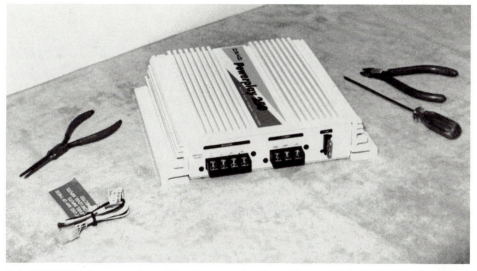

4-3 A 250-W high-powered amplifier to connect to the radio receiver, cassette, and CD player.

Varactor tuners

The varactor tuner is used to select broadcast stations. When the tuning system supplies voltage to the varactor diode, a different station is tuned in. Just by checking the different voltages on the varactor diode, you can determine if the tuning system or the RF, oscillator, and mixer stages contain the defective parts (Fig. 4-4).

In the Audiovox AV-215 with an FM station tuned in at 92.1 MHz, the control voltage was 2.1 V. This voltage increases as stations are tuned higher in the FM band. At 88 MHz, the voltage was 1.5 V, and at 108 MHz, the voltage was 5.5 V. In the AM band, at 530 kHz, the voltage was 0.99 V, and at 1400 kHz, the tuning voltage was 5.83 V. This voltage will vary in different car radios.

When no tuning voltage is found at the varactor diodes or controlled transistors, check the CPU PLL processor. Measure the supply voltage (V_{CC}) on the supply pin. If voltage is supplied to the varactor diodes without any stations tuned in, suspect problems within the front-end circuits instead of the CPU-quartz or digital synthesized circuits.

There she blows

If the car radio is dead, check the fuse. If the "A" lead to the fuse holder and receiver are burned, suspect shorted audio output transistors. The audio output transistor might be located upon a separate heat sink or metal receiver framework. The output transistors in the high-powered amplifier are located on heat sinks with transistors bolted to the large metal flange-type body (Fig. 4-5). Check for a piece of mica insulation between the transistor and the heat sink.

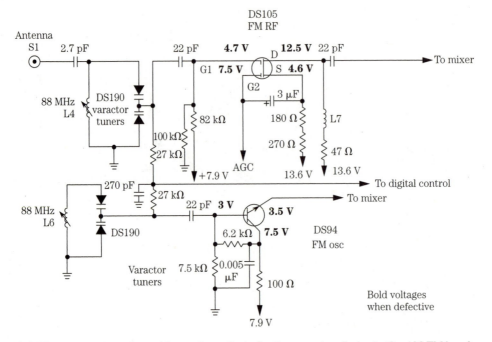

4-4 The system control provides tuning voltage for the varactor diodes in the AM-FM bands.

Locate the leaky transistor with in-circuit transistor tests. Remove the suspected transistor, and check it out of circuit. While the output transistors are out of the circuit, check the bias resistors and diodes for leaky conditions. Double-check the directly coupled driver transistor for open or leakage conditions.

Too-hot-to-handle audio output IC components can cause the fuse to blow. When an oversize fuse has been inserted, the leaky output IC might burn the connecting foil of PC wiring and "A" lead harness wires. Often, the too-hot output IC might have gray or white burned marks upon the body area. Check the IC circuits for a change in resistance or leaky bypass capacitor tied to the output IC terminals (Fig. 4-6).

Keeps blowing the fuse

When the fuse blows right after replacement, suspect a grounded feed-through capacitor (C213) or polarity diode (Fig. 4-7). Sometimes these feedthrough units break down and short the battery voltage against the metal radio case. Check for a leaky polarity diode (D217) connected to the power voltage source. This diode is installed to prevent wrong polarity "A" voltage from damaging solid-state devices in the auto receiver. Sometimes the auto battery will get low and charged up backwards and can damage semiconductors.

Check for leaky or shorted output IC or transistors. Leaky high-powered transistors can keep blowing the "A" line fuse. Remove the large heavy metal case from the high-powered amp, and be careful if the power output transistors are bolted to

4-5 Check for leaky output transistors that can blow the line fuse.

the metal case (Fig. 4-8). Check each transistor in circuit for leakage. Test all output transistors and ICs when the fuse will not hold right after replacement.

Common man

Most failures within the auto receiver are common to other units. The most common failures are the audio output transistors and ICs. Worn or scratchy volume controls must be replaced after several years of wear and tear. Distorted, intermittent, and weak reception often occurs in the audio circuits. Fuse failure is caused by a leaky audio output component. The intermittent cassette audio is caused by leads torn or worn off of the tape head connections.

Mechanical problems within the auto-cassette player are speed, play, and eject problems. Speed problems result from worn or loose motor belts (Fig. 4-9). Hold the capstan wheel, and if the motor stops rotating, the belt is good. Replace the drive belt when the motor pulley rotates inside the drive belt. A grinding noise can result from the capstan-flywheel dragging against the bottom brace area. Install a new flywheel, or remove and tap the flywheel up into the correct position. Apply epoxy ce-

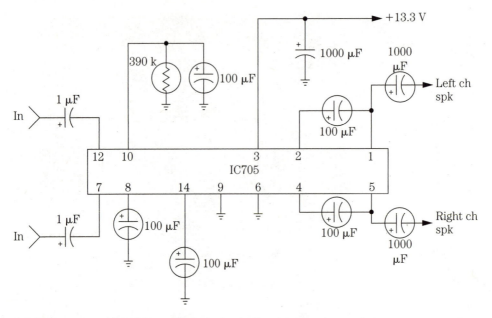

4-6 If the power output IC runs too hot, check all resistors and capacitors.

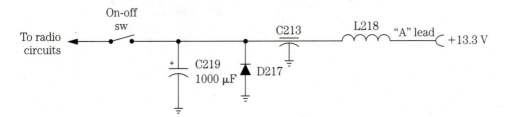

4-7 If the fuse keeps blowing, suspect a leaky polarity diode (D217) and feedthrough capacitor (C213).

ment around the bottom capstan shaft and the body of the flywheel. However, clean up the area with alcohol and a rag before applying the cement.

Always clean up the moving components in the cassette player after making repairs. Clean up all parts with alcohol and a cloth or with cleaning fluid. Clean the motor pulley, belt, capstan, pressure rollers, and reels (Fig. 4-10). Check the wiping tongs or magnetic switch mounted upon the supply reel when the player keeps ejecting the cassette. Clean off the rotating switch-eject surface and each tong to ensure a clean switching contact.

No AM

If the FM is normal, but there is no AM reception, the defective part must be in the AM front-end circuits because both AM and FM use the same IF circuits (Fig. 4-

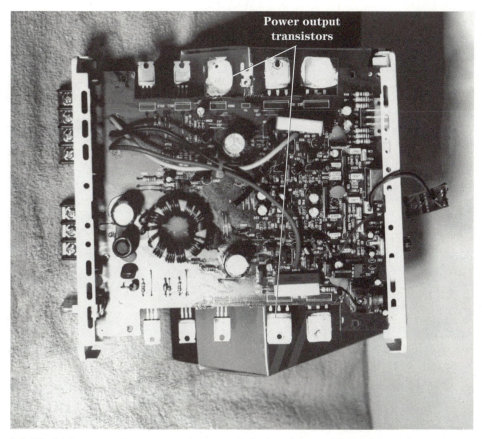

4-8 The high-powered output transistors are bolted to the metal flange heat sink.

11). Locate the RF transistors or ICs near the ferrite antenna coil. The coil might be mounted on the board close to the RF transistor. Trace out the RF and oscillator circuits near the main tuning capacitor. Look for varactor diodes near each coil in the digital or quartz tuning.

When the RF stage is not functioning, placing the positive probe on the collector terminal of the RF transistor might bring in a local broadcast station. This means that both the mixer and oscillator circuits are normal. Check the voltages of the transistors or IC components in the mixer and oscillator circuits to determine if the circuits are functioning. You might find that both AM and FM operate from the same component. Clean the AM-FM switch with cleaning spray.

The FM circuits consist of an RF, oscillator, and mixer stage like the AM section, except at a much higher frequency (88 to 108 MHz). Most of the problems that occur in the FM stages are caused by leaky transistors, burned bias resistors, and improper applied voltage. Check all three transistors within the FM circuits (Fig. 4-12). Take critical voltage and resistance measurements.

If the AM section is normal, the FM problem must be in either the RF, oscillator, or mixer stages. Locate the RF FM transistor or IC from the antenna switch and

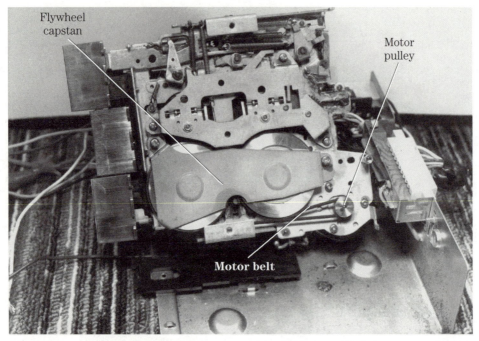

Flywheel
capstan

Motor
pulley

Motor belt

4-9 For slow-speed problems, check the motor drive belt to see if it is loose, stretched, or worn or if there is oil on the belt.

4-10 For slippage, clean up the loading and take-up reels with alcohol and cloth.

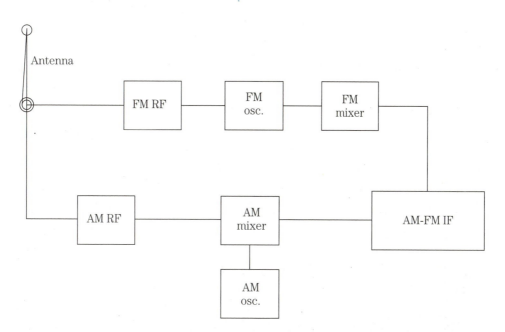

4-11 A block diagram of the early AM-FM RF, oscillator, and mixer circuits.

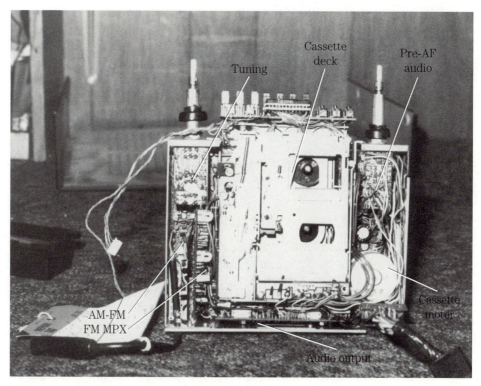

4-12 Locate the various sections of the radio before tests are made.

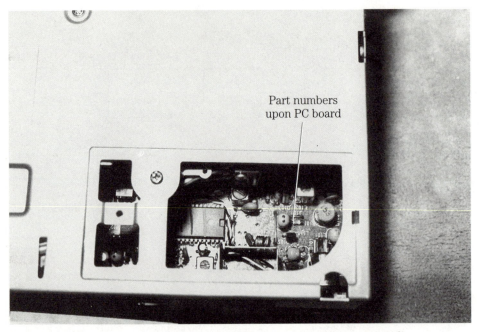

Part numbers
upon PC board

4-13 Notice the parts are mounted upon the PC board in the Audiovox car radio.

small coils of wire. Notice that the FM coils are only a few turns of bare wire, while the AM coils have many turns on a form or rod. Parts in the front end can be located with markings on the PC board (Fig. 4-13).

A loud FM rush with the volume turned high can indicate a defective mixer or oscillator circuit. Sometimes a local station can be tuned in with a meter probe on the collector terminal of the RF FM transistor. Don't overlook a defective AM-FM switch.

After locating a defective transistor, replace it with a universal replacement. Universal replacement transistors will function nicely in AM-FM and other car radio circuits. Cut the replacement terminal leads the same length, and mount the transistors in the very same place to prevent FM alignment.

Weak reception

Go directly to the antenna and RF circuits when the signal is weak for either AM or FM reception. Locate the RF transistor tied to the coils and the AM-FM function switch. Take critical voltage measurements. If a schematic is not available or you cannot find a transistor number, use the operating voltage and find it in the manual of universal replacement transistors.

Determine if the transistor is an NPN or PNP type. If the radio is fairly new, you can assume it is an NPN transistor. With a positive voltage measurement on the collector terminal and the emitter tied to ground, the transistor is an NPN type.

Improper supply voltage to either the AM or FM circuits can produce weak reception. If the supply voltage is low, visually trace the voltage back to the isolation resistors and zener diode circuit. Usually the voltage is regulated by a zener diode or transistor. A defective regulator prevents a normal power source (Fig. 4-14).

Don't overlook a dirty distant switch with weak FM reception. Besides checking for a weak AM-FM transistor, double-check the ferrite rod for breaks. Erratic AM and FM reception can be caused by a dirty AM-FM function switch.

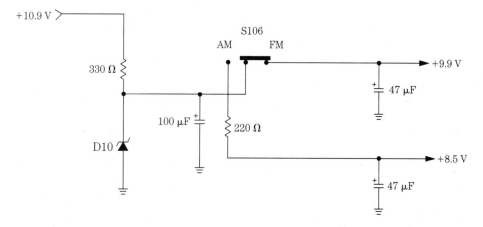

4-14 Leaky zener diode D10 caused weak AM-FM reception.

Intermittent front-end reception

Try to determine if the intermittent is in the audio or front end of the radio by clipping an external amp to the top of the volume control. Turn the receiver volume down. Notice if the receiver is cutting up or down with signal from the front end of the radio. Isolate the front end to determine what stage the intermittent is in. Likewise, isolate the audio section.

If the AM and FM reception is intermittent and the cassette player is normal, you know the trouble must exist from the function switch to the IF section common to both bands. Usually when the AM, FM, and cassette player are intermittent, the trouble lies in the audio AF to the speaker. With both AM and FM stages intermittent, either the IF circuits or the supply voltage contains the defective component (Fig. 4-15). Don't overlook broken or poor outside antenna connections.

Poorly soldered connections, cracked boards, bad transistors or ICs, poorly soldered terminal leads or eyelet connections, poor antenna connections, a defective antenna jack, or intermittent voltage applied to the front-end circuits can cause intermittent problems. Suspect a dirty or worn volume control with intermittent reception.

Surface-mounted components (SMD)

The latest cassette/CD player might have both conventional and surface-mounted components. The surface-mounted parts can be mounted on one side of

IF
section

Output
IC

Volume control
board

4-15 Locate the IF transformers upon the chassis, and check the supply voltage to these circuits with a DMM.

the board, with regular parts on the other. Several different boards are found within the cassette-tuner board (Fig. 4-16). Notice the very fine PC wiring that connects the surface-mounted processor to the circuit.

Surface-mounted transistors and IC components are used throughout the circuits. Besides surface-mounted ICs and transistors, diodes, capacitors, and resistors are used in many circuits. Notice that diodes can have two active tabs, and three-legged transistors can have different lead identifications (Fig. 4-17). Power output ICs should be bolted to heavy heat sinks.

Digital transistors can be found in the audio pre-amp and the mechanism and system control. The digital transistor can have an internal resistor in the base circuit or another bias resistor between the base and emitter terminal (Fig. 4-18). When these digital resistors are checked with a transistor tester or the diode test of the DMM, the resistance is higher in the base terminal. Likewise, the measurement from base to emitter, with the internal base-to-emitter resistor, is different. Compare another similar digital transistor with the low base-to-resistor test before discarding the suspected leaky transistor.

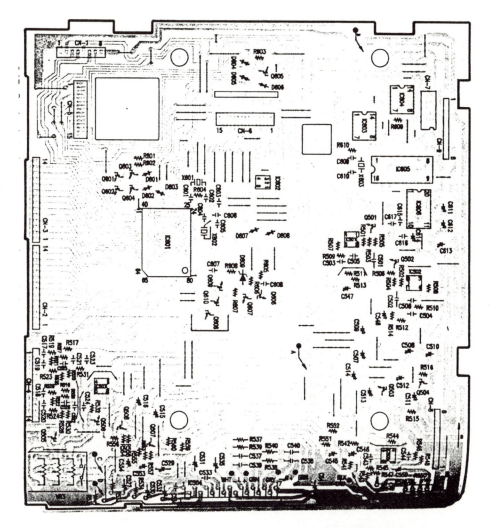

4-16 SMD components are found mounted upon a trace PC board layout in a Radio Shack auto radio. (Radio Shack)

Weak audio

Weak sound can be caused by any component in the audio circuits. Try to isolate the weak section to the left or right channel. If both channels are weak, suspect an improper supply voltage on a common audio output IC. Weak audio can be caused by an open ground within the volume control. If sound is coming into the top and not at the center terminal, suspect a defective control (Fig. 4-19).

Most weak audio conditions are caused by transistors, ICs, and coupling or by-pass capacitors. If the left side is weak but the right is not, isolate the weak stage with an external amp. Go from base to base on transistors or IC input and output ter-

IC AND TRANSISTOR LEAD IDENTIFICATION

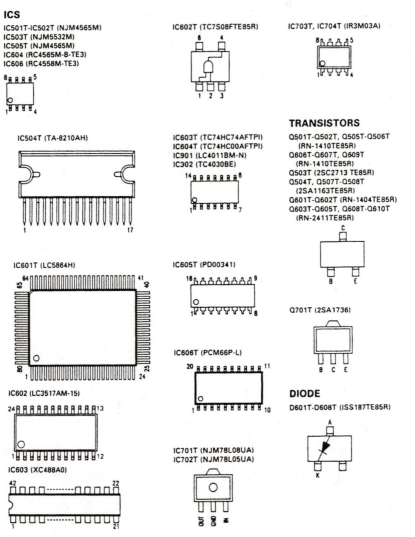

ICS

IC501T-IC502T (NJM4565M)
IC503T (NJM5532M)
IC505T (NJM4565M)
IC604 (RC4565M-B-TE3)
IC606 (RC4558M-TE3)

IC504T (TA-8210AH)

IC601T (LC5864H)

IC602 (LC3517AM-15)

IC603 (XC488A0)

IC602T (TC7S08FTE85R)

IC603T (TC74HC74AFTPI)
IC604T (TC74HC00AFTPI)
IC901 (LC4011BM-N)
IC302 (TC4030BE)

IC605T (PD00341)

IC606T (PCM66P-L)

IC701T (NJM78L08UA)
IC702T (NJM78L05UA)

IC703T, IC704T (IR3M03A)

TRANSISTORS

Q501T-Q502T, Q505T-Q506T
 (RN-1410TE85R)
Q606T-Q607T, Q609T
 (RN-1410TE85R)
Q503T (2SC2713 TE85R)
Q504T, Q507T-Q508T
 (2SA1163TE85R)
Q601T-Q602T (RN-1404TE85R)
Q603T-Q605T, Q608T-Q610T
 (RN-2411TE85R)

Q701T (2SA1736)

DIODE

D601T-D608T (ISS187TE85R)

4-17 Notice the different terminal layouts upon transistors, resistors, diodes, and IC components. (Radio Shack)

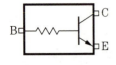

4-18
Digital transistors have resistors in the base and emitter terminals inside the SMD part.

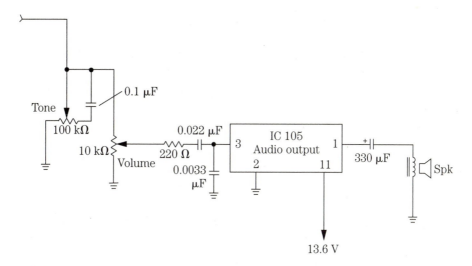

4-19 Check the volume with an external amp at the top and center of the control for defective volume control.

minals to determine where the signal is lost. Take critical voltage and resistance measurements when a suspected component is located.

Go directly to the audio output transistors or ICs with a weak or distorted signal. Check if the transistor or IC is running excessively warm. In push-pull transistor output circuits, you might find one transistor open and the other leaky. Open or leaky driver transistors can cause weak and distorted music. Suspect a dried-up electrolytic capacitor, especially those of 1 μF or 3.3 μF, for a loss in capacity.

Distorted sound

Distortion in the audio circuits can occur from the AF transistor through the audio output circuits. An open AF or driver transistor can cause weak reception and distortion in the speaker. Cracking noises and distortion can result from a defective output IC or transistor. If, after the chassis runs for a few minutes, the music becomes mushy, suspect an overheated transistor or output IC component (Fig. 4-20).

When the left channel is distorted and the right channel is normal, signal trace the left channel. Check for distortion at the volume control. Signal trace the rest of the audio circuit with an external amp to locate the distorted stage. Injection of a sine- or square-wave signal will indicate which stage is defective, with the scope as indicator. Clipped sine waves or rounded square waves indicate distortion.

Improper voltage found on the transistor or IC can be caused by a leaky zener diode or transistor regulator. Don't overlook shorted or leaky bypass capacitors and leaky decoupling electrolytic capacitors for insufficient voltage that can produce distortion. Overheated or red-hot output ICs definitely cause excessive distortion. Suspect the common output IC when both channels are distorted.

4-20 Hot output transistors or diodes can damage the "A" lead to the radio/cassette player.

Dead left channel

Locating the cause of dead audio in either channel is easy. If both channels are dead, suspect something common, such as the power supply or dual output IC. Isolate the dead channel to the left or right with audio from the speakers. Locate the dead left channel starting with the speaker or volume control. Signal trace the audio from stage to stage with an external amp or scope.

Most dead channels are caused by open transistors, ICs, and coupling capacitors. Check for an open speaker. Of course, if only one speaker is defective in a high-powered car radio, the speaker itself is defective. However, if all speakers on the left side are dead, check the amp circuits.

With no rush in the sound, quickly rotate the volume control to determine if the audio circuits are functioning. If the audio is normal for a few minutes, makes a noise, then goes dead, check for an open AF or driver transistor or IC. Don't forget the leaky decoupling capacitor feeding voltages to the AF or driver circuits. Suspect a dual output IC when only a hum can be heard from both channels (Fig. 4-21). Look for leaky or open bypass capacitors connected to the output IC for dead conditions. Check for a leaky output IC or transistor when there is no audio and repeatedly blown fuses.

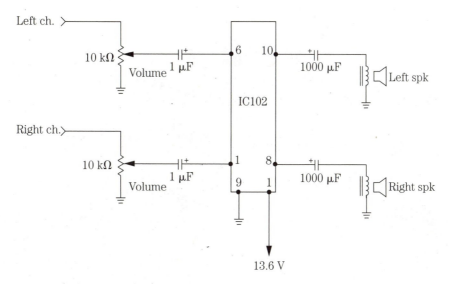

4-21 Suspect IC102 when you hear hum in both audio channels.

Intermittent audio

When the radio warms up and then cuts out, suspect a leaky power IC. An excessively worn volume control can cause the audio to cut out. A shorted volume control can also produce intermittent sound. Broken coupling capacitor leads or an intermittent break inside can produce intermittent reception. Bad board connections, terminal leads, and speaker connections can cause intermittent audio.

Try to isolate the intermittent audio to the left or right channel. Locate the correct intermittent channel with an external amp connected to the volume control of the intermittent speaker. Sometimes, when signal tracing, if the probe is touched to a transistor terminal, the sound returns to normal. Carefully go from stage to stage to locate the intermittent sound problem.

Suspect the dual audio IC when both channels are intermittent. If the IC seems normal, check for a voltage change at the low-voltage power supply. Check for a broken voice coil wire in one speaker. Loose speaker wires and cables can cause intermittent sound problems. Check all speakers for intermittent open or loose clip connections (Fig. 4-22).

Speaker problems

The auto speaker might sound distorted or mushy as a result of extreme weather conditions. The cone begins to warp and the voice coil rides against the center magnet. Excessive dirt and dust can fall into the speaker cone and cause distortion or noisy reproduction. Speakers that are mounted upwards in the front dash or rear deck have a tendency to warp and mush up. Small holes can be punched through the speaker cone or the cone can be torn, resulting in vibrating music (Fig. 4-23).

4-22 Check for intermittent speaker connections, voice coil, fader control, and cables for intermittent reception.

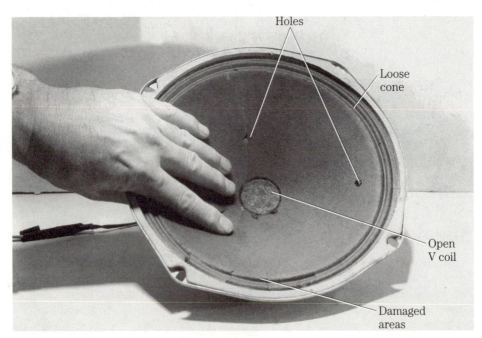

4-23 Check the speaker for holes, cone damage, dropped voice coil, and open speaker winding.

An intermittent speaker might be caused by too much applied power, which damages the voice coil. Sometimes the flexible wire that is soldered to the voice coil will break and cause intermittent audio. Simply remove the speaker from the mounting, connect the speaker wires, and lightly press down on the cone of the speaker. Move the cone up and down as the music cuts in and out. Replace the speaker if it is defective. Check the speaker terminals and wires for intermittent audio. Substitute another speaker to see if the wiring is broken or there is a poor connection.

Because at least four speakers are in the new auto hookups, you might have four times more speaker damage (Fig. 4-24). The voice coil could be blown or torn loose from the cone of the speaker as a result of excessive volume. Most speakers damaged by excessive volume are large woofers. Check the suspected dead speaker with an

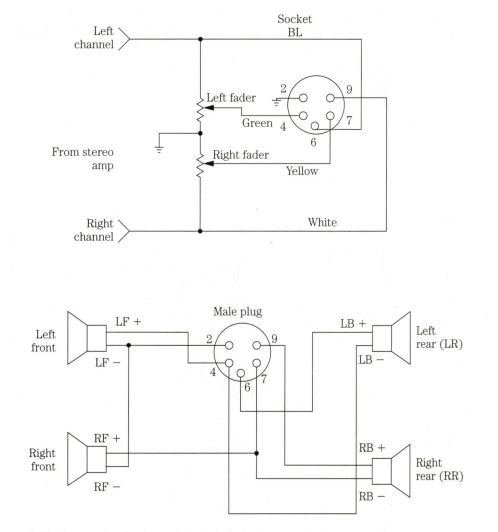

4-24 The speaker hookup might include fader controls, harness, and two or more PM speakers.

RX1 ohmmeter range. Replace the defective speakers with universal PM types that have the same wattage, size, magnet, and mounting holes.

Noisy reception

Check the output transistors and IC components for popping and crackling noises. If you hear a loud pop and the channel goes dead, suspect the output IC. Sometimes, when the stations are tuned in or the unit is first turned on, a noisy output IC will cause a crackling noise. Low frying noises can be caused by noisy transistors or ICs.

A motorboating noise found in either channel can be caused by the power output IC. Suspect the dual output IC when very loud motorboating is heard with the volume turned down. A chirping noise can be caused by open or blown speaker coupling capacitors. Check for poor grounds in the output circuits or on the IC heat sink when motorboating and buzzing noises are heard. Suspect the first AF transistor for microphonic noise.

Check the large filter capacitors or output ICs when you hear only a hum in the speaker. Overheated transistors or IC output components can produce hum in the audio. Low pickup hum can be caused by decoupling capacitors in the AF or driver circuits. Improper grounds in the volume control or AF transistor circuits can cause a pickup hum noise. Suspect a defective volume control for open or noisy conditions (Fig. 4-25).

4-25 Replace the erratic or noisy volume control when it will not clean up with cleaning spray.

Lower the volume control to determine if the noise or hum sound is in the audio or front-end section. Suspect a red-hot output IC for a loud howling noise. Shunt the main filter capacitor for a loud or high-pitched noise. When stations are tuned in and you hear a screeching noise, check the small decoupling capacitors in the AF and driver circuits. Inspect shields and look for loose IC mounting screws when you hear a squawking noise with the volume turned down after the set warms up.

Noisy outboard

Don't overlook car engine noise picked up by a poor grounding of the top cowl antenna. Remove the lead-in plug, and see if the noise disappears. A broken or grounded lead-in or water inside the lead-in can cause intermittent or excessive pickup noise (Fig. 4-26).

Stacked boards

Besides the main circuit board, you might find audio, volume control, front-end, and FM MPX boards in the auto receiver. Some of these boards are difficult to get at. It seems that the defective component is always under the last board. Sometimes metal screws and soldered chassis tabs must be removed before the board can be removed (Fig. 4-27). Before removing any board, look over all possible methods of holding the board in place. Make sure the defective component is located on a cer-

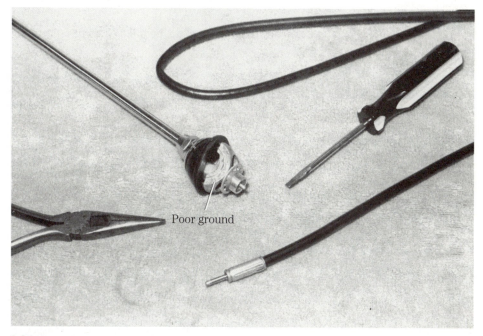

Poor ground

4-26 A poor antenna ground can pick up motor noise at the base of the antenna.

4-27 Check for a defective part on one of the stacked boards before trying to remove it.

tain board before removing it. Remember, there might be a lot of wires to remove. Write down each wire connection.

Look outside the PC board

Many car radio problems are caused by components mounted on the metal chassis off the PC board. Check for engine or antenna pickup noise instead of noisy components in the radio. Noisy or intermittent speakers can result from a dragging voice coil or broken coil leads. Connect another speaker and test.

Poor speaker switch connections or a bad cable harness can result in intermittent operation. Suspect a leaky output transistor or IC when there are burned cable wires. Check the fuse for incorrect amperage or tin foil wrapped around the fuse.

Auto receiver hookup

The auto radio has a set of cables that connect to the ignition, fuse holder, chassis ground, and several speakers. Usually this external cable plugs into the rear of the auto radio. When one or more components are connected together like a cassette/tuner, cassette/receiver, or under-dash CD player, several cables and plugs are required.

The cassette/tuner or cassette/receiver with CD controls operates on a CD changer within the auto trunk. The line output of the CD player connects to the line out of the cassette/tuner. Both the cassette/tuner and the CD player "A" leads must be connected to the 13.3-V battery source. If the CD player has internal stereo power output stages, they are connected to the speakers (Fig. 4-28). When neither unit has power-output amps, the cassette/receiver must be connected to the external power amplifier and speakers.

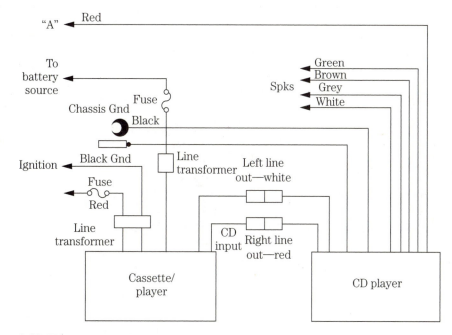

4-28 The cassette/radio and CD player hookup cable connects to the ignition, battery, and speakers.

Various symptoms

The following is a list of a few different car radio symptoms that might help you locate a defective component. How to test for and locate the component is given in each case, as well as how to repair the different symptoms.

No AM or FM reception

Isolation Look for something common to both the AM and FM circuits like the AM-FM switch, antenna, or low-voltage power source.

Location The AM or FM switch might be located on the front panel. Trace the wires to a possible leaky diode or transistor.

Repair Substitute another antenna into the radio while it is in the car. Plug the lead into the antenna jack, and hold it outside the window. Spray cleaning fluid inside the AM-FM switch area. If the supply voltage source is low, suspect a leaky zener diode, transistor regulator, or filter capacitor (Fig. 4-29).

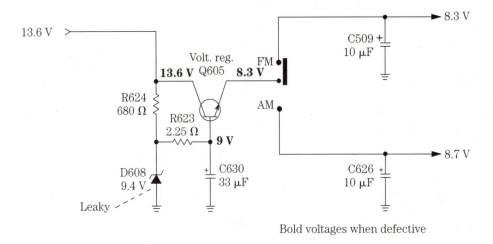

4-29 Leaky D608 prevented AM and FM reception with a low dc supply voltage source.

Weak reception

Isolation If all stations can be heard over the entire AM band with weak reception, suspect an open RF stage or converter.

Location Locate the AM antenna jack. The RF and converter transistor should be nearby.

Repair Take voltage measurements on the RF and converter transistor or IC component. A leaky RF or converter transistor can cause the weak signal problem. If not, check the RF coil and antenna connections (Fig. 4-30).

Dead

Isolation Touch the center terminal on the volume control with a test probe or screwdriver blade to see if you hear a click or noise in the speaker. If audio is present, suspect the front-end section.

Location Start at the volume control with touch tests.

Repair With normal audio, inject a signal from a noise generator at the volume control, detector, and IF stages working towards the front-end section. Touch the base of each IF transistor until the noise disappears. Take critical voltage and resistance measurements (Fig. 4-31). Most AM-FM IF coils have the colors green, yellow, and black on the top.

Noisy AM reception

Isolation After a couple of hours, the AM stations become noisy. FM is normal. Rotate the volume control clear down to see if the problem is in the audio or front-end. The noise was eliminated with the volume lowered in a Chrysler 4048367.

Location Start at the base of each IF stage and ground out with a 10-μF capacitor. When the base convertor was shorted, the noise disappeared; however, when the capacitor was placed on the base of Q1, the noise was still there.

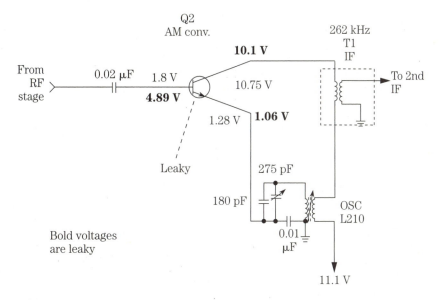

4-30 Check the RF and oscillator transistor for leaky or open conditions.

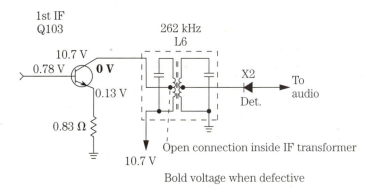

4-31 The dead AM symptom was caused by an open or broken connection inside the IF transformer.

Repair Noisy transistors or ICs must be replaced. Testing the transistor does not help. After Q1 was replaced, the noise was still there. Voltage measurements on the collector indicated low voltage. Resistance measurements on R3, R4, and R2 indicated a change in resistance (Fig. 4-32). The resistor module (3597230) must be ordered from Chrysler or a supply house.

Only local AM stations

Isolation If the FM reception is normal and AM reception is weak, suspect a defective RF stage.

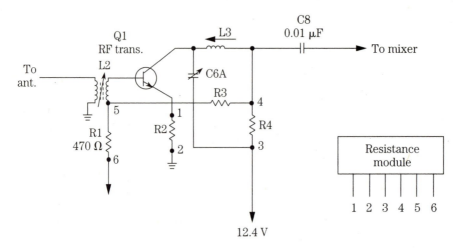

4-32 Noise in the Chrysler car radio was caused by a changing resistance in a resistor module.

Location   Locate the antenna jack and trace the circuit to the RF transistor.

Repair   Check the RF transistor in-circuit. Take critical voltage and resistance tests.

Dead, only hum

Isolation   Isolate the front end from the audio at the volume control. In this case, the audio was dead.

Location   Locate the sound board, and trace the center lead of the control to the electrolytic coupling capacitor.

Repair   Take in-circuit transistor and voltage tests (Fig. 4-33).

No AM, normal FM

Isolation   The AM reception is very weak AM station, but FM is good. Check RF or converter transistor or IC.

Location   Trace from the antenna jack to the antenna coil and RF transistor.

Repair   Test the transistor in-circuit. Remove the transistor if there is leakage between the base and emitter. A positive voltage of 8 to 10 V indicates the collector terminal, and the emitter has a leakage less than 470 Ω.

Intermittent FM stereo

Isolation   Try to locate an intermittent by pushing and prodding the chassis. Good reception on monaural FM but intermittent on stereo indicates the MPX section.

Location   Try to locate the MPX section.

Isolation Monitor the signal with an external amp. Take critical voltage and resistance tests. Suspect the MPX IC. Look the IC number up in the universal replacement manual.

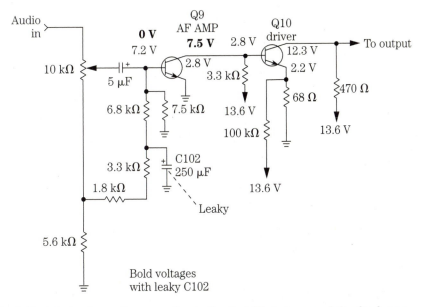

4-33 A leaky decoupling capacitor in the first IF stage caused the dead symptom.

Repair Don't overlook open or leaky bypass capacitors connected to the IC. If one stereo channel is dead with signal coming out of the MPX IC, replace it.

No FM, AM normal

Isolation If AM is normal, look in the FM front-end section.

Location After taking low-voltage measurements, go directly to the voltage source. Check the AM-FM switch.

Repair R501 was operating red hot. Q12 was normal. Low FM voltage source. C122 was leaky (Fig. 4-34).

No AM-FM rush, dead AM, no cassette

Isolation All systems dead. Check the power supply.

Location Locate the large filter capacitor terminals.

Repair Take voltage checks. Check for leaky filter capacitors.

Dead left channel

Isolation Check the audio in the left channel.

Location Locate the left channel on the board with a speaker hookup. Also use an external amp.

Repair Signal trace the left channel audio with an external speaker. Take critical voltage and resistance measurements.

Dead, hum only in speaker

Isolation Check the audio at the volume control. If it is normal, proceed to the audio circuits. Check the output IC.

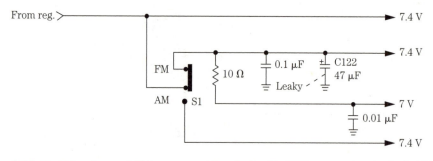

4-34 No AM and normal FM was caused by a leaky 47 µF filter capacitor in the 7.4-V source.

Location   Start at the volume control, and signal trace.

Repair   Check the signal in and out on the IC. Take critical voltage measurements on the IC. Check components tied to the IC. Replace the IC.

No cassette right channel

Isolation   Check the audio at the volume control. If the left cassette audio is normal, suspect the head or pre-amp IC. Check for wires off of the tape head.

Location   Start at the right cassette tape head connections, and signal trace signal in and out of the pre-amp IC. Locate the pre-amp IC.

Repair   Suspect IC201 with normal cassette input and no output. Check the signal at pin 8 and pin 6. Replace IC201 when the left channel is normal. Suspect D202 if there is signal out at pin 6 and no signal at volume control (Fig. 4-35).

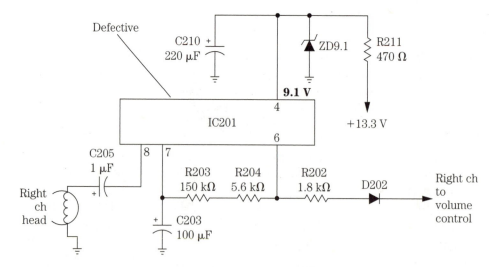

4-35 Defective pre-amp IC201 caused a dead right channel in the cassette player.

Weak, distorted sound

Isolation Check for distortion in the output stages.
Location Locate the output transistors or ICs.
Repair Check the transistors with in-circuit tests. Take critical voltage and resistance tests. Check the bias resistors.

Left channel distorted, only on tape

Isolation Check the left channel with an audio amp. Check for distortion in the outputs.
Location Locate the output IC or transistors.
Repair Check the signal in and out. Take critical voltage tests. Check the speaker above ground in some Delco models (Fig. 4-36).

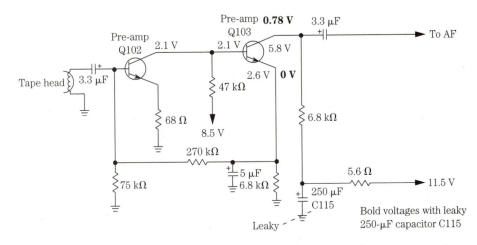

4-36 Leaky C115 produced a distorted left channel in the cassette player with normal radio reception.

Loud hum on the left channel

Isolation If the right channel is normal, go to the left channel.
Location Locate any hot ICs or transistors.
Repair Check for distorted input. Replace the output. Check for leaky coupling capacitors.

Intermittent right channel

Isolation Check the audio on the volume control. Monitor the right channel.
Location Locate the right channel with a speaker, and signal trace with an external amp.
Repair Monitor the right channel with an external amp, scope, and voltage tests. Check for poor board connections. Check for intermittent transistors, ICs, and coupling capacitors (Fig. 4-37).

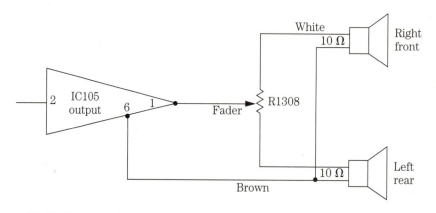

4-37 IC105 caused the left channel to be distorted in a Delco 80BFM1 car radio.

No cassette rotation

Isolation Locate the cassette motor within the cassette player.

Location After locating the motor, check the voltage upon the motor terminals. If voltage is present, measure the continuity of the motor for open windings.

Repair No voltage was found upon the motor terminals, indicating R173 or D130 open (Fig. 4-38). The 2.2-Ω resistor was found open. Replace resistor with a 2-W metal-type resistor.

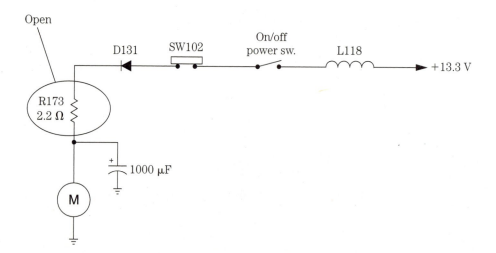

4-38 No cassette rotation resulted from open resistor R173 in the motor circuit.

Noisy after being on 5 minutes

Isolation Does lowering the volume control cut out the noise? If noise is present, suspect noise in the front end.

Location   Determine if it is one channel or both. If both, suspect the outputs. Voltage or resistance tests are no good. Locate the noise with an external amp. Short out noise with a 5-μF capacitor.

Repair   Remove the collector pin of each transistor and see if the noise disappears. White water rings were found around VC2. When pushing on the trimmer, the noise comes and goes. Replace with the exact part number (Fig. 4-39).

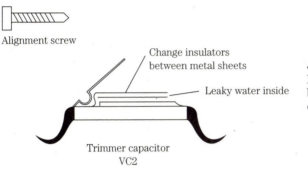

Alignment screw

Change insulators between metal sheets

Leaky water inside

Trimmer capacitor VC2

4-39 The cracking noise in the RF converter stage was caused by water inside the trimmer capacitor VC2.

Stations not tuning in on digital tuner

Isolation   Either tuning control or varactor-tuned circuits are defective.

Location   Locate the PLL processor on the board. Check it with the replacement manual.

Repair   Take critical voltage measurements. Check the VDD voltage. If it is low, check the voltage in the power supply source. With a normal supply, check the voltage on the varactor diodes while rotating the dial assembly. If the voltage changes, check the tuned circuits.

Audio radio troubleshooting chart

Check the troubleshooting chart in Table 4-1 for additional auto radio problems.

Table 4-1. Radio reciver troubleshooting chart

Symptom	Circuit	Remedy
Cannot tune in stations	RF oscillator mixer circuits	Check the oscillator and mixer transistors and ICs. Check the voltage.
No tuning lights or digital channel	Synthesizer controller	Determine if the controller is tuning stations. Check for applied voltage on the varactor diodes.
No AM	AM converter	Test the AM converter transistor or IC. Take voltage tests.

Table 4-1. Continued.

Symptom	Circuit	Remedy
Weak AM	AM antenna coil	Check for a broken core or winding.
	AM RF amp	Check the RF FET transistor. Perform critical voltage tests.
Distorted AM	AGC circuits	Check the resistors and capacitors in AGC circuit.
Noisy AM	Transistors and ICs	Determine if there is noise in the radio. Remove the outside antennal plug. Check the AM converter and IF transistors.
No FM—Normal AM	FM-RF oscillator & mixer circuits	Check the FM voltage source. Check the FM FET transistor.
	Dirty AM-FM SW	Spray down into the switch area, and clean the contacts.
Weak FM	Check RF-FM transistors and ICS	Test the RF FET transistor. Perform critical voltage tests . Perform a complete FM alignment.
FM distortion	IF agc circuits	Test the AGC transistor. Perform critical voltage test. Perform critical voltage tests. Replace the matrix IC.
	Matrix IC	Perform critical voltage tests. Replace the matrix IC
Noisy RM	IF-IC and FM matrix IC	Apply coolant and heat on the ICs and transistors.
FM hum	Power supply	Check the filter capacitors. Sub the filter capacitors.
	Voltage regulators	Test the transistor and zener diode regulators.
No stereo operation	MPX circuits	Check the voltages on the MPX IC. Signal the stereo channels a scope or external audio amp.
Excessive stereo	MPX IC	Take critical voltage on the IC. Replace the MPX IC.
Stereo light does not light or stays on all the time.	MPX IC or transistor light indicator	Check for a leaky transistor or IC. Check the LED or indicator light.

5
CHAPTER

Troubleshooting cassette players

Cassette players come in many sizes and forms: miniature portables, tag-alongs, professional, boom-box, cassette deck, and in the car radio (Fig. 5-1). Mono sound is found in some small cassette players, while stereo channels exist in large portables, table models, and car radios. Besides being used for playback, cassette players are also popular because they can be used for recording of messages and music.

Although many service problems are the same in most models, the larger deluxe units and car tape players have additional features. When mechanical and electronic components work side by side, trouble surfaces. Most mechanical problems can be seen or easily located, while the audio channels must be isolated and located with test instruments.

Required test equipment

The following equipment is needed to service cassette players:
- DMM or voltmeter
- Oscilloscope
- Audio signal generator
- Sine- and square-wave generator
- Capacitor tester
- External amp
- Audio analyzer (optional)
- Wow and flutter meter (optional)
- Test cassettes

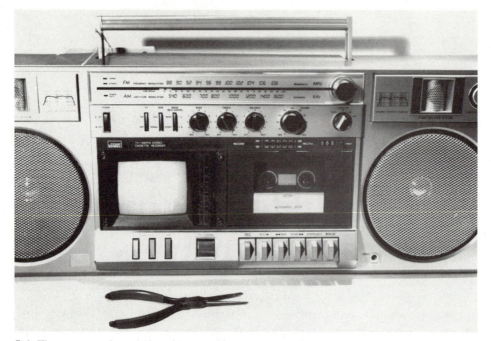

5-1 The cassette player is found in portables, tag-alongs, boom-boxes, cassette decks, car radios, and portable TVs.

Dragging your feet

Slow speed can result from a dry capstan bearing, dry pinch roller, tape wrapped around the pinch roller, a loose or oily belt, or a defective motor. Check for gummed up or dry capstan and flywheel bearings. Simply remove the capstan, clean it with alcohol and a rag, lightly oil it with phono lube or light oil, and replace it (Fig. 5-2). Replace all oily, cracked, loose, or shiny belts.

Check the batteries when the tape slows down during battery operation. Try the machine using ac operation to determine if weak batteries are causing the slow speed. Sometimes batteries that were left in the player too long will pull down the motor speed during ac operation. Remove the batteries, and check them in a battery tester or with a DMM.

Erratic or intermittent speeds can be caused by a dirty leaf switch, belt slippage, or a clogged pinch roller. Don't overlook a defective motor when slow-down or dragging occurs. Sometimes tapping the motor while it is operating will cause it to cut up and down. Replace a defective motor with the original part number.

Dead, no operation

Suspect a dirty or bent leaf switch, dirty function switch, defective play/record switch, dead batteries, defective motor, or a mechanism that will not seat properly when the player will not operate. Listen for a low hum noise from the motor. If there

5-2 Clean all moving parts with alcohol and cloth.

is no noise, check the batteries or the power supply. Clean the leaf switch if the batteries are normal.

Look for the function and leaf switch assemblies (Fig. 5-3). Sometimes push buttons are used for all functions, while one large rotary switch selects the different operations. Spray cleaning fluid inside the switch area and rotate the switch back and forth to clean the contacts. Clean the leaf switch by placing cardboard between the contacts, squeezing the points together, and pulling back and forth to clean the contacts.

Measure the voltage on the tape motor when the batteries and on/off switch are normal. The voltage should equal the battery voltage or dc supply source. If the cassette player operates from four C batteries, the total voltage is 6 V. If it is operated from a power line, the dc voltage should be around 6 V. Check for weak batteries, rusty battery springs, and poor battery connections when the unit operates on ac but not during battery operation (Fig. 5-4).

No tape motion

Check for a loose, oily, cracked, or slipped belt when the motor works but there is no tape motion. Look for a frozen capstan bearing when the motor pulley rotates inside the belt. Take voltage measurements on the motor terminals to determine if the motor is dead. Don't overlook a dirty leaf switch with no voltage applied to the motor or sound circuits.

5-3 Check for a dirty leaf switch in the small cassette player with an intermittent or dead symptom.

Sometimes you can hear the motor spinning, but nothing happens in play, record, fast forward, or rewind mode. Remove the bottom cover and see if the capstan/flywheel is rotating. Check for a missing or broken belt. Replace the capstan belt if it is loose or especially greasy. Make sure the power is off (Fig. 5-5).

No fast forward

In most direct drive mechanisms, the idler wheel is pressed against and rotates the take-up reel (Fig. 5-6). The idler wheel is rotated by a friction drive against a wheel that is attached to the capstan/flywheel shaft. Suspect slippage if the player operates in play but is slow in fast-forward operation. Clean all drive surfaces with alcohol. When the fast forward is belt-driven, clean the belt and drive pulley. If both play and fast forward are slow, clean the motor belt and flywheel surfaces.

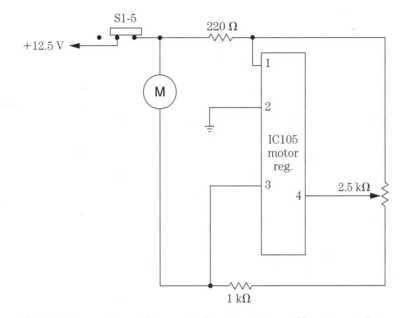

5-4 The tape motor might operate from transistor or IC motor regulator circuit.

5-5 Check the motor belt to see if it is too loose, stretched, or broken or if there is oil on the belt.

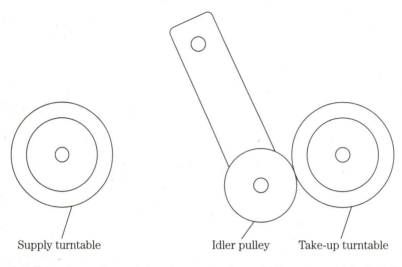

Supply turntable   Idler pulley   Take-up turntable

5-6 Clean the surfaces of the take-up, supply, and idler pulleys with alcohol and cloth.

Some small tape players have plastic gears that mesh in fast-forward mode. Check for broken teeth or gears that are not in line. Look for a missing C washer if the gears are out of line.

Will not rewind

Poor or no rewinding can be caused by slick surfaces on the idler or turntable areas. Clean these areas with alcohol. Rewind and fast forward should run faster than the play/record mode. In older and lower-priced players, in rewind mode, the shifting idler wheel is shifted when the rewind button is pushed against the turntable and reel assembly. Check for a dry or worn idler pulley. Remember, the pinch roller does not rotate in either rewind or fast forward.

Tape pulling

Eating or pulling tape can be caused by no movement or erratic movement of the take-up reel. Shut off the machine at once, or the tape might wind around the capstan and pinch roller.

Try operating the tape player in play mode without a cassette loaded. Notice if the take-up reel stops or rotates unevenly (Fig. 5-7). If so, clean the rubber drive surface on the bottom of the reel and idler drive pulley. Make sure the player is running smoothly before trying another cassette.

Suspect a worn or uneven pinch roller if the take-up reel is normal. Clean the pinch roller with alcohol and a cloth. Sometimes excessive tape oxide collects on the pinch roller.

5-7 Check the take-up reel for pulling of tape from the cassette.

Jammed cassette

Suspect that the tape has wrapped around the capstan when the cassette door will not open or the tape will not rotate. Don't pry open the door. Remove the back cover and try to locate the capstan/flywheel. Rotate the flywheel backwards by hand to loosen up the tightly wound tape, then try to open the door (Fig. 5-8).

The cassette tape might be damaged to the point where it cannot be used. However, sometimes the tape can be saved if not too much tape is wound around the capstan. Cut the tape loose from the capstan and pinch roller if it is wound inside the pinch roller assembly. Clean out all of the excess tape. Clean the capstan, pinch roller, and spindles after removing the tape.

Try to release the door from the back side if the catch or release is bent or jammed. Sometimes you can remove the front-panel bezel or plastic lid. Look for the door release lever. A jammed gear or lever in the eject mechanism can prevent the door from opening.

The squealer

The speaker in the cassette player might let out a loud squeal or warbling noise when the cassette function switch is pressed. The large cassette deck or player might have push buttons that operate many different functions. When the cassette switch is pressed and locked in, with a loud squealing noise, suspect a dirty function switch. This switch can be a rotary or push-button type that contains a stiff wire moving a flat function switch with many contacts (Fig. 5-9).

5-8 Suspect a jammed cassette when the door will not open.

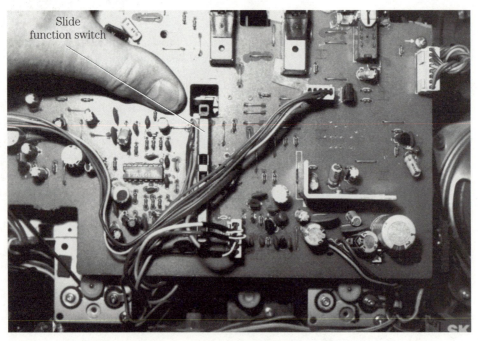

Slide
function switch

5-9 Clean the function switch by spraying cleaning fluid down into the switch area.

If there is extreme noise in the speakers, clean up the function switch. Spray cleaning fluid down in the dirty contacts, and rotate the switch back and forth to help clean up contacts. Push one button and then the cassette button to clean up switch contacts. The silver contacts upon the sliding switch become tarnished and make poor switching contacts.

The slush pile

Excess oil found upon the motor, drive belt, take-up, and supply reels can destroy the tape motion instead of helping. Do not oil any components within the cassette player unless the bearing is frozen or squealing. Light oil sprayed upon gears, take-up reels, and plastic gears can leak down upon moving parts that can cause extra slippage (Fig. 5-10). All oil spots must be wiped and cleaned up.

5-10 Do not spray on the take-up or supply reels, drive pulley, and plastic gears for faster speeds.

Do not apply oil or grease on top of the bearing of the capstan/flywheel if it is suspected of rotating at a slow speed. Remove the flywheel and capstan. Clean off the old grease with alcohol and a rag. Apply a dab of phono or light grease at the capstan/flywheel bearing. Remount the capstan, and wipe off the capstan area with alcohol. Likewise a drop of oil upon the pinch roller is adequate. Wipe off any oil or excess oxide that is on the rubber roller.

Tape head motor

5-11 Motors with worn or loose bearings should be replaced.

Most cassette motors do not need lubrication. They are lubricated at the factory. If one of the motor bearings begins to make a noise, a drop of light oil down into the bearing might help stop the noise. Motors with worn bearings must be replaced (Fig. 5-11).

Piling on

Blobs of solder upon a PC board can cause intermittent operations. Sometimes, when the boards are run through a solder bath, several areas might contain excess solder. In replacing a component, the technician can apply too much solder, resulting in shorting out two or more traces or wiring. Extra care must be added when replacing ICs or processors with gull-wing terminals. The PC wiring is very thin around these components (Fig. 5-12).

Blobs of solder piled up at each end of an SMD part can result in a cold soldered connection. Clean up the old connection so that the SMD component can lay flat, and melt solder to flow into the connection. Remember, too much soldering iron heat can destroy the SMD device.

Dead right channel

If one channel is dead, suspect an open tape head, AF transistors, output ICs or transistors, or an open speaker. Notice if the tape is rotating. Check to see if the tape lies against the tape head. Look for a missing mounting screw that lets the tape head back so that it does not engage the tape. Does the tape rotate between the take-up and supply reel? Check the audio with an earphone.

5-12 Check large blobs of solder for poor terminal contacts with an intermittent chassis.

Rotate the volume control up and down rapidly to determine if a noise or hum is heard in the speaker. Place a finger or screwdriver blade on the center terminal of the volume control and listen for a low hum (Fig. 5-13). Inject external sound at the volume control to see if the audio is normal. If the sound is good in the speaker, check for a dead front end or tape head. A loud rushing noise in one channel can indicate an open tape head or broken terminals.

If the audio output stages are normal, insert a recorded cassette and trace the audio with an external audio amp. Trace the audio from the tape head to the input terminals of the pre-amp IC or base of the first transistor. The signal is quite weak at this point.

In IC pre-amps, check the audio at the input and output terminals. If you find adequate volume at the volume control and none at the speaker, the output circuits are defective. When you find weak or no sound at the volume control, the defective stage is in the input circuits. Check the voltage supply terminal of the IC. No voltage or low voltage can indicate a defective power supply or IC.

Weak left channel

Clean the tape heads with alcohol. Determine if both channels are weak, or just the left channel. Signal trace the left channel from the tape head to the pre-amp and AF circuits. Weak audio can be caused by a dirty tape head, open or leaky transistors or ICs, open or dried-up coupling capacitors, or an improper voltage source (Fig. 5-14).

Signal trace the left channel with an external amp until the signal appears weak. If the signal going into the IC is normal but is weak at the output terminal, suspect a

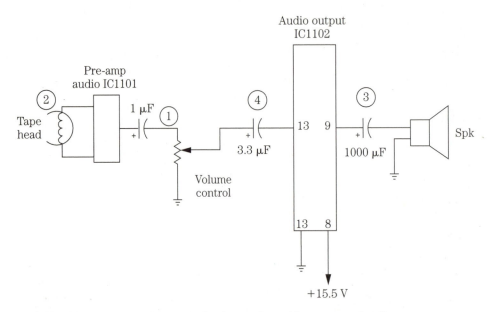

5-13 Troubleshoot the audio circuits by the number with an external audio amp.

5-14 Weak and distorted audio can be caused by a dirty tape head.

leaky or open IC. Measure the supply voltage. This voltage should be close to the battery or ac power supply voltage at the supply pin of the output IC.

Check all components tied to the IC terminals on the left channel. Take voltage measurements on each pin. Shut off the machine, and take low resistance measurements (Fig. 5-15). Compare these readings with the normal right channel. Suspect a leaky capacitor or a change in resistance of a bias resistor with a low measurement. Test all capacitors with a capacitor tester.

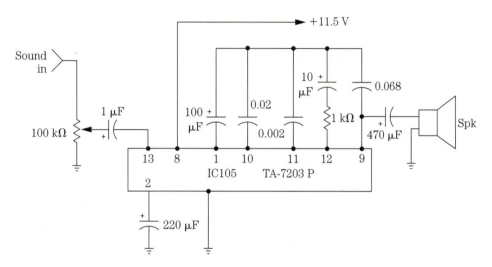

5-15 Take critical voltage and resistance measurements to determine if the IC is defective.

Locate the left channel with the speaker or an external audio amp. Transistors and IC parts often are found in a line on either the left or right sides in the audio circuits. Look for audio electrolytic coupling capacitors between the output IC and speaker (Fig. 5-16). Don't overlook a grounded tone control when weak volume and lower frequency in the left channel is a problem. Trace the weak channel from the volume control to the small coupling capacitor to the base of the transistor or input terminal of the IC.

Intermittent right channel

Locate the intermittent channel with the speaker or an external amp. Monitor the signal at the volume control to determine if the intermittent is in the front-end or output circuits. Most audio intermittents are caused by transistors, ICs, coupling capacitors, bad terminal connections, or poor board connections. Intermittent audio is the most difficult problem to locate and solve in the audio circuits and requires a lot of service time.

Monitor the intermittent audio with an external amp, voltage, and scope waveforms. Place a test cassette in the machine, and let it play. When the sound cuts out, locate the defective stage with amp and voltage tests. Sometimes the voltage on all

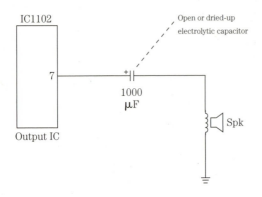

5-16
Suspect that the speaker coupling capacitor is open or dried up for weak and distorted audio.

IC terminals might be quite normal with an intermittent IC. Check the components tied to the IC pins.

Push up and down on the PC board to make the sound act up. Cracked wiring and poor board and terminal connections will show up with this method. Probing parts might turn up an intermittent component or bad terminal. Resolder all the terminals in a section that appears to be intermittent to locate a bad connection (Fig. 5-17).

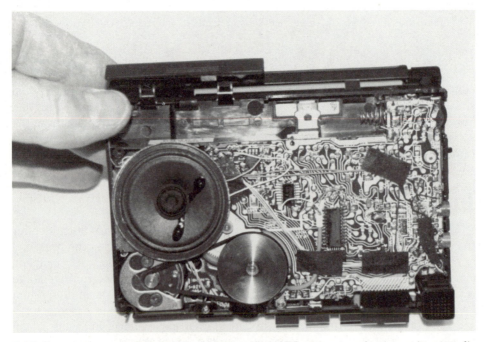

5-17 Sometimes you need to solder a section of the PC wiring to solve intermittent audio problems.

Eye contact

Sometimes it takes very sharp eyes and a lot of patience to trace out the PC wiring or cables to a given component. For example, if one stereo channel is dead or weak and you do not know which components tie into that defective channel, start at the speaker terminals. Trace the speaker wires to a plug or soldered connection upon the PC board. Because small stereo amplifiers have a speaker electrolytic coupling capacitor between the speakers and the output transistor or IC, trace it out (Fig. 5-18).

5-18 Broken PC wiring can be located with an ohmmeter and a strong light.

The defective channel can be circuit-traced right up to the audio output IC upon a heat sink. If both channels are fed from one dual IC and you don't have a schematic, take critical voltage and resistance measurements upon the IC terminals. Remember, stereo circuits are identical with one another, and voltage measurements should be within a fraction of a volt with normal channel.

After circuit-tracing the PC wiring and locating large output ICs or power transistors, signal trace the defective channel with the external audio amp or scope. After locating the defective channel with IC or transistor output parts, signal trace the

IC or transistors on the same side as the output component. The dead or weak channel can be compared to the normal one at several test points upon the chassis.

Transistor replacement

By locating the transistor number on the body of the part and looking in the universal replacement manual, you can determine what type it is, where it works, and its correct replacement (Fig. 5-19). Look for terminal markings on the bottom of the PC board to locate where each terminal connects. Mark each down on a separate piece of paper for reference.

Transistor styles include small, medium, and large power and special regulator types. The transistor number tells you what type of transistor it is and whether it is low frequency, high frequency, FET, SCR, triac, MOSFET, or opto device:

- A—Denotes the type of transistor and the number indicates the active electrical connections plus one (Fig. 5-20)

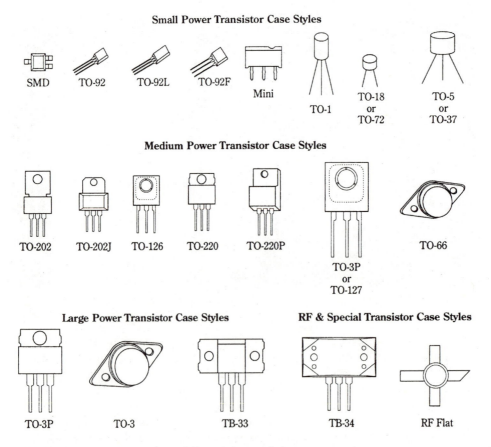

5-19 Transistors appear in many different sizes and shapes.

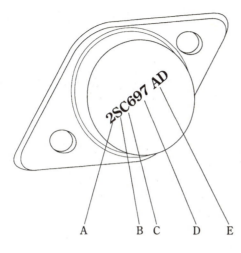

5-20
The 2SC indicates that the transistor is a high-frequency NPN type.

A   B C   D   E

- B—Japanese (EIAJ) or American (EIA)
- C—The transistor application and whether it is an NPN or PNP type
- 2N—Transistor, FET, SCR, or triac
- 2SA—Transistor, high frequency, PNP type
- 2CB—Transistor, low frequency, PNP type
- 2SC—Transistor, high frequency, NPN type
- 2SD—Transistor, low frequency, NPN type
- 2SJ—FET, P channel
- 2SK—FET, N channel
- 3SK—MOSFET, N channel
- 3N—MOSFET, dual triacs
- 4N—Opto devices

By looking at the transistor number, you can determine if it operates at a certain frequency and whether it is a PNP or NPN type. The suffix letter after the last number indicates an improvement over another transistor. Note that a B or C suffix letter after the number indicates a much superior transistor than the letter A.

If no numbers are found on the transistor, measure the supply voltage on the transistor, determine what type of a circuit the transistor works in, whether it is high or low frequency, and if it is small, medium, or large power by the size of the transistor case. Look up the data in the universal replacement manual, and replace the questionable transistor. You can also look at a schematic of the same brand or a similar circuit to get the working transistor number.

IC replacement

Locate the defective IC, and check the part number found on top of the IC. Look up the replacement in the replacement manual. Locate the input and output terminal pins. Signal trace the audio on the input terminal and then the output pin. The signal should be much greater on the output if the IC is normal.

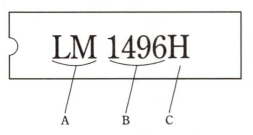

5-21
Locate the letters and numbers
on top of the IC, and look it up in
the universal semiconductor
manual.

The top numbers and letters on the IC indicate the manufacturer, the type of device, whether it is American or Japanese, improvements, and package style. Most manufacturers go by the following guidelines (Fig. 5-21):

- A—Denotes the manufacturer of the device when registered with the Japanese or American EIA numbers (Fig. 5-21)
- B—Indicates the type of device, root number, and that it is used primarily with American manufacturers
- C—Suffix letter indicates an improvement. The letter N is superior to the letter B. The number can indicate the package style.
- C—Can or metal
- D—Dual in-line package (DIP)
- E—Flat package (SMD)
- K—TO-3 transistor package
- L—Single in-line package (SIP) (Fig. 5-22)

5-22 Only one or two small ICs are found in the hand-held portable cassette player.

- P—Plastic package
- R—Reversed pin configuration
- S—Single in-line package (SIP)
- T—TO-220 transistor package
- V—Mounts vertically (ZIP)

Recording problems

Check the tab on the rear of the cassette when the cassette player will not record. The recording button cannot be engaged if this tab is removed. Try another cassette with the tab in place, or put tape over the opening. Clean the play/record head with alcohol and a cotton swab (Fig. 5-23).

Tape head

5-23 Clean the oxide from the erase and R/P tape heads for weak and distorted audio.

Does the cassette play in play mode? If so, the playback head and recording switch are normal in the path of the audio signal. Suspect a broken or dirty switch when the tape will play but not record. The switch contacts might not leave the play position. A weak recording can be caused by an oxide-coated tape head on either channel or no bias signal on the tape head. Check the oscillator bias waveform with the scope. A worn tape head can prevent recording in one channel.

Intermittent recording with a warbling sound can be caused by a dirty play/record switch. Suspect a dirty switch when you get a loud hum or buzz. Clean the play/record switch if a loud crackling noise is heard when it is switched to the record mode.

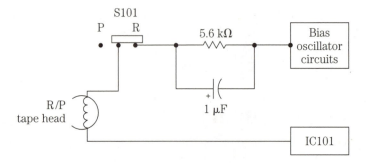

5-24 Trace the bias oscillator waveform with the scope at the R/P tape head.

Suspect a defective bias oscillator circuit when there is no waveform on one side of the recording head terminal (Fig. 5-24). Check for voltage on the bias oscillator transistor, an open transformer winding, and no supply voltage in record mode.

Poor recording

When both channels have a messed up, garbled, or jumbled recording, suspect the erase head circuits. Clean the erase head with alcohol. Sometimes the small gap can become packed with oxide dust, which results in no recording. Try another cassette. Check that the ground wire is connected to the erase head.

Check the bias oscillator for erratic or intermittent operation with poor recording symptoms. Poor oscillator transformer connections can produce intermittent recordings. Weak recording can be caused by a dirty record head or low supply voltage to the bias oscillator. Clean the play/record switch. If both record and play are weak or distorted, check the audio output and pre-amp circuits. If the record time indicator will not move, suspect a broken belt or jammed assembly (Fig. 5-25).

Noisy operation

Check the capstan/flywheel bearings, idler pulley, and motor for squeaks or screeching noises. Lubricate with a light oil. A noisy cassette hub can be checked by changing to another cassette. An open head connection can cause a loud howling in the speaker with the volume wide open. A dirty play/record button can also cause a howling noise.

When both channels have a low hum or frying noise, replace the output IC. If the noise continues with the volume turned down, suspect a noisy output IC or transistor. When the noise can be turned down with the volume control, check the pre-amp transistors, ICs, and bypass capacitors. Short the base to the emitter terminal of the transistors to locate the noisy stage. Don't overlook a defective motor (Fig. 5-26).

5-25 Suspect a jammed gear or broken belt when the numbers will not rotate.

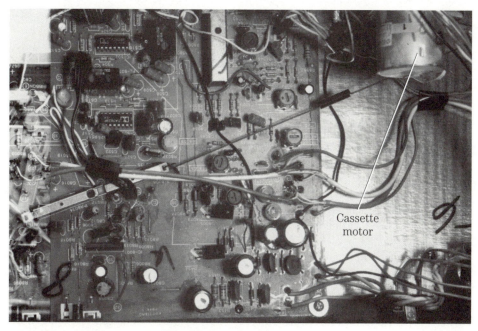

5-26 Locate the noisy cassette motor and replace it.

The runaway

High-speed tape movement can result from a defective regulator circuit, a belt riding high upon the motor flange of a pulley, and a defective motor. Check the cassette motor speed by inserting a 3-kHz tone cassette and connecting the frequency counter at the speaker terminals. Make sure the belt is riding in the motor pulley groove and not upon the edge of the pulley. Place the cassette player in play mode. Connect a 10-Ω resistor at the earphone jack or speaker terminals (Fig. 5-27).

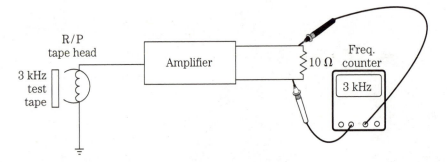

5-27 Check the cassette motor speed with a 3-kHz cassette and a frequency counter.

Measure the frequency at the frequency counter. If the reading is 3 kHz, the speed is normal. A higher reading indicates faster speeds, and a lower measurement indicates slower speeds. If the speed is rather low, suspect belt slippage, a loose belt, or a defective motor. Oil deposits on the capstan/flywheel can produce low speeds. Check at the end of the motor, and see if there is a screw speed adjustment. If not, check the motor speed regulator for a control or screwdriver adjustment. Adjust the speed on the frequency counter at 3 kHz.

Motor resistance

The small cassette motor might be intermittent, erratic, slow, or dead. The intermittent motor might start one time and be dead the next time. Then, when the motor shaft is rotated, the motor will begin to run. Erratic motor rotation is often caused by a worn or dirty commutator. Some of these motors are too small to take apart, but sometimes with patience and care, the armature and wire tongs can be cleaned (Fig. 5-28). Other times, you can tap the outside metal belt of the motor and it will resume speed.

Check the motor winding with a quick RX1 resistance measurement. No or high resistance measurements indicates an open winding or dirty commutator. Most small dc motors will have a resistance of less than 10 Ω. Check for correct voltage applied across the motor terminals with slow speeds. The small motor operating from the ac power supply might operate from 12 to 18 V, while in the dc battery operated cassette player, the voltage equals the total battery voltage.

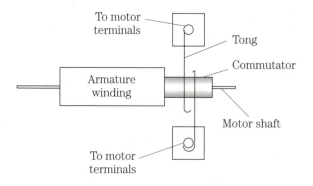

5-28
The construction of a small dc motor with tongs and armature.

Tape head resistance

Besides collecting oxide dust, the R/P tape head can become open, intermittent, or cause distortion in the speakers. The worn tape head can cause a loss of high frequencies. A magnetized tape head can cause extra noise in recording. An open head winding will produce a dead channel. Poorly soldered connections or broken internal connections can cause intermittent music. A grounded tape head winding to the outside metal shell can cause a dead or weak audio.

Check the open or high-resistance head winding with the ohmmeter (Table 5-1). Inspect the cable wires and resolder for intermittent conditions. These tape head wires seem to break where the cable connects to the head-winding terminals. Make sure the tape head is engaging the tape. Check for a loose screw, which would let the head swing out of line.

**Table 5-1. Typical tape head resistance
of several different cassette players.**

Model	Tape-head resistance
GE-3-5808KA	225 Ω actual measurement
Panasonic RQ-L315	315 Ω actual measurement
Sony M440V	348 Ω actual measurement
Sony TCS 430	512 Ω actual measurement

Typical R/P tape head resistance: 200 to 830 Ω
Typical erase head resistance: 200 to 1K Ω

The erase head is mounted ahead of the R/P head to erase any previous recording. Often the erase head has two leads and is excited by a dc voltage or by a bias oscillator circuit. Suspect the erase head when garbled or two different recordings are heard in the speakers. Measure the dc voltage across the tape head. If there is none, check the bias oscillator circuits. Simply trace the wiring from tape heads to the dc source or bias oscillator circuits. Check the open erase head with the low range (RX10) ohmmeter range (Fig. 5-29). The erase head resistance can vary from 200 to 1000 Ω.

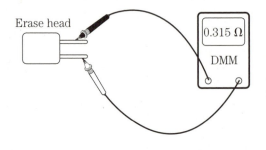

5-29
Check the erase head resistance with the RX10 DMM ohmmeter measurement.

Broken plastic door

In many of the dual-cassette decks, the front door will come off. Lightly pry open the front door. The front door slips into a four-sided track that clips the plastic door to the front-loading assembly. In other doors, the plastic door pulls out of the round track (Fig. 5-30). Sometimes, one or two of these plastic tips breaks and the door falls off. Apply cement to the broken plastic ends, and slip the door back into position.

Hinged doors on some models will unclip at the bottom. Lightly pull upon the door assembly, and see if the door will loosen. Inspect the bottom shaft that holds the door, indicating correct removal. Hinged doors that have long threaded rods must be replaced. On expensive models, two etcheon-type screws hold the plastic door in place. The plastic door or cassette chassis should be removed to easily and properly clean heads and remove jammed cassettes.

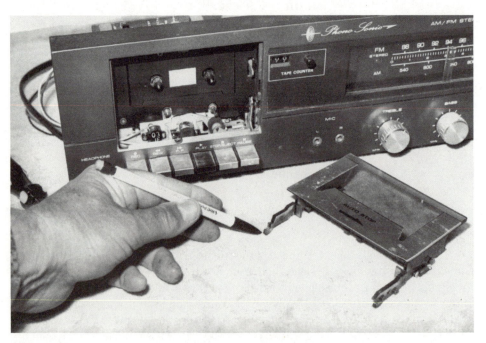

5-30 Replace the broken door when it will not lock in place.

Cassette player symptoms

The various cassette player symptoms and isolation, location, and repair information can help you solve cassette player problems. These symptoms can be applied to most any tape deck or player. Several case histories are given for both cassette and car cassette players:

Squeals when switching

Isolation   Check for a dirty function switch, and locate the switch upon the cassette PC board.

Location   Check behind push buttons or mounted on the chassis.

Repair   Spray cleaning fluid down inside the sliding switch area. Rotate the switch back and forth to help clean the silver contacts. Several spray applications might be needed.

Will not erase

Isolation   A defective erase head, broken cable, defective bias oscillator, or loose erase head mounting can cause poor erasing of the previous recording.

Location   If the play/record head plays a cassette normally, suspect a broken cable or erase head. Locate the erase head to the left side of the play/record head.

Repair   Check for a broken lead or an open erase head. Take continuity test with a DMM. Clean the play/record switch assembly and the erase head.

Cracking noise at turn on

Isolation   When the ac switch is turned on, a frying and crackling is heard in the cassette deck.

Location   Determine if the on/off switch has dirty or worn contacts. Excessive arching of the on/off switch should be replaced.

Repair   If the on/off switch has no resistance in closed position and is only noisy at turn on, place a 0.22-μF 150-V capacitor across the switch contacts (Fig. 5-31).

No left channel

Isolation   If the right channel is normal, signal trace the audio circuit from the tape head to the speaker with an external amp on the left channel.

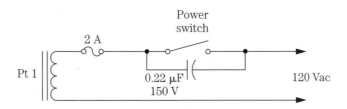

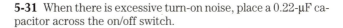

5-31 When there is excessive turn-on noise, place a 0.22-μF capacitor across the on/off switch.

Location Locate the pre-amp IC and input and output transistors. Check for audio at the left channel volume control.

Repair Trace the center terminal of the audio coupling capacitor to the input terminal of the output IC. Normal signal in and no signal out at pin 8 of the IC indicates a defective IC (Fig. 5-32). Check supply pin 8 and the voltage and resistance on all terminals.

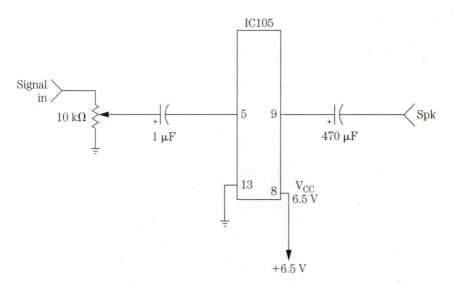

5-32 Check the supply pin voltage with input sound and no audio at the output pin 9.

Cassette will not eject

Isolation Suspect the door latch, a defective plunger, or a loose pin in the bottom of the cassette player.

Location Located on the door latch or plunger.

Repair Replace the pin, and make sure the plunger seats properly. Place rubber silicon cement over the pin so that it will not work out of position.

Motor does not rotate

Isolation Check for a defective leaf switch, function switch, or motor or improper voltage to the motor terminals.

Location Locate the leaf switch under or near the play/record button. Locate the small motor terminals at the opposite end of the motor shaft or wires coming out of the motor.

Repair Clean the leaf or function switch. Replace it if it is broken. Check the voltage at the motor terminals. Replace the tape motor if the voltage is normal (Fig. 5-33).

5-33 Check the voltage across the cassette motor terminals to determine if the motor is defective.

Keeps blowing main fuse

Isolation After replacing the 2-A fuse in a J.C. Penney 683-3207 player, the new fuse would blow. Output transistors, ICs, and leaky components in low-voltage source can blow the fuse.

Location Check the silicon diodes in the low-voltage power supply. Locate the power output transistors on the heat sinks.

Repair Test and replace the leaky Q210 and Q212 output transistors.

Distorted right channel

Isolation In a J.C. Penney 3245 AM-FM-MPX player, the right channel was distorted. Go directly to the output transistors because most distortion problems are found in the output circuits.

Location Locate the output transistors.

Repair Check each transistor in the circuit for open or leaky conditions. The distortion was caused by an open between the base and emitter terminals within output transistor Q921 (Fig. 5-34).

No tape motion

Isolation In a Pioneer KP-3500E car radio, the motor would not rotate. Check the voltage on the motor terminals.

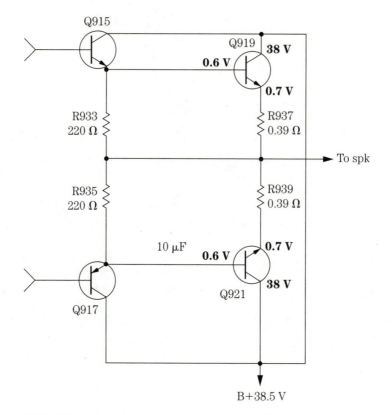

5-34 Q921 was open between the base and emitter in a J.C. Penney cassette player.

Location Locate the motor and terminals. If there is no voltage found on the motor terminals, look for a defective voltage regulator transistor or zener diode.

Repair Trace from the motor terminals to the voltage regulator transistor. Q2 was tested in the circuit and found to be open (Fig. 5-35). Q2 was replaced with an SK9042 universal replacement.

Shuts off after 5 minutes

Isolation Remove the motor belt to see if the motor quits rotating, and monitor the voltage at the motor terminals.

Location Locate the motor and circuits.

Repair After the motor ran for a few minutes, no voltage was found at the red motor terminal. The motor regulator transistor was open.

Keeps kicking out cassettes

Isolation Check if the mechanism is electronically or mechanically operated.

Location Locate the revolving switch, which is usually located off the take-up reel or belt drive.

Repair Clean the contacts of the revolving switch and shorting tongs.

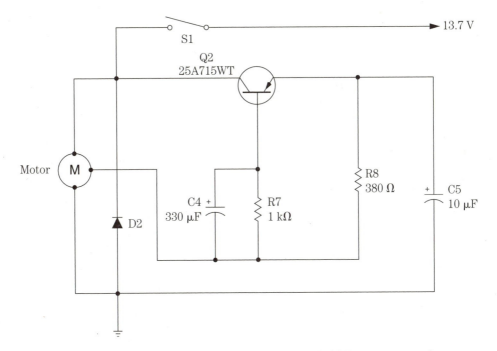

5-35 Open Q2 prevented motor operation in a Pioneer KP-3500 E auto cassette player.

Keeps changing directions

Isolation Check for dirty or worn rotating contact switch connections.

Location Look for round switch contacts on top of the turntable.

Repair These long shorting tongs can get bent out of line or can become dirty. Clean the silver contacts on the turntable and the contacts of the shorting tongs (Fig. 5-36).

Automatic reverse keeps reversing

Isolation Check for a contact switch on the spindle or turntable.

Location Locate the shorting contacts or magnetic switch under the turntable.

Repair Both (red and blue) switch wires were broken off the magnetic switch (Fig. 5-37).

Rush noise on left channel

Isolation If a rush occurs when the volume is up but there is no music, suspect a broken head wire.

Location Check all the wires and cables connected to the tape head.

Repair Resolder the broken wire on the left channel tape head.

Low volume on the right channel

Isolation Check the tape head or amp circuits.

Location Look for the play/record head.

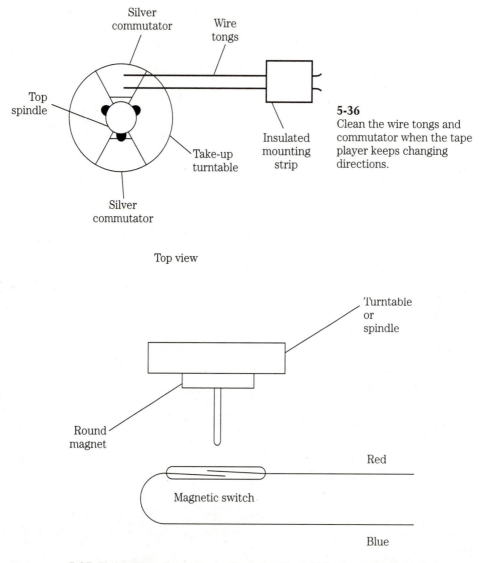

Silver commutator

Wire tongs

Top spindle

Take-up turntable

Insulated mounting strip

Silver commutator

Top view

5-36
Clean the wire tongs and commutator when the tape player keeps changing directions.

Turntable or spindle

Round magnet

Red

Magnetic switch

Blue

5-37 Check for broken wires to the magnetic switch when automatic reverse keeps reversing.

Repair Clean the tape head. The right channel might be packed with oxide dust.

Sound in only one direction

Isolation The unit will play normally in one direction, but when reversed, there is no sound.

Location Look for two different heads or two capstans and two pinch rollers.

Repair Found a broken wire on the left tape head.

Keeps pulling tapes

Isolation The capstan or pinch roller is dirty, or the take-up reel is not operating.

Location Look at the pinch roller, and notice if the tape is wrapped around the capstan.

Repair Noticed the take-up reel was rotating intermittently. Clean the idler pulley and contact surfaces in the play and record modes.

Normal speed, then high-speed reverse

Isolation Check for a defective motor.

Location Locate the motor and two sets of capstan/flywheels.

Repair When the end of the cassette was reached, the end switch changed the direction of the motor and capstan/flywheels. Found excess tape on one of the capstan shafts. Remove and clean the capstan.

Intermittent left front speaker

Isolation Checked amp with another speaker. Also checked wiring to the speaker.

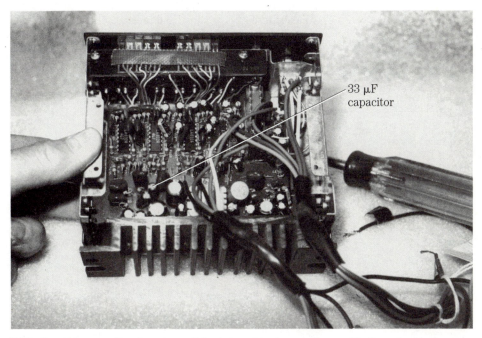

33 μF capacitor

5-38 Check the equalizer booster unit for intermittent or weak sound in the auto speaker system.

Location   Located in the amp and left front speaker.

Repair   Substituted another left front speaker, and after hours of isolation, pulled the unit from the car. Found a defective equalizer booster unit. Replaced an intermittent 3.3-μF coupling capacitor (Fig. 5-38). Sometimes you have to check outside the unit for sound problems.

Cassette player troubleshooting chart

For additional troubleshooting tips check Table 5-2.

Table 5-2. Typical cassette troubleshooting chart.

Symptom	Cause	Remedy
Cassette cannot be inserted	Inserted improperly	Check the cassette for damage. Check for foreign material inside. Notice if the play or record button is depressed.
Record button cannot be depressed	No cassette loaded Cassette tab removed	Reload the cassette. Check for removed tab at the back. Place tape over cutout if you want to record on this cassette.
Playback button cannot be locked in	Check for completely wound tape.	If the tape is completely wound toward arrow direction, rewind the tape with the rewind button.
Tape does not move	Incorrect batteries Weak batteries No ac Check motor	The batteries are in backwards. Test, and replace below 1.2 V. The power adapter is not connected. Check the voltage across the motor. Continuity test the motor winding to see if it is open.
No sound from speakers	Are the headphones plugged in? Check the location of the volume control.	Remove the headphone plug. The volume is turned down.
Fast tape speed	Check the setting of the speed control.	Readjust the speed control.
Weak or distorted sound	Weak batteries Dirty heads	Test or replace. Clean the heads with alcohol and cleaning stick.
Poor recording	Weak batteries Dirty R/P tape heads	Test or replace. Clean up with alcohol and cloth.
Poor erase	Improper connection Dirty erase head No dc voltage No oscillator waveform	Check all the head connections. Clean the erase head. Check the dc voltage on the head. Take a scope waveform on the head terminals.

6
CHAPTER

Repairing the black-and-white TV chassis

Although most problems with the black-and-white TV chassis are the same as those found in color chassis, they can be repaired more quickly and easily. Fewer components and lower voltages are found in the black-and-white TV chassis, and the various parts are easy to locate and remove. Look for a solid-state manual tuner in these chassis (Fig. 6-1).

The pocket black-and-white TV

Troubleshooting and repairing the small hand-held portable TV can be difficult without a schematic (Fig. 6-2). This small portable has the TV input, video, and sweep circuits on one board and the audio components upon the other. Miniature regular parts are found upon the top side of the chassis, and SMD components are on the PC wiring.

In Fig. 6-3, notice that the top side has white outlines of SMD parts located on the bottom side of the PC board. Special parts must be used in these small TV receivers. Components like electrolytic and bypass capacitors, resistors, and batteries can easily be replaced. Special parts—such as the small tuner, IF coils, digital display, and variable controls—should be ordered from the manufacturer.

Dead, nothing

Because 85 percent of black-and-white TV troubles exist in the horizontal and vertical circuits, check these circuits first. The dead symptom can be caused by no voltage from the low-voltage power supply or horizontal circuits. Take a voltage test on the horizontal output transistor. Remember, these voltages might be low if a step-down transformer is found in the power supply. Circuits working directly from the power line will have higher dc voltages through the chassis.

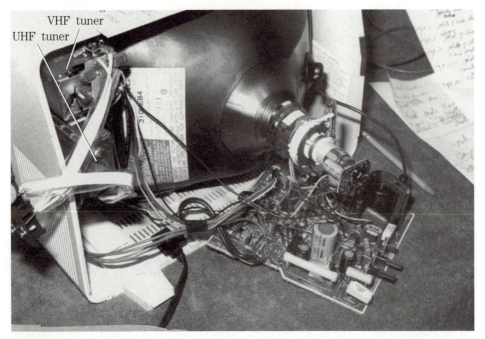

6-1 The solid-state VHF tuner is at the top and UHF at the bottom in the black-and-white chassis.

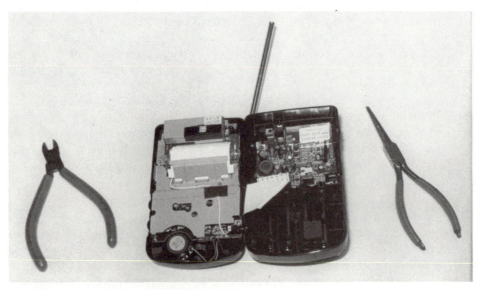

6-2 The small hand-held portable TV has LCD display and audio on one board with video-IF and sweep circuits on the right board.

6-3 Notice that the white lines represent the SMD parts underneath on the PC wiring.

Next, check the dc voltage at the fuse supplying voltage to the horizontal output circuits. A waveform test on the horizontal output transistor will indicate if the horizontal circuits are functioning (Fig. 6-4). If they are not, inject a horizontal signal at the base of the horizontal output transistor. Usually this transistor is located on a separate heat sink on the PC board. Look for a white raster with injected signal, indicating that the horizontal circuits, flyback, and picture tube are normal.

Keeps blowing fuses

When both the line and B+ fuses open, suspect a defective component within the horizontal output circuits. If the line fuse blows and the B+ fuse does not, check the parts in the low-voltage power supply. Go directly to the horizontal output transistor and measure the resistance from the body to common ground. A low resistance measurement under 100 Ω indicates a leaky horizontal output transistor or damper diode (Fig. 6-5).

Check the silicon diodes, filter capacitors, and regulator circuits in the power supply circuits when the main fuse opens under any conditions. Lightning or flashover of parts can cause the fuse to open. Replace the fuse and let the chassis run for 8 to 24 hours to see if additional parts are needed.

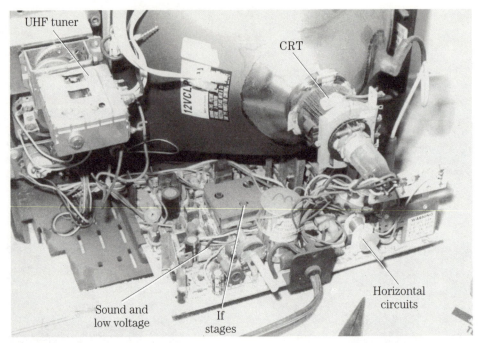

6-4 Locate the various components on the heat sinks and letters or numbers on top of ICs.

6-5 Replacing a leaky horizontal output transistor on the black-and-white TV chassis.

Intermittent raster and sound

Monitor the low-voltage power supply and waveforms on the collector terminal of the horizontal driver transistor. Connect a voltmeter to the B+ fuse or low-voltage output and a scope to the driver transistor located near the driver and output transistor on the heat sink. Check the horizontal driver and oscillator transistor when voltage lowers and improper waveforms are found at the collector of the driver transistor.

If possible check the horizontal driver and oscillator transistors in-circuit. Sometimes they will test good, and the raster will return to normal. Most intermittent horizontal problems are caused by an intermittent driver transistor, transformer, or horizontal oscillator transistor. Don't overlook poor board connections or damaged horizontal hold coil connections.

No raster, no high voltage

Check the voltage at the B+ fuse and waveforms at the horizontal output transistor. Determine if the lack of high voltage is caused by the horizontal, power supply, or high-voltage circuits. No waveform at the horizontal output and lower voltages indicates defective horizontal circuits.

Inspect the horizontal hold control coil for broken connections. This control sticks out the back of the chassis and is easily broken (Fig. 6-6). Inject a horizontal

6-6 The vertical and horizontal hold, with vertical size controls at the rear of the black-and-white chassis.

signal to the base of the horizontal output transistor to determine if the output circuits are normal. Check the flyback and high-voltage circuits with no raster and injected signal.

If there is no waveform at the base of the horizontal output transistor, scope the driver and oscillator circuits. Test the horizontal oscillator and driver transistors in the circuit with a transistor tester or DMM diode test. Take critical voltages on the oscillator and driver transistors. Don't overlook a shorted primary winding of the driver transformer or poor soldered terminals.

Sides pull in

Check the B+ regulator, B+ adjustment, and low-voltage power supply for low voltage to the horizontal circuits. The power regulator transistor might be located on a separate heat sink with the B+ adjustment nearby. Check the voltage at the silicon diodes or bridge circuit and at the collector terminal of the power regulator transistor.

A defective output or driver transistor can cause poor width problems (Fig. 6-7). Burning or low resistance of the primary winding of the driver transformer can cause the sides to pull in. Check for low high voltage at the anode socket of the CRT. Most black-and-white portables have a high-voltage measurement from 6 kV to 10.5 kV. A white line down the middle indicates high voltage and vertical sweep with no horizontal sweep caused by an open horizontal yoke winding or an open capacitor in the return circuit.

6-7 The defective output transistor and HV diode rectifier in the black-and-white chassis.

Horizontal drifting

Check the horizontal oscillator components for horizontal drifting. Spray small capacitors within the oscillator circuits with coolant or heat from a hair dryer. Horizontal drifting might go out of sync after the chassis heats up. Replace both the capacitors and oscillator transistor in the horizontal hold oscillator circuits. Monitor the voltage at the collector terminal of the horizontal oscillator. Locate the hold control coil on the back of the chassis (Fig. 6-8).

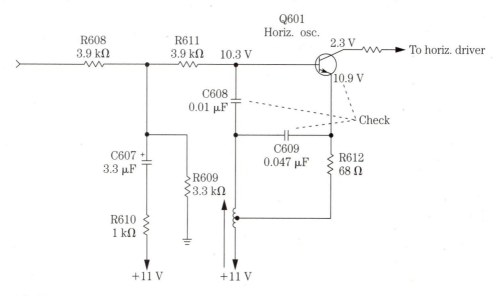

6-8 Check the following parts for horizontal drifting.

Only a horizontal white line

A white line with normal sound indicates no vertical sweep with normal high voltage. The horizontal, high-voltage, low-voltage, and picture tube circuits are normal. Most vertical sweep problems are caused in the vertical output circuits. A scope test at the vertical output capacitor and oscillator transistor or IC will indicate vertical sweep conditions (Fig. 6-9).

No vertical sweep is caused by defective output transistors, ICs, or capacitors or low or improper voltage. Check the power supply source terminal (V_{CC}) for correct voltages. Measure the voltage at the collector (metal) terminal of the vertical output terminals instead of the output IC. Test each transistor with in-circuit tests. Replace both transistors if one is found to be leaky.

Insufficient vertical sweep

Look for leaky or open output transistors or ICs when there is only a few inches of vertical sweep. Replace both output transistors when one is found to be leaky (Fig. 6-

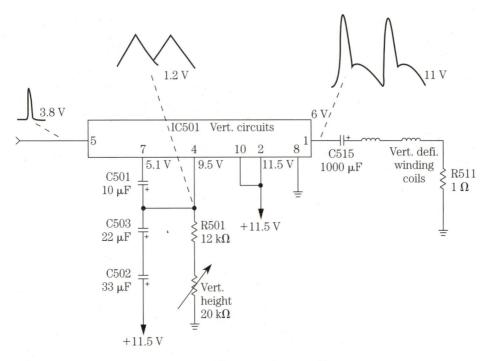

6-9 Take vertical scope waveforms upon the vertical output IC.

10). Check for burned bias resistors and leaky diodes in the output circuits. Locate the vertical output transistors and ICs on separate heat sinks. Take critical voltage and in-circuit tests. Problems can also be caused by improper power supply voltage, burned isolation resistors, or broken connections in the supply voltage wiring.

Six inches deep

Insufficient height in a Sylvania black-and-white portable (BWE150SL01) was caused by improper supply voltage at the vertical output transistors. The output transistors are usually mounted upon a heat sink and were mistaken for two regulator transistors upon the same heat sink. Both vertical output transistors were marked and located upon the curve of the PC chassis (Fig. 6-11). TS423 and TS422 were the vertical output transistors and TS426 driver transistor. These were traced out with the various resistor and diode components.

Voltage measurements were made upon the collector terminal of all three transistors and appeared quite low. Most vertical output circuits including output ICs have a 20- to 25-V supply voltage. All three transistors tested normal in the circuit. A power transformer was located off of the chassis with a bridge rectifier network. Here the voltage source was around 15 V, which indicated another voltage source was used for the vertical output circuits. No doubt, the supply voltage was taken from the secondary-derived power source.

6-10 The vertical output transistors might be located upon a separate heat sink.

The vertical output voltage source was traced from the collector of TS423 with the low-resistance range of DMM to a common source in the flyback circuits. A close inspection of all diodes upon and around the flyback indicated that a resistor showed signs of overheating in series with a zener diode. The zener diode (D474) had a resistance of 0.17 Ω (Fig. 6-12). R411 should have a resistance of 390 Ω, and tested less than 10 Ω. Replacing R471 and D474 returned the 24.40-V source to the two vertical output transistors. Poor vertical linearity and fold-over can be caused by defec-

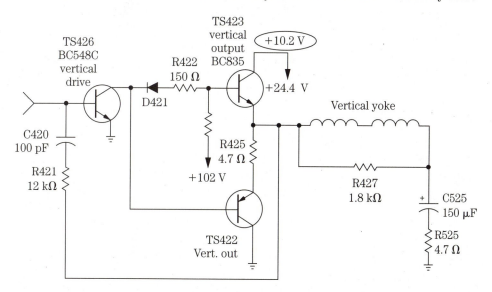

6-11 Improper voltage supply source causes a 6" vertical deflection upon the CRT.

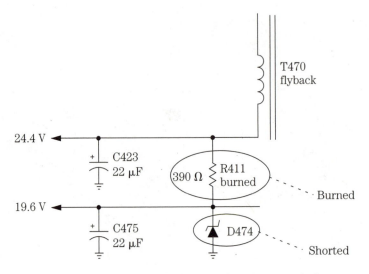

6-12 Leaky D474 and burned R401 caused improper vertical sweep.

tive feedback circuits, C524 and C525. Also, check vertical output transistors for insufficient vertical sweep.

Rollaway

Notice if the picture has black bars in the raster when it is rolling or flipping. Check for a leaky B+ regulator or open or dried-up capacitors in the low-voltage power supply. Check for hot or overheated ICs or transistors. Defective or open electrolytic capacitors in the vertical output circuits can cause a flipping or rolling raster. Make sure vertical sync is present at the input of the vertical oscillator transistor or IC.

Vertical crawling can be caused by an open or dried-up filter capacitor in the main power supply. Check for poor filtering in scan-derived voltage circuits of the flyback feeding vertical circuits.

Vertical rolling

The picture had vertical fold-over and would not stop rolling in a Goldstar KMA-0401 model. The picture was only 5" in height. Because many of the vertical output transistors are not mounted upon a heat sink, check for a large IC for vertical oscillator and driver or countdown circuits. Scope the many terminals and trace the vertical waveforms to the base of a vertical driver or output transistor.

All three vertical output transistors were found in a cluster with fairly normal input and output waveforms. A bias resistor of 2.2 Ω was found between Q652 and Q653 with a fairly normal output waveform (Fig. 6-13). Voltage measurement of 9.2 V was found on collector Q652. This voltage is normal because the 47-μF capacitor is rated at 10 V. Very little waveform was found on the input to the vertical yoke winding. A continuity measurement between yoke and common ground was less

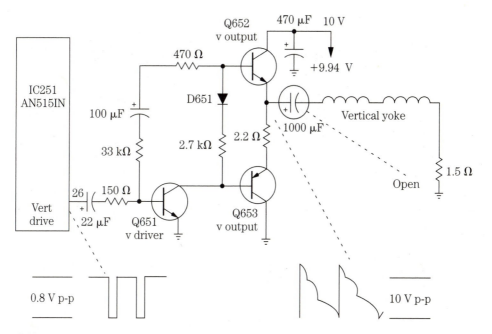

6-13 An open 1000-µF coupling capacitor produced a vertical rolling and improper sweep.

than 30 Ω. When the 1000-µF coupling capacitor was shunted, the raster returned to normal height.

Vertical fold-over

Check for leaky vertical output transistors when vertical fold-over or insufficient height is a problem. Take critical voltage tests on the output transistors or ICs (Fig. 6-14). Remove one lead on the vertical output coupling capacitor. Clip another capacitor of the same or higher capacity from the vertical output sweep transistor or IC to the yoke winding. Notice if fold-over and insufficient height disappears.

Dead, no sound

From volume control to speaker, a defective component can cause the no-audio symptom. Signal trace the sound circuits from the volume control toward the speaker with an external amp. Locate the volume control, and check the audio at the center terminal. If it is normal, proceed toward the speaker. Go from base to base of the audio transistors or to input and then output terminals of suspected IC.

When the sound quits, test transistors or ICs in that stage. Take critical voltage measurements on the output circuits. Suspect a leaky IC or an open coil in the B+ circuits when there is no sound and just a hum in the speaker. Locate the sound circuits on the PC board starting at the speaker, output IC, or transistors (Fig. 6-15).

6-14 Locate the vertical output transistor on a heat sink.

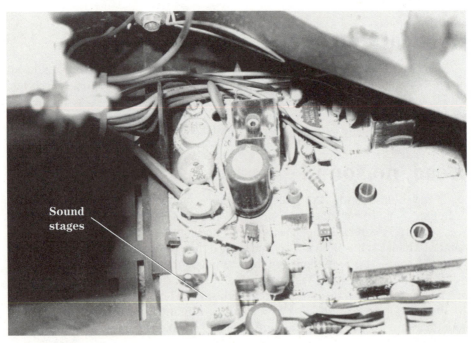

6-15 Locate the sound circuit with IF coils and output IC or transistors.

Weak sound

Signal trace the audio circuits with an external amp when weak sound is a problem. Weak audio can result from open AF transistors, coupling capacitors, or speaker capacitors. Look for dried-up coupling capacitors between stages with loss of audio. Weak sound can be caused by open or dried-up capacitors across emitter resistors in the audio output circuits. A dropped cone or shorted voice coil to the center magnet of the speaker can produce weak sound.

Distorted sound

Often, distorted audio is found in output circuits with leaky transistors, ICs, or capacitors. Check the speaker or leaky output transistor for mushy sound. Check the speaker cone to see if it is dragging against the center pole piece. Garbled sound can be caused by open, burned, or broken foil wiring in the base circuit of the audio output transistors. Replace the output single-ended transistor if popping noises occur after the set warms up.

Extreme tunable distortion can be caused by a change in frequency in the discriminator or sound coil. Touch the coil with an insulated tool with a station tuned in. This coil is eliminated in some of the newer black-and-white chassis and is replaced with a fixed 4.5-MHz ceramic filter (Fig. 6-16).

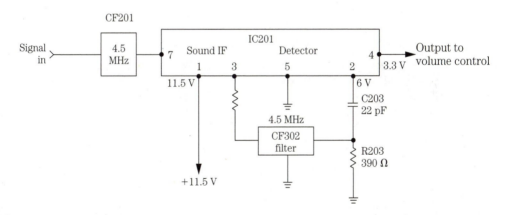

6-16 A ceramic filter might be found in some sound circuits instead of the discriminator coil.

Humbug

Hum in the sound and black bars upon the screen can be caused by dried up or open filter capacitors (Fig. 6-17). Shunt each large filter capacitor in the low-voltage power supply. Clip the electrolytic capacitor across the suspected one with the power off. Also, check the voltage regulator circuits in the older black-and-white chassis for hum and black bars in the raster. A leaky or shorted regulator transistor can cause dark bars in the picture.

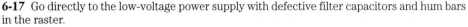

6-17 Go directly to the low-voltage power supply with defective filter capacitors and hum bars in the raster.

Defective filter capacitors feeding the vertical circuits can cause vertical crawling. Check the voltage source at the vertical output transistors. Suspect the low-voltage power supply with a low-voltage source. In a J.C. Penney 685-2064 black-and-white chassis, vertical crawling was noticed in the raster. This voltage source was traced back to a transformer half-wave rectifier source (Fig. 6-18). The voltage measured at C909 and C729 was 19.1 V. When the 100-µF capacitor was shunted, the picture returned to normal.

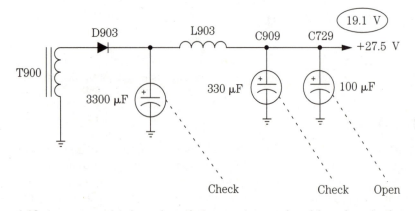

6-18 An open or dried-up electrolytic capacitor produced hum bars in the raster and picture.

Intermittent video

Monitor the video with a scope at the video amp output. Most video problems result from leaky video transistors or ICs, improper voltage, open peaking coils, or a change in resistance of collector load resistor. Take critical voltage tests.

If you have white retrace lines and cannot turn down the brightness, check the video amp for leaky or open conditions. Sometimes the transistor will open and test open or leaky, and sometimes the transistor will open after a voltage is applied under load. Spray coolant on the transistor to make it act up. Check for poor coil and terminal connections within the video circuits.

Fuzzy/dark picture

The picture was not clear with a dark raster in a Sanyo 21T68 black-and-white portable. With the brightness control wide open, insufficient brightness was noted. The dark screen appeared like a defective picture tube. Voltage measurements were taken upon the CRT, and very little voltage was noted upon pins 6 and 7 (Fig. 6-19). The screen grids should have a high voltage from the boost power source.

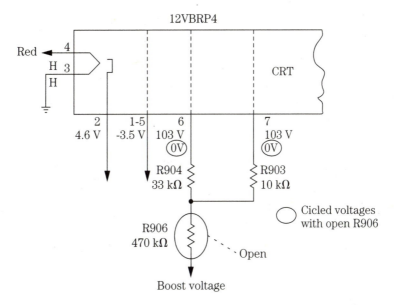

6-19 R906 produced a fuzzy and dark picture in a J.C. Penney 685-2064 portable.

Pins 6 and 7 were traced back to voltage-dropping resistors, R903 and R904. No voltage was found at either end of the resistors. R906 (470KΩ) had high-boost voltage on one side and zero on the other. A quick resistance test of the 470K resistor was found open. Check all voltages upon the picture tube before replacing the CRT.

Bright sunlight

Excessive brightness and an all-white screen can be caused by defective components in the video or picture tube circuits. Extreme brightness with no control over the brightness with normal sound was found in a Goldstar KMB220G chassis. Go directly to the video output stage for no control of brightness. The video output transistor was located by tracing back from the cathode terminal of the CRT (Fig. 6-20).

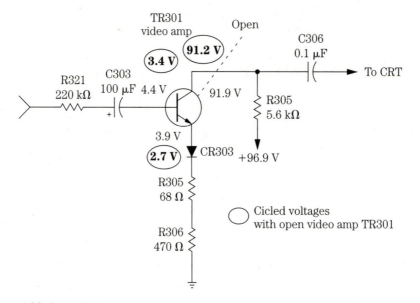

6-20 Open TR301 caused excessive brightness in a Goldstar portable TV.

A quick waveform test of the broadcast signal was found upon the base terminal of the video amp, and very little signal was found at the collector terminal. TR301 was suspected of being open; however, when it was checked in the circuit, it tested normal. Voltage measurement upon the collector terminal was quite high (91.2 V) compared to the base (3.4 V) and emitter (2.75 V). The video output transistor was replaced and restored a normal picture. No doubt the transistor broke down when the higher voltage was applied.

Poor brightness

Poor or improper brightness can be caused by defective components in the video and picture tube circuits. Check the high voltage on the picture tube. Take critical voltages on the CRT socket (Fig. 6-21). Low brightness can be caused by improper voltage on the grid terminal with open boost resistors or circuits. Low or no boost voltage can be caused by a leaky boost diode, open resistor, or dried-up electrolytic capacitor in the boost voltage source.

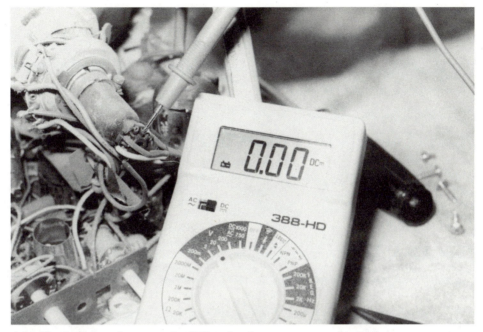

6-21 Check all the voltages on the CRT socket for poor or excessive brightness.

A gassy or shorted picture-tube gun assembly can cause scanning lines and no control over brightness even with normal CRT voltages. Lightly tap the end of the picture tube if you have a flashing raster. Don't overlook a defective heater element inside the gun assembly. A picture tube with a heater element over 45-Ω resistance should be replaced. Normal heater resistance in black-and-white tubes is less than 25 Ω. The high resistance heater will have a weak and negative picture.

Isolate and locate the suspected component on the PC board. Take transistor and IC numbers and letters to locate the replacement, type, and circuit that it performs in. Look up semiconductor parts in a universal replacement manual. Locate defective circuits with components found on separate heat sinks. Take critical voltage and resistance measurements and waveforms, and use an external audio amp to locate the defective part.

Unusual picture

The picture was all dark except a faint outline around the picture with light at the outside edges in a Goldstar VR-230 portable. At first, a picture tube was suspected, but it tested good. Voltage measurements upon pins 6 and 7 should be around 106 to 110 V; however, the Goldstar measured only 17 V. The boost voltage originates in the horizontal output and flyback circuits (Fig. 6-22).

Boost rectifier CR302 and R308 tested normal. A 10-μF electrolytic 150-V capacitor was clipped around C302. The chassis was fired up, and the brightness came up with 110 V of boost voltage. C302 was replaced with a 10-μF, 160-V replacement capacitor.

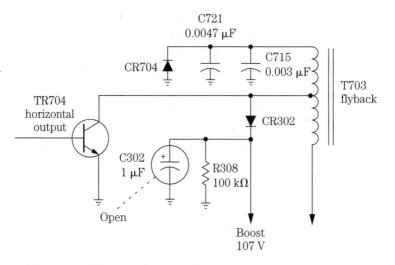

6-22 An unusually dark picture with a faint outline was caused by an open C302 in the boost voltage circuits.

On a diet

Suspect the horizontal or low-voltage power supply when the sides are pulled in. Hum might be heard in the sound with width problems. Narrow width with 60 cycles in the picture indicates power supply problems. Intermittent voltage regulator transistors can cause intermittent width. Improper adjustment of the B+ control can cause the sides to pull inward.

In a Goldstar KMC1910G portable, the sides were pulled in and the raster appeared like a barrel. Low voltage was found at the collector terminal of the horizontal output transistor. The B+ adjust would not change the dc voltage at the collector terminal. The voltage regulator (TR904) was located with the B+ control (Fig. 6-23). No voltage was found upon the emitter terminal with high negative voltage at the collector terminal. A 6.8-V zener diode was leaky, and a voltage-dropping resistor appear burned. Replacing R908 and D906 solved the barrel-type symptom. Always, adjust the B+ control for correct voltage on the horizontal output transistor.

Storm watch

The dead black-and-white chassis might be a victim of a lightning storm or power outage. Check the input terminals of the ac cord and antenna-tuner cable for burned and damaged wiring. Parts in the low-voltage power supply can be blown apart, including the PC wiring. When the PC wiring is stripped from the board, the TV chassis should be totaled out. Sometimes lightning will occur several blocks away and only damage the antenna wires and tuner cable. When the TV chassis has a direct hit in the power line, suspect damage to the low-voltage power supply.

Check and repair the ac input terminals. Test the step-down transformer for open primary winding (Fig. 6-24). Test each silicon diode in the bridge circuit. Check the

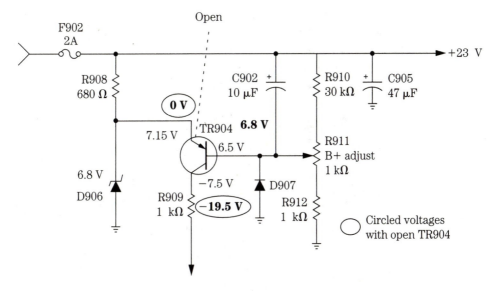

6-23 The narrow raster resulted from an open TR904 regulator in the low-voltage power supply.

on/off switch and connecting wires. Test all regulator transistors and diodes found in the regulator circuits. Replace all transistors in the regulator circuit if there is extreme lightning damage. Check all PC wiring in the low-voltage power supply for stripped copper or beads of copper. Repair damaged PC board with regular hookup wire.

Black-and-white TV symptoms

These various symptoms might help you isolate, locate, and repair the black-and-white TV chassis. The case histories can be applied to any black-and-white TV set.

Sides pulled in

Isolation   You cannot adjust the B+ control up on a J.C. Penney 685-2064 TV.
Location   Locate the horizontal output transistor mounted upon a heat sink.
Repair   Replace TR801, R802, and R805. R802 and R805 were found burned and a change in resistance. TR801 was found open (Fig. 6-25).

No raster, no picture, no sound

Isolation   Check the low-voltage power supply or horizontal output circuits.
Location   Locate the horizontal output transistor on a separate heat sink close to flyback.
Repair   Take a voltage test on the horizontal output and driver transistors. If there is low or no voltage, suspect a leaky output or poor voltage source. Test both transistors in-circuit. Replace them if they are leaky or open.

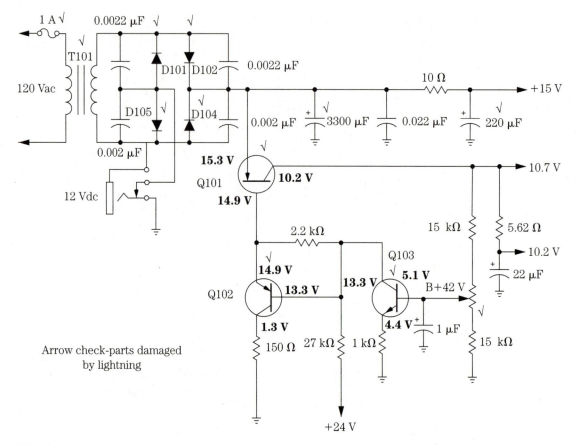

6-24 Check the following components when lightning damages the black-and-white TV chassis.

Dead, no sound

Isolation If voltage is low on the horizontal or driver transistor, scope the driver and oscillator waveforms. No waveform on the base of the driver indicates a defective horizontal oscillator circuit.

Location Locate the horizontal hold coil control near the rear of the chassis. The oscillator transistor should be near the coil and horizontal output transistor.

Repair Take critical voltage and resistance measurements on the horizontal oscillator transistor. Inspect the oscillator coil for broken connections. Take in-circuit tests of the oscillator transistor and continuity tests of the oscillator coil.

Intermittent raster and sound

Isolation Monitor the horizontal output and driver transistors with a scope and DMM.

Location Locate the horizontal output on a heat sink, and the driver transistor will be nearby.

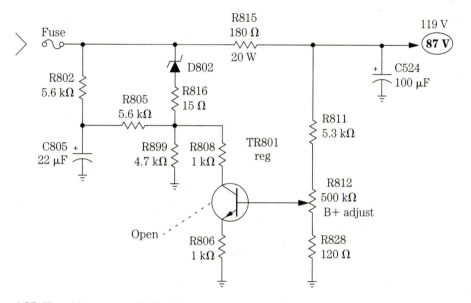

6-25 The sides were pulled in because of an open TR801 regulator in a J.C. Penney black-and-white portable.

Repair Where waveforms cease, check the driver and oscillator transistors. Take critical voltage tests of the horizontal oscillator and driver transistors. In a Quasar 5TS-629 chassis, both waveforms and low voltage were found on the driver transistor (Fig. 6-26).

Overscan lines

Isolation In a G.E. 12X13 chassis, the raster had overscan lines and no control over the brightness (Fig. 6-27).

Location Locate the CRT socket, and take voltage measurements.

Repair The voltages upon the picture tube were fairly normal. When the end of the CRT was tapped, the raster would come and go. The picture tube must be replaced with a shorted gun assembly.

Horizontal drift off-frequency

Isolation Usually horizontal drifting occurs in the oscillator transistor, coil, capacitors, or circuit.

Location Locate the oscillator coil and transistor.

Repair Replace the oscillator transistor and capacitors in the tuning circuit. Spray coolant on the transistor and capacitors to see if the unit drifts off-frequency.

Drifts off after 5 minutes

Isolation Monitor voltages on the horizontal oscillator. Use coolant on each transistor and capacitor.

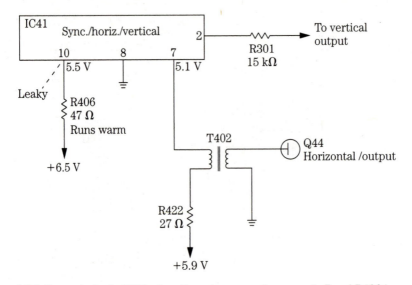

6-26 Suspect a leaky IC41 when there is no waveform on pin 7 and R406 is warm.

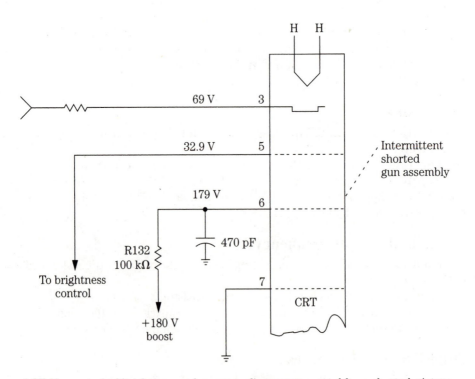

6-27 No control of brightness and overscan lines were caused by a shorted picture tube.

Location Locate the oscillator coil and transistor. They should be near the driver and transformer.

Repair Q601, C607, and C606 were replaced to solve the horizontal drifting problem in a Samsung BT-2397TR (Fig. 6-28).

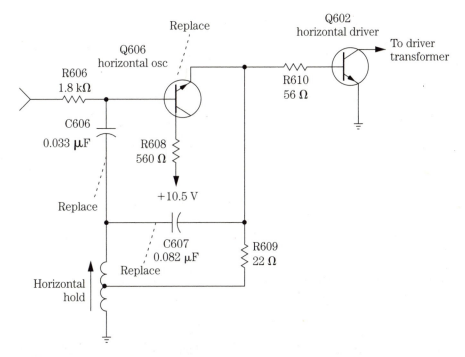

6-28 Replace Q606, C606, and C607 in a Samsung portable for horizontal drifting.

No high voltage, no sound, no raster

Isolation Check waveforms on the base of the horizontal output and driver transistors.

Location The driver transistor is near the driver transformer and output on a separate heat sink.

Repair There was low voltage on Q602 and suspected leakage. R612 was running hot. The driver primary winding was found to have only 0.2 Ω. Replaced T601 and Q602 in a Sears 401.51060050 (Fig. 6-29).

Horizontal white line

Isolation No vertical sweep.

Location Look for the vertical output transistors on separate heat sinks or an IC on a heat sink.

Repair Check the metal part of the transistors for voltage. One transistor will have a positive high voltage, and the other a low voltage, tied to ground through a resistor. Check the transistors in-circuit. Measure the supply voltage on suspected ICs.

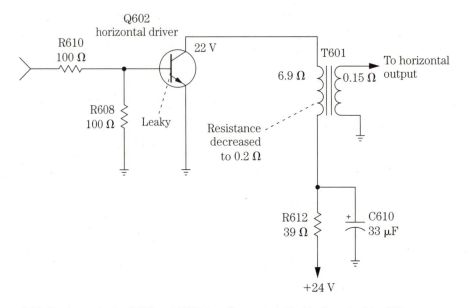

6-29 Replace a leaky Q602 and T601 in a Sears portable black-and-white TV.

Vertical fold-over, keeps flipping

Isolation   In a Citek 2213, the picture was only 5" high. Usually fold-over occurs in the vertical output circuits.

Location   Locate the vertical output IC. All vertical circuits are found in this IC.

Repair   IC501 was running red hot. A vertical waveform was found on the input and output terminals of the IC. Yoke assembly C511 (1000 μF) was open (Fig. 6-30).

Intermittent vertical sweep

Isolation   Intermittents can occur in just about any part of the vertical circuits. Suspect transistors, ICs, coupling capacitors, bad wiring, poor terminal connections, and improper low voltage.

Location   Locate the output transistors on the heat sinks and the vertical hold control on the rear apron.

Repair   Test the transistors in-circuit. Take voltage measurements. Check for burned resistors in the power supply source. Monitor the supply source for intermittent voltage (Fig. 6-31).

Insufficient vertical sweep

Isolation   Check the output transistor or IC and the voltage supply source.
Location   Locate the heat sinks.
Repair   Check the output transistors in-circuit for open or leaky conditions. Replace both when one is found to be leaky.

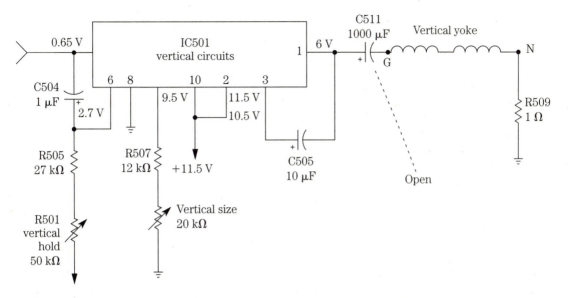

6-30 The picture was only 6" high because of an open coupling capacitor (C511).

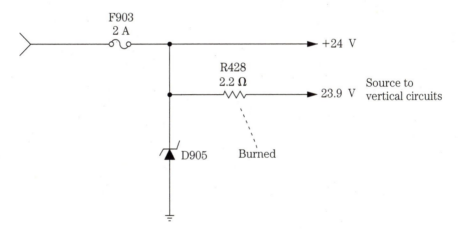

6-31 R428 produced intermittent vertical sweep in a Sears black-and-white TV.

Vertical bouncing

Isolation Vertical jitters, bouncing, rolling, and crawling can be caused by bad electrolytic capacitors in the supply voltage or vertical circuits.

Location Locate the vertical output transistors.

Repair Shunt electrolytic capacitors in the vertical circuits and low-voltage source (Fig. 6-32).

Dead, no sound

Isolation Locate the dead section with an external audio amp starting at the volume control.

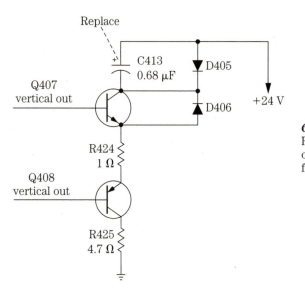

6-32
Replace C413 in the vertical output circuit of a J.C. Penney TV for vertical bouncing.

Location Locate the volume control, and trace to the first AF transistor or IC.

Repair The no-audio problem was found in a Zenith 19GB1Z with signal traced up to the connection of R1013 and R1014, but there was no speaker sound. C1018 was found to be open (Fig. 6-33).

No audio, hum in the speaker

Isolation Check the speaker, volume control, and AF transistors or ICs with an external audio amp.

Location Locate the speaker, volume control, and AF transistors or ICs or the transistor pre-amp.

Repair Signal trace the audio by the numbers. There was no signal at the sound IF amp (IC301) in a Samsung BT-239TK. Voltage tests indicated that the IC was leaky. Replaced IC301 with ECG712 universal replacement (Fig. 6-34).

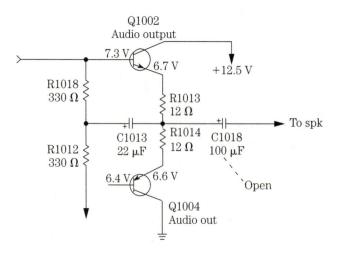

6-33
An open C1018 produced no sound in a Zenith portable.

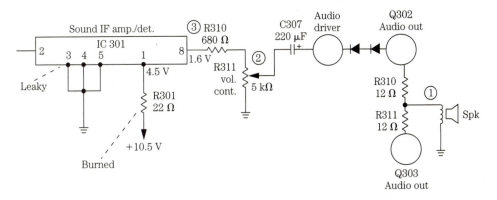

6-34 Check by the numbers with an external amp to signal trace audio circuits.

Brightness comes and goes

Isolation   Check the video output transistor or IC and the picture tube circuits.

Location   Locate the CRT socket.

Repair   Check the high voltage on the CRT (2 kV). Check voltages on other pins with low voltage at pin 6. Should be 110 V. Resistor R261 was found to be open (Fig. 6-35).

Cannot turn down the brightness

Isolation   Check the CRT circuits, boost voltage, and video output transistors or ICs.

Location   Take voltages on the CRT socket. Trace the circuit back to the video output transistors.

Repair   Found low voltages on video amp Q202 in a Philco CHUWA chassis (Fig. 6-36).

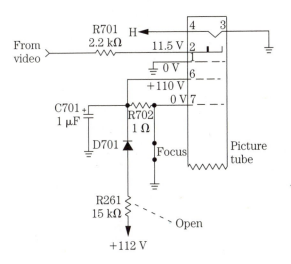

6-35
An open R261 in the boost source caused brightness to come and go in a Technika chassis.

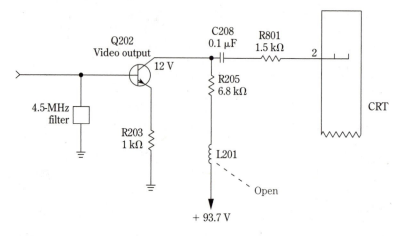

6-36 Brightness could not be turned down with open L201 in a Philco CHUWA chassis.

Black-and-white TV troubleshooting chart

See troubleshooting chart (Table 6-1) for additional black-and-white TV problems.

Table 6-1. Troubleshooting the B&W TV chassis chart.

Symptom	Cause	Repair
Tuner		
Snowy picture	Dirty tuner Antenna input	Clean the manual tuner. Check the antenna balumn coils.
Intermittent picture	Dirty tuner	Spray cleaning fluid into the tuner switching areas.
No picture—rush in sound	Defective tuner or IF circuits	Check the voltage at the tuner. Check the AGC voltage at the tuner. Sub the tuner to see if the tuner or IF or AGC circuits are defective.
Video		
Really bright screen	Video amp Picture tube	Test the video amp in-circuit. Check the picture tube voltages. Test the CRT with a tube tester. Check all the voltages on the CRT. Check the boost voltage.
Cannot turn down the brightness	Open video transistor	Test the video amp. Check the voltages on the video amp. Scope the video circuits.
	CRT	Test the picture tube. Check the voltages on the CRT socket.
Dark screen	Weak CRT	Test the picture tube.

Symptom	Cause	Repair
	No HV voltage	Check the anode with an HV probe.
	Check video circuits	Scope the video circuits.
	No CRT heater	Test the continuity of the CRT heaters.
Vertical		
No vertical sweep	Check vertical out	Test the output transistors. Scope the signal in and out on the IC.
	Electrolytic capacitor	Shunt the yoke coupling capacitor.
	Yoke	Test the continuity for open.
Insufficient vertical sweep	Low-voltage source	Test on the vertical outputs.
	Vertical output	Test the transistors. Scope the output circuits. Check the voltages on the output IC. Check the bias resistors and diodes.
Rolling pictures	Low-voltage supply	Test the voltage at the output transistors and ICs.
	Feedback circuits	Check the electrolytic capacitors.
	Improper hold	Check the resistance of the hold control.
Vertical crawling	Low-voltage source	Shunt the filter capacitors in the voltage source.
	Vertical output transistors	Look for leaky or open transistors.
Vertical foldover	Check vertical feedback circuits vert. coupling capacitor. Leaky IC output	Shunt each electrolytic capacitor in the vertical circuits. Shunt the capacitor between the yoke and output. Scope the signal in and out. Take critical voltage and resistance measurements.
Horizontal		
No horiz. sweep	Horiz. output	Test the horiz. output transistor. Check the damper diode. Check the flyback.
	Horiz. deflection IC	Scope the output waveform. Check the supply voltage source. Scope the drive input terminal.
	Horiz. driver	Test the driver transistor. Check the supply voltage.
	Flyback	Look for leaky winding or HV diodes. Overheats check the drive signal.
Horiz. output runs hot	Horiz. output	Check the amplitude of the drive waveform. Test the horiz. output transistor. Replace the horiz. output. Check the low-voltage source. Replace the electrolytic capacitor on the primary winding of the driver transformer. Replace the driver transformer. Solder the

Table 6-1. Continued.

Symptom	Cause	Repair
		driver transformer pc lead connections.
Sound		
No sound	Audio output	Test the signal in and out. Scope the audio signal. Test the transistors. Test audio output IC. Check the low-voltage source. Check the coupling capacitor. Test the speaker and wires. Sub the speaker.
Weak sound	Output circuits	Test the transistors. Signal trace the audio in and out. Check the output IC.
	Coupling capacitor	Look for an open or dried-up driver or AF capacitor. Look for an open speaker coupling capacitor.
Distorted sound	Output circuits	Test for leaky or open transistors or IC. Scope the signal. Check the burned bias resistors. Check the low-voltage source. Look for a defective speaker.
	Discriminator circuits	Readjust the discriminator coil. Check the capacitors in the discriminator transformer circuits.
Low-voltage power supply		
No voltage source	Regulators	Test the transistor or zener diodes. Test the regulator transistors. Look for a leaky zener diode. Take voltage and resistance measurements on the regulator IC.
Low-voltage source	Filters	Shunt the filter capacitors in the voltage source. Check for burned resistors. Check regulators.
Hum in the audio	Capacitors	Shunt all the filter capacitors. Check for a leaky regulator transistor or leaky zener diodes.

7
CHAPTER

Servicing the compact disc player

Like the video cassette recorder, the compact disc (CD) player is difficult to service without a schematic, so try to locate a diagram for the player or one similar to service the player (Fig. 7-1). There are many tests that can be made without a schematic. A block diagram will help in isolating the various circuits (Fig. 7-2). The block diagram of each circuit might help you understand how CD circuits perform.

Critical waveforms, voltage, and audio tests might locate the defective circuit. Checking the laser diode, RF, and EFM signal will indicate if the laser pickup assembly is functioning. Voltage measurements and signal waveforms on the loading, slide, and disc motors can determine if motor circuits are normal.

Critical waveforms throughout the signal circuits up to the digital-to-analog (D/A) converter indicate that these signal circuits are working. An external audio amp can signal trace the audio from the D/A stereo output to the line output and headphone jacks.

Laser circuits

The laser diode output can be checked with a laser light meter or infrared indicator. Measure the dc source applied to the laser diode. This might be a direct voltage or voltage applied from a laser driver transistor or IC. The laser diode output is fed directly to the RF amp IC or processor.

The eight-to-fourteen (EFM) signal is a very complex encoding scheme used to transfer digital data to a form that can be placed on a disc. The EFM waveform is found at the output of the RF IC and is fed to the digital signal processor (Fig. 7-3). When the error signal is not present at the servo IC, the chassis might shut down at once.

The RF, RFO, or EFM signal can be scoped at pins 15, 14, and 8 of IC4. When no output waveforms are present, check the voltage supply source (V_{CC}) on pin 11. If

7-1 The CD player might be found in the auto, home, tag-along units and inside a CD changer.

the laser assembly is normal but there is no RF or EFM voltage, suspect the RF amp (IC4). The EFM signal appears in a diamond-shaped waveform. A blurry diamond pattern is no good.

Eyes have it

When the eye pattern is not found at the RF amplifier IC or transistors, the CD player will automatically shut down. The laser optical assembly must be operating with a normal RF amplifier, and there must be an adequate low-voltage source feeding the front-end circuits. Scope the eye pattern (EFM) at the RF amp output (Fig. 7-4). This RF signal feeds the digital signal processor IC and servo circuits.

Play it safe

The laser optical beam cannot be seen by the naked eye. The service technician should avoid looking directly at the laser beam (Fig. 7-5). Keep a CD loaded on the disc platform at all times. Keep your eyes at least 25" from the optical laser beam. Place a conductive mat under the test equipment and player while servicing it. Wear a wrist strap to leak off static charges from your body to the chassis and ground. Do not forget to remove shorting or interlock devices after repairing the CD player. Take critical leakage tests. Replace the critical parts with originals.

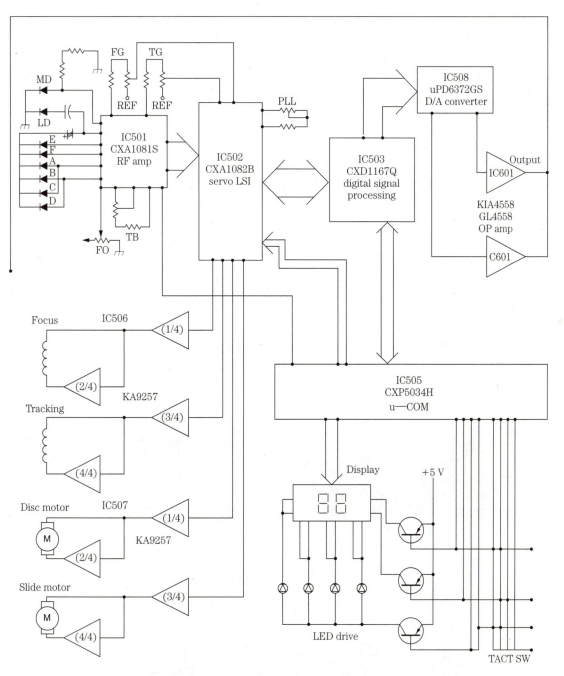

7-2 A block diagram of a portable CD player.

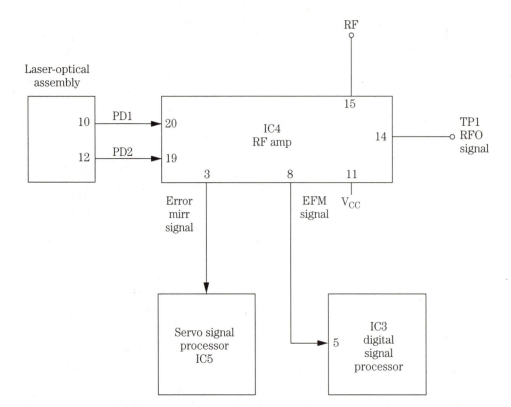

7-3 A block diagram of the laser optical assembly and RF amp with RF and EFM signal.

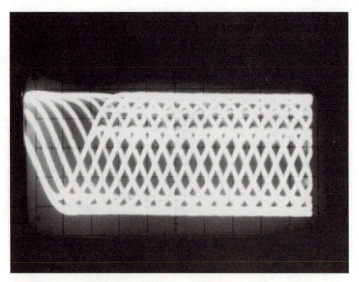

7-4 The eye pattern found at the output of the RF amplifier.

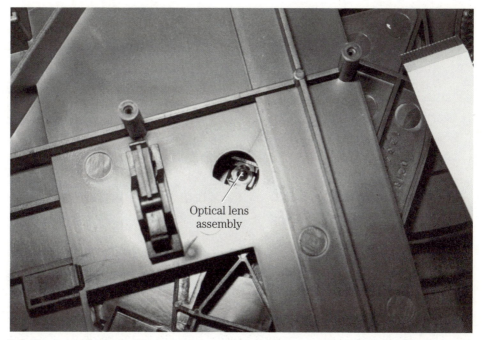

7-5 The laser lens assembly that shines against the bottom side of a CD disc found in a CD changer.

Digital signal processor

The EFM comparator changes RF signal into a binary value. The binary-coded signal is then fed to pin 5 of the digital signal processor (IC3), where it is demodulated, a bit clock is generated, errors are detected and corrected, lost data is interpolated, and the sub code for track number/elapsed time is demodulated. A bit clock must be derived from the EFM signal by a variable crystal oscillator (VCO) so that the data in the EFM signal can be read.

The demodulated EFM signal is converted to digital data and stored in a random-access memory (RAM) (IC3). The data is read from the RAM and fed to the digital filter and then to a D/A converter (IC11) (Fig. 7-6).

IC3 contains a digital low-pass filter with a cutoff frequency that is the first harmonic of the sampling frequency. Some CD signal circuits have a separate low-pass filter between the digital signal processor and the D/A converter. An analog filter in the output stage filters out frequencies above 20 kHz. However, eliminating these frequencies completely would require a very complex analog filter because they are very close to the derived output frequencies. By removing higher frequency components related to the sampling frequency using a digital filter, the specifications of the analog low-pass filter at the output stage can be simplified.

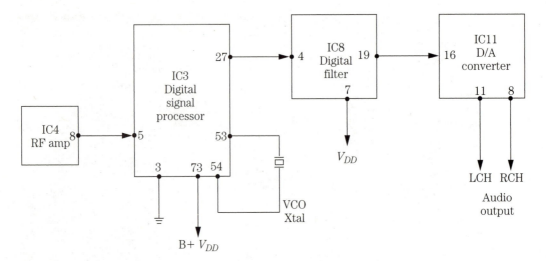

7-6 A block diagram of the signal processing circuits.

Digital/analog converter

The D/A converter (IC11) converts a digital signal to an analog signal (Fig. 7-7). The audio signal is then fed to the final output amplifier. One or two stages of amplification are found in most CD players. This audio signal is fed to a stereo line output jack or to a separate headphone amp. The line output voltage of most CD players is around 2 V.

Headphone amplifier

The headphone driver (IC1) is fed the stereo audio signal from the D/A converter audio output through a dual volume control to input pins 3 and 30 (Fig. 7-8). The amplified audio is taken from pins 10 and 13 to the stereo earphone jack. A muting transistor is found in each audio stereo channel at the earphone terminals (Fig. 7-9).

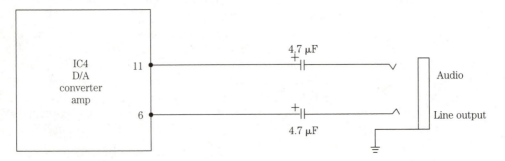

7-7 A typical block diagram with stereo line output jack.

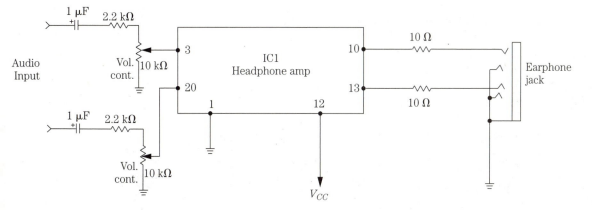

7-8 A block diagram of the headphone amplifier in the portable CD player.

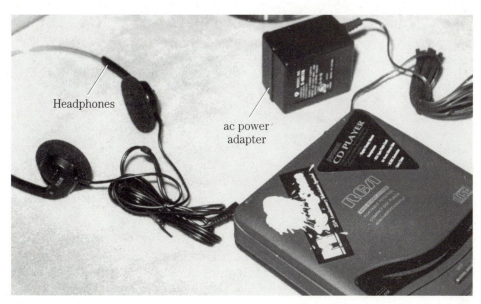

7-9 A pair of headphones plugged into an RCA portable CD player.

Low-voltage power supply

The low-voltage power supply within the portable CD player consists of a dc-dc converter supplying a higher dc voltage (5 V) from two or three batteries. The dc-dc converter provides a higher operating voltage with transistor output filtering to eliminate ripple.

The low-voltage circuits in a table model CD player or changer might have a small power transformer, bridge rectifiers, electrolytic filter capacitors, transistors, and zener diode regulators (Fig. 7-10). Several different voltage sources are needed to operate the various different circuits. Check the voltage on the main filter capacitors for correct voltages. Often, small electrolytic capacitors are found at each output voltage source in the CD player.

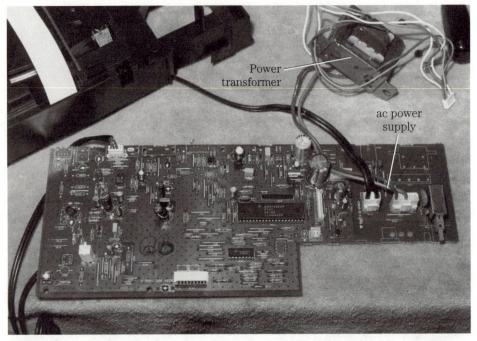

7-10 The step-down power transformer with bridge rectifier, regulator, and filter capacitor in the ac-operated CD player.

In the boom-box cassette/CD player the low-voltage power circuits are quite simple, with bridge rectification and high filter input (Fig. 7-11). The cassette motor and other circuits have a transistor-regulated voltage source, while the CD circuit has one or more transistor-zener diode regulators. The dc-dc converter in the CD section provides a 5-V and 10-V source for the CD circuits (Fig. 7-12).

The CD player operated by an ac power transformer has transistors, ICs, or both in the low-voltage power supply. Here a +12-V, –12-V, and –22-V regulators are found in the table-model CD players. The +12-V source is fed to regulator IC501 with +5-V output, IC502 with a –5-V output, and IC503, which provides voltage to the audio power output and capstan reset circuits (Fig. 7-13). Bridge rectifiers with high input filtering provide power to most CD regulator circuits.

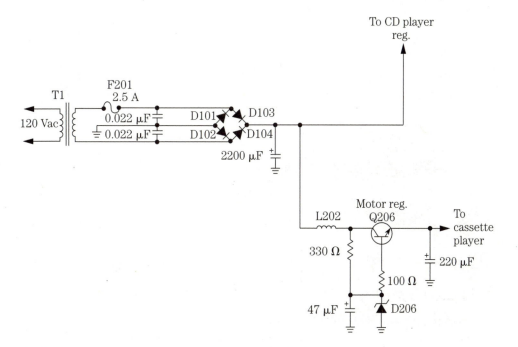

7-11 The low-voltage power supply in a boom-box CD player.

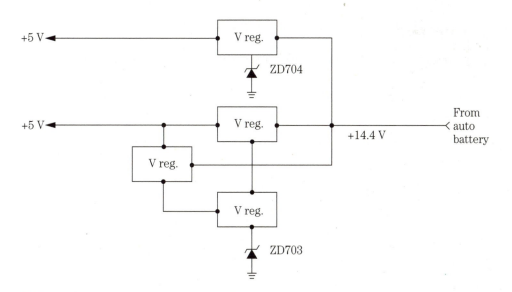

7-12 A typical block diagram of an auto CD player voltage-regulator system.

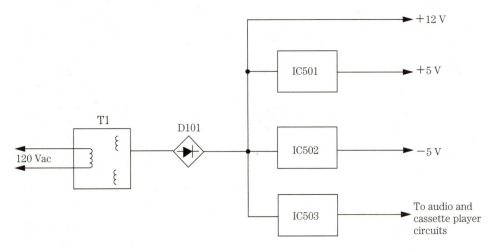

7-13 A typical block diagram of a table-top low-voltage power supply.

Focus error amp

The focus error amp is the difference between the RF 1-V amplifier output ($A +$ C) and the RF 1-V amplifier output ($B + D$). (The output becomes $A + C - B - D$.) The focus error (FE) is fed to the servo LSI, which controls the focus coil with the IC focus coil driver (Fig. 7-14). Likewise, the tracking error (TE) amp signal is fed to the servo LSI or IC to control tracking with the tracking coil driver IC.

Drag and sag

The loading motor might be referred to as a *tray motor*. The tray or loading motor moves the loading tray in and out for loading and unloading the disc (Fig. 7-15).

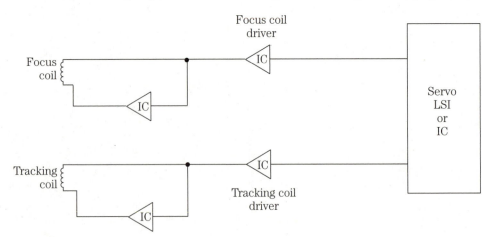

7-14 A block diagram of a focus error (FE) and tracking error (TE) signals from servo LSI or IC.

7-15 The loading motor pulls and pushes the loading tray in and out of the CD player.

The loading motor might drive a plastic gear that moves the tray assembly. This same plastic gear might raise and lower the clamper assembly. If the tray moves outward, the clamper raises, and after loading, it adds pressure on the loaded CD and holds it in position.

Intermittent operation of the loading tray can result from foreign material within the track area, a slipping drive belt, and a jammed mechanism. A dirty open/close switch can cause intermittent operation. Check the tray switch by shorting across its terminals. Notice if the tray rack or gears are binding. When the loading motor drives a belt to load and unload, check the belt for oil spots. Check the drive belt for worn or cracked areas. Inspect the loading pulley or plastic gear assembly for stripped or broken teeth. Check the CD changer loading tray for wires caught on the bottom of the moving tray (Fig. 7-16).

Loading motor

Usually the loading motor is controlled directly from the digital or control system IC. When the tray is open, a zero voltage is applied, and when closed, +5 V are applied from the control IC. This control voltage is fed to a driver IC or transistor circuit and feeds voltage to the loading motor winding. When measuring the voltage applied to the motor, you might find +4 V to open the tray and –4 V to close the tray and pull the CD into the player (Fig. 7-17).

7-16 The loading tray of a CD changer can hold 5 or 6 different discs.

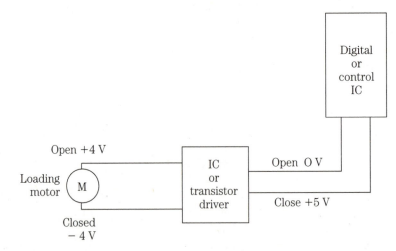

7-17 A block diagram of the loading motor circuits controlled by a digital-control IC.

Disc and slide motors

The disc motor rotates the CD. This motor is also called a *spindle motor*. The slide, SLED, or feed motor moves the optical pickup assembly from the center of the CD toward the outside edge during playing. It is just the opposite of the phono player. Both of these motors are controlled from the servo IC or processor (Fig. 7-18).

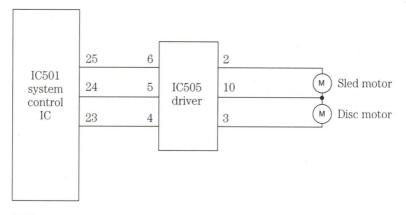

7-18 The SLED and disc motors are controlled by driver IC505 and IC501.

The spindle or disc motor is usually controlled by a constant linear velocity (CLV) servo circuit. Because the pits on the disc must be read at a constant rate, the linear velocity of the track must be constant. Therefore the angular velocity of the disc must be changed, depending on the position of the track being read to maintain a constant linear velocity. The angular velocity must increase as the pickup moves toward the center of the disc. The disc platform is directly coupled to the shaft of the spindle or disc motor assembly (Fig. 7-19).

7-19 The disc platform is mounted on the top of the disc motor shaft with the SLED motor moving the optical assembly outward to the edge of the disc.

The slide or SLED motor brings the laser pickup within the fine tracking control range. A tracking servo signal is used to move the pickup horizontally. The tracking coil keeps the pickup assembly on the track.

Important waveforms

RF-EFM waveforms

Check for an RF or EFM waveform at the RF test point or on the RF amp IC. This EFM eye pattern indicates that the laser pickup and RF amp circuits are normal (Fig. 7-20). If the waveform is missing or there is low voltage, suspect a defective laser diode or RF amplifier IC.

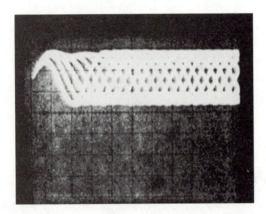

7-20
The eye pattern or EFM waveform at RF amp.

Test the laser diode with an infrared indicator to determine if light is being emitted from the laser diode. Check all voltages on the RF amplifier. The highest voltage should be the supply voltage pin terminal. The RF amp can be traced from the laser pickup wires to the PC board. Check all voltages on the laser driver transistor. Replace the RF amp before scrapping the laser assembly, as it is rather expensive to replace.

FM signal

The EFM or FM waveform indicates that both the RF amp IC and laser pickup assembly are functioning. Often the waveform signal will be on the high terminal numbers of the RF IC. Remember, the voltage found on the supply pin will be around +5 V and on the terminals will be around +2.5 V. This waveform is taken during normal disc play operations (Fig. 7-21).

Focus coil drive waveform

The focus coil (FO) waveform from the pins of the focus driver IC or transistor indicate that signal is applied to the focus coil (Fig. 7-22). The focus coil movement can be seen while it is searching by looking at the movement of the focus coil assembly. When the waveform is missing, check the focus signal from the servo IC or processor.

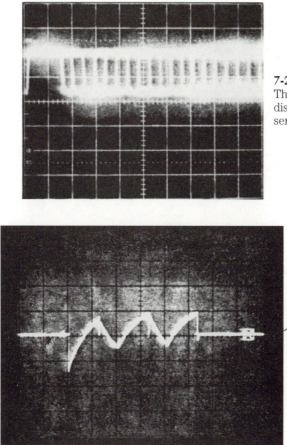

7-21
The FM signal waveform with the disc playing, fed to the signal and servo ICs.

7-22
The focus coil drive waveform when focus search failed or there was no disc on the tray.

0 V

The focus coil continuity can be checked with the low ohmmeter scale with the unit shut down. Sometimes the focus coil will move with the ohmmeter leads attached, indicating a normal focus coil assembly.

In the waveform in Fig. 7-23, the focus drive signal in CH1 of the scope indicates when the focus search is accomplished. The focus okay (FOK) signal is shown in CH2 from the RF amplifier. This circuit generates a signal to determine when the laser spot is in the reflecting surface of the disc. Often the FOK signal is high when the laser is in focus. When the waveform is missing at the transistor or drive IC, suspect the servo or RF processors.

Tracking coil waveform

The tracking coil waveform indicates that the tracking driver IC or transistor is normal (Fig. 7-24). The tracking coil waveform can be traced from the driver IC up to the tracking coil assembly. Locate the tracking coil assembly wires on the terminals of the pickup assembly. If the waveform is missing, take voltage measurements on the driver IC or transistor. Trace from the tracking coil winding to the driver IC. In some later models, the tracking and focus coil driver circuits are found in one IC component.

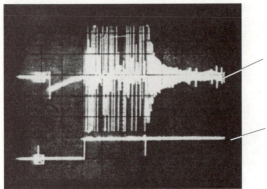

7-23
The focus drive waveform when focus search is accomplished.

0 V (CH1)

0 V (CH2)

CH1: Focus coil drive signal
    2 V/div.
CH2: FOK

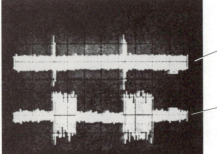

0 V (CH1)

0 V (CH2)

7-24
The tracking coil drive signal of CH2.

CH1:TEO
    1 V/div.
CH2:Tracking coil drive signal
    2 V/div.

In Fig. 7-25, during forward track traverse, the CH2 coil drive signal is at the bottom when time division is 0.5 mS/div. During backward track traverse, the tracking coil drive signal is at 2 V/div. (Fig. 7-26).

Feed motor drive waveform

During normal play, the feed motor drive waveform should look like Fig. 7-27. This waveform indicates that the feed motor IC and servo processor are normal. Suspect a defective driver IC or transistor when no signal is applied to the feed or slide motor. Check the supply voltage source. Locate the feed motor on the chassis, and trace out the motor wires to the driver IC.

The portable CD player

The portable CD player was designed for the ardent, on-the-go, music lover (Fig. 7-28). The portable player operates from batteries or a battery pack. The portable

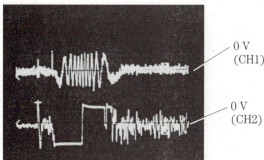

0 V
(CH1)

0 V
(CH2)

7-25
The tracking coil drive waveform
from the tracking coil driver IC.

CH1:TEO
1 V/div.
CH2:Tracking coil drive signal
2 V/div.

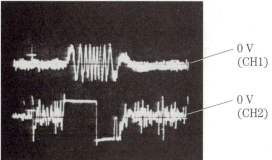

0 V
(CH1)

0 V
(CH2)

7-26
The tracking coil drive waveform
during backward track traverse.

CH1: TEO
1 V/div.
CH2: Tracking coil drive signal
2 V/div.

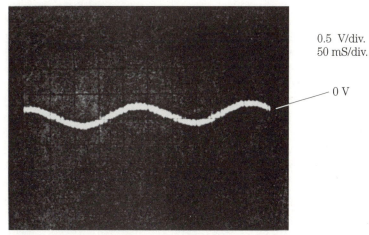

0.5 V/div.
50 mS/div.

0 V

7-27 A feed motor waveforms directly from the feed motor IC.

Disc
motor platform

7-28 The small RCA CD portable with a lid interlock.

CD player has circuits similar to the large table-top or changer player. Of course, there are only two motors, slide and disc motors, because the player has top loading. The motor circuits might be controlled by one IC. You can listen through a pair of headphones with either battery or ac operation. The portable player has two line output jacks to play through an external amp.

All portable compact disc players have a top-lid interlock switch system that provides protection for the operator when the top lid is opened for loading of disc. The lid interlock switch disables the laser signal. Remember, the optical lens assembly shines upward toward your eyes so that the power to the laser diodes is removed when the door is open (Fig. 7-29).

The focus, tracking, and motor circuits in a Realistic CD-3370 player are controlled by IC6. The focus coils are connected to pins 43 and 44, while the tracking coil connects to pins 41 and 42. The focus coil test points are TP10 (F–) and TP12 (F+) (Fig. 7-30).

IC6 also provides drive for the spindle and SLED motors. The spindle motor connects to pins 36 and 37. The SLED motor terminals connect to pins 34 and 35. In this portable CD player, the coils and motors are controlled with one IC6 driver. The slide motor moves the laser assembly out on sliding rods, and the disc motor rotates the playing disc (Fig. 7-31).

The portable CD player contains many SMD components, gull-wing IC and microprocessors, and several PC boards. These small components, PC wiring, and parts make servicing the portable CD player more difficult. There are more IC components found on the portable player because of the physical size. The suspected IC might control two or more different circuits. Check the supply voltage, and scope the sig-

7-29 The pen points at the optical lens assembly.

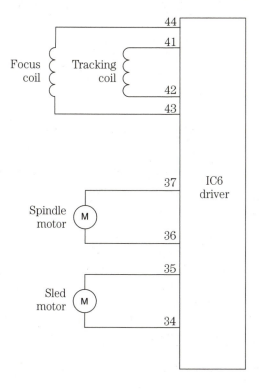

7-30
IC6 drives the tracking and focus coils, spindle, and SLED motors in a Realistic CD-3370 player.

7-31 The SLED motor moves the optical assembly on the guide rails in a portable CD player.

nal in and out of the suspected IC. Take voltage and resistance measurements upon each terminal pin before replacing IC.

The CD changer

The table-top or auto CD changer might contain five or six CD discs. The turntable or roulette motor rotates the tray holding the disc and stops for the correct selection. The roulette sensor circuits provide correct start and stop indicators of the roulette motor. The typical roulette motor might be controlled by transistors, ICs, or a combination of both (Fig. 7-32). A signal from pin 21 of system processor IC201 controls the transistors and ICs in the roulette motor circuits.

The roulette right and left input is applied to pins 5 and 6 of roulette motor driver IC103. IC103 provides a right and left voltage to the base of driver transistors TR103 and TR104. The reverse voltage applied to the base of each transistor can control the rotation of motor to the left or right rotation. Voltage at the emitter terminals of TR103 and TR104 is applied directly to the roulette motor terminals through CB111 and CB103 (Fig. 7-33).

Most electronic circuits found in the CD changer are similar to those found in any large table-top CD player. The big difference is the four to six different motors found in the changer CD player. A slide or SLED motor moves the optical assembly along the rails from inside to the outside edge (Fig. 7-34). The loading motor loads

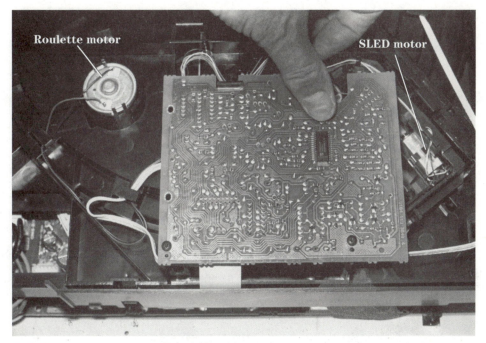

7-32 The roulette motor rotates the turntable or carousel in the CD changer.

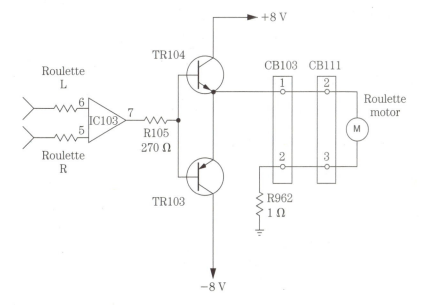

7-33 Driver transistors TR103 and TR104 produce voltage to rotate the roulette motor.

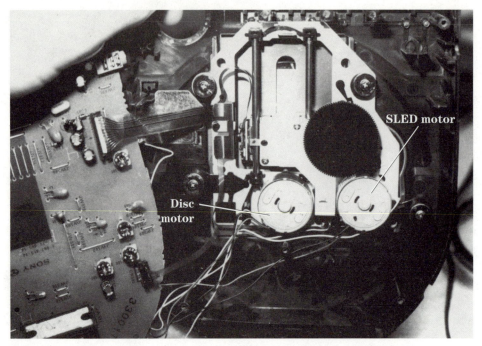

7-34 The SLED motor moves the optical assembly on the rails of a CD player.

the tray by moving it in and out. The roulette or select motor rotates the five or six discs in a turntable for selection. A disc or spindle motor rotates the compact disc at 500 to 200 rpm. The up/down motor assists in loading and playing the disc while the magazine or carousel motor rotates the turntable. In some changers, the carousel, tray or loading, and turntable motors operate from the same microprocessor (system control) (Fig. 7-35).

You will find several different PC boards found in the CD changer. A large board contains the ac power supply with voltage regulators, signal processors, and system control circuits (Fig. 7-36). While the different motor circuits are tied to a servo control board, the servo control board has motor transistors and ICs that make the different motors operate. Removing the bottom cover might uncover the servo board and motors (Fig. 7-37). The resistance of the SLED motors can range between 10 and 20 Ω, roulette motors 15 to 18 Ω, and loading motors 5 to 20 Ω. Check the continuity or motor resistance when voltage is applied at the motor terminals and the motor doesn't rotate.

Boom-box CD circuits

The boom-box with CD might have an AM-FM-MPX radio and cassette player within the same unit. Most boom-boxes have top-loading CDs, without a loading motor (Fig. 7-38). A rotating keeper at the top lid holds the disc in playing position.

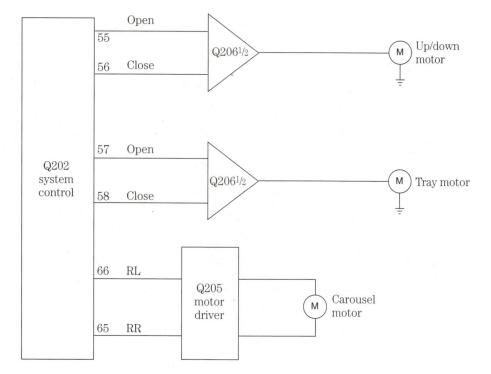

7-35 Q202 controls the up/down, tray, and carousel motors in a CD changer.

7-36 The ac power supply, voltage regulator, signal processor, and system control are found on one large board in a Sharp CD changer.

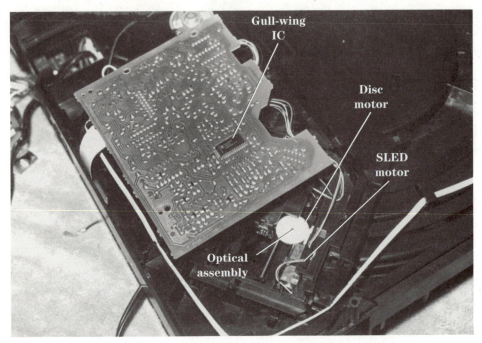

7-37 Here one gull-wing IC controls the various motors in the CD changer.

7-38 The CD is top-loaded in the boom-box combination player.

Block diagrams of boom-box CD circuits are quite similar to any CD player, except that you might find several ICs serving double duty. Only a disc and slide motor are found in the boom-box player. The low-voltage power supply circuits are rather simple, providing dc voltage for all circuits within the CD player.

Instead of a line output or headphone jacks, the boom-box output is fed into a function switch that selects cassette head amp, radio signal, or CD audio (Fig. 7-39). These signals are fed to the stereo audio output IC. A volume and tone control are found between the function switch and IC amp. The CD section operates, like the radio and cassette circuits, from several "D" batteries or an external dc source.

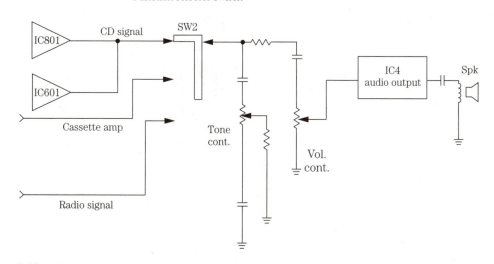

7-39 A block diagram of the CD function switch in a boom-box CD player.

Similar schematics

If possible try to locate a similar schematic when the correct schematic is not available. A block diagram of a boom-box, car CD player, or portable CD player might help in isolating the problem and locating the parts. Often universal ICs and processors are not provided for CD players. Use the original part numbers.

Checking SMD components

Be very careful when taking voltage and resistance measurements on SMD ICs, processors, and transistors. Use a magnifying glass to get on the correct terminal. Sometimes these gull-like leads are so close together that it's very easy to short two terminals together with regular test probes. Either sharpen the test probe to a fine point or purchase a pair of thin probe tips with leads.

The SMD signal processor might have 80 terminals. Usually the signal processor IC can be located as the one having the most terminals. The servo signal processor might have around 50 terminals. In portable CD players, you might find the focus, tracking, spindle, and slide motor drive circuits on one large IC having 44 or more terminals. Of course, this processor IC can be located by its PC board wiring connected to the motors, focus, and tracking coils (Fig. 7-40).

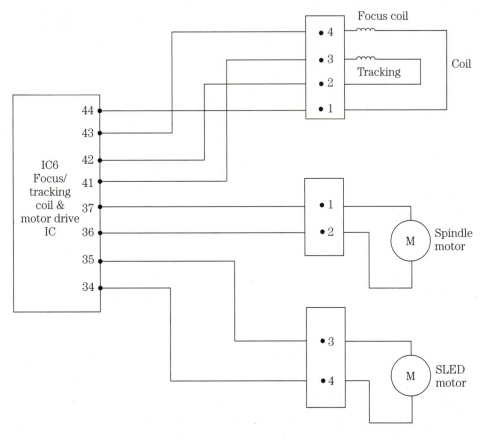

7-40 IC6 provides drive for a spindle and SLED motor, with focus and tracking coils.

Troubleshooting RF amp and pickup

Take an RF or EFM waveform on the terminals of the RF amp IC or transistor. If there is no RF signal, suspect a defective laser diode or RF amplifier. Take critical voltage measurements on the RF amp IC. If voltages are very low, suspect a leaky IC or power source. Locate the RF amp IC by its leads coming from the laser pickup. Check the voltage supplied to the photo detectors in the pickup assembly. Replace the RF amp IC that has poor voltage and little or no waveform.

Troubleshooting the digital signal processor

Check for EFM waveforms from the RF amp IC or transistor to the input terminal of the digital processor. Does the loading motor operate the disc tray? Usually the loading motor is operated with signal from the digital signal processor. Take a waveform to see if the crystal circuits are oscillating. Locate the digital signal processing IC. Measure the supply source voltage (V_{CC}), which is often the highest voltage on the IC. If these measurements are normal, the digital signal processor is probably okay. If not, take voltage measurements on all other terminals.

Troubleshooting the D/A converter

Check for analog or audio at the left and right output channels of the D/A converter with an external audio amp. No or low audio signal can indicate that signal is not getting to the D/A converter IC or that there is an improper supply voltage source. Measure the supply voltage (V_{CC}) from the low-voltage source. Don't overlook a defective mute transistor system when one channel is normal and the other is low in volume. Proceed to the mute or headphone circuits when there is low or no audio.

Servicing the audio mute system

Low or weak audio can be found in a D/A converter with a defective mute circuit. If both channels are weak, suspect a defective mute system in the digital signal processor. Disconnect either collector terminal of the mute transistor from one of the channels, and notice if the sound is normal. Suspect a defective mute transistor when the audio in one channel is very low. The audio can be signal traced from the D/A converter to the audio amp IC to the line output jack with an external audio amp.

Servicing headphone circuits

The headphone amplifier circuits can be signal traced with an external audio amp. Often the headphone amplifier connects to the audio signal before the line output transistors. The audio is amplified by one or two separate IC amp drivers. You might find one large IC for both channels in portable CD players.

The audio is fed to a dual volume control and input circuits of the headphone driver amp. The amplified audio is capacity coupled to the earphone jack with two muting transistors in the stereo channels (Fig. 7-41).

Signal trace the audio in and out of the headphone driver IC. Suspect the driver IC or defective mute circuits when both channels are dead. With one defective audio channel, suspect a mute transistor, coupling capacitor, or driver IC. Take voltage measurements on the driver IC to determine if the IC is defective. Remove the col-

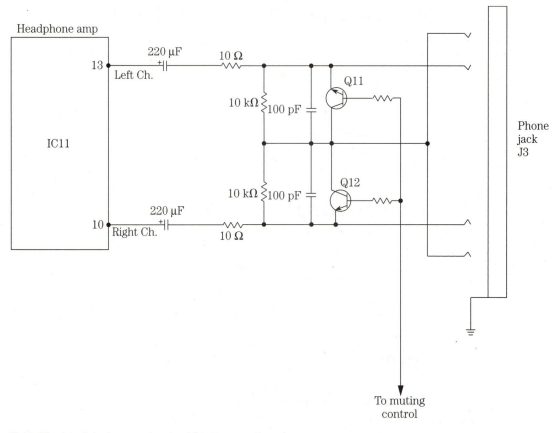

7-41 The headphone amp circuit with mute transistors.

lector terminal of the dead or weak channel's mute transistor. Locate the driver IC by tracing the PC board wiring from the jack to the first component.

Servicing servo circuits

The servo processor provides power to the tracking and focus coil, spindle, and slide motors through respective IC or transistor devices. The servo IC must receive tracking error (TE), focus error (FE), mirror, focus okay (FOK) data, interface data, system control, and clock data from the system control IC before the various circuits can function. The data latch, phase comparator input, VCO, spindle motor drive, and spindle motor on/off control signal from the digital processor to the input servo processor.

Improper operation of these functions might point to a defective servo control IC. Take critical voltage tests on each terminal pin. Locate the servo processor IC as the one with 40 or more pin terminals.

Check the focus and tracking coils with continuity ohmmeter tests when there is no or improper focus and tracking action. Measure the voltage on each driver transistor or IC. Take critical waveforms at each focus or driver IC.

When either the spindle or SLED motor will not function, take motor continuity tests with an ohmmeter. Measure the voltage at the motor terminals. Take critical voltage measurements on each driver IC or transistor. Scope the motor control and drive IC. Locate the driver and control IC by tracing the motor PC board wiring back to the first driver IC.

Poor loading

Poor or intermittent loading can result from a binding of gears and tray. Dress all cables and wires so that the turntable or roulette motor will rotate. Make sure no cables or wires are holding up the loading tray. Likewise check the belts and gears that are rotated by the loading motor. A jammed or broken gear might not let the motor rotate.

Locate the loading motor close to the front of the unit or belt and gears tied to the loading mechanism. Often the loading motor rotates a plastic gear or clamper assembly (Fig. 7-42). Trace out the wires from loading motor to the PC board to locate the driver IC or transistors.

Monitor the various voltages and waveform signals of the motor control circuit. Clip a scope probe to the servo IC output that drives the motor driver IC or transistors. Monitor the voltage across the motor terminals. Replace the defective motor when it stops and voltage is applied to the motor terminals. Remove belt or gear train if suspected of loading or binding up the motor.

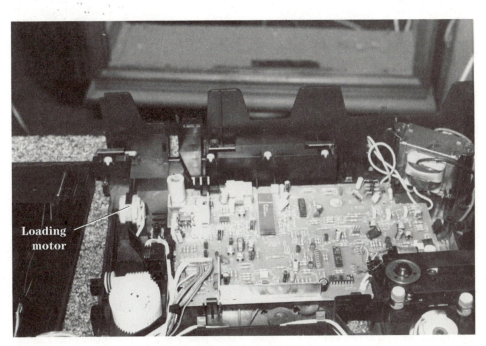

7-42 A loading motor found in a Magnavox CD changer.

No disc rotation

The disc motor is easy to locate right under the platform that rotates the disc. The compact disc is recorded at a constant linear velocity. This means that the disc rotation must be redirected as the laser pickup tracks to the disc's outer edge with a revolution varying from 500 to 200 rpm.

In a portable CD player, the motor speed is controlled by signals from pin 14 (SLD) of servo IC2 that are fed to pin 2 of the driver IC6 (Fig. 7-43). D1– and D2+ (17 and 18) MV voltage is fed to the disc motor terminals through connector CN3.

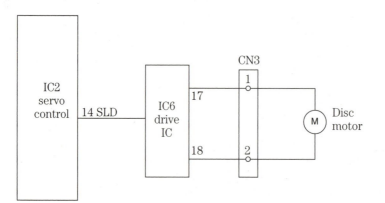

7-43 IC2 controls driver IC6 and the disc motor in a Radio Shack CD-3380 portable player.

Locate the disc driver IC, and scope the terminals fed to the disc motor. If there is no signal, scope the waveform from the servo control IC. Measure the voltage across the disc motor terminals. If voltage is present, take a quick continuity test of the motor winding. The disc motor can be checked to run by applying a 1.5-V battery across the motor terminals. By applying external voltage, you know the motor operates. Remove wires from motor terminals before applying voltage. Then check the voltage source supplied to the driver IC. Often, a positive and negative supply voltage is fed to the driver disc motor IC.

Symptoms

These various symptoms might help you locate and isolate the defective circuits in other machines.

Dead, no disc rotation in a Sony D-14 portable CD player

Isolation   The laser beam was not lit, and there was no disc motor rotation or indicator lights. Check the batteries or insert ac adapter.

Location   If there is still no operation, suspect a defective interlock (Fig. 7-44).

7-44 The open lid breaks the circuit of an interlock switch in the portable CD player.

Repair   Where the top lid extension point is inserted into the top interlock, the shorting switch terminals were out of line and dirty. Clean the contacts and prongs.

A brand new CD player came into the shop without any disc movement

Isolation   Had the customer removed the shipping screws?

Location   Turning the unit upside down showed the screws were still in place (Fig. 7-45).

Repair   Removing the shipping screws provided correct disc operation.

Intermittent and no CD function with normal radio and cassette operations in a Realistic 14-529 boom-box CD player

Isolation   All other functions were normal except the CD player, so the dc voltage source might be the problem.

Location   Locate the voltage regulator and electrolytic capacitors.

Repair   Take critical voltage measurements on the capacitor terminals. No voltage was found at the CD regulator or fuse F202. Cleaning the function switch solved the problem (Fig. 7-46).

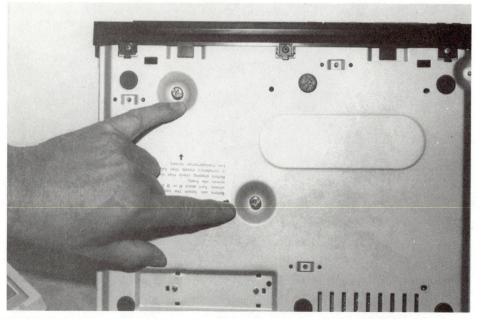

7-45 Remove the shipping screws in the bottom of the CD player before plugging in the ac cord.

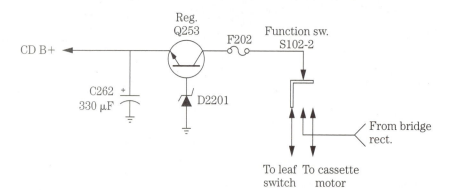

7-46 A dirty function switch S102-2 in a Realistic CD player caused intermittent operation.

A JVC VL-V400B CD player would start then shut down

Isolation First determine if the laser pickup and RF amp are supplying a signal to the signal processor and servo controller IC.

Location Locate the two RF amp transistors, Q201 and Q202, from the photo diode leads of the laser pickup.

Repair No RF or EFM waveform was found at the signal processor, Q201, or Q202. You should be able to scope the RF signal, even for a very short duration, before shutdown. RF amp Q202 was found to be leaky and replaced.

An Alpine 5900 car CD player wanted to run, then shut down

Isolation   The CD player was removed and connected to a 13.8-V dc source. An infrared test on the laser diode appeared normal. Either the RF amp (IC652) or optical pickup assembly was dead or there were defective signal processing or servo processing.

Location   RF amp IC652 was found with leads from the pickup assembly (Fig. 7-47).

Repair   A waveform (EFM) was missing at the RF test point. Voltage tests on the pins were quite low, below 3 V. Usually a 5-V source feeds pins 23 and 24. Suspected and replaced the RF amp (IC652).

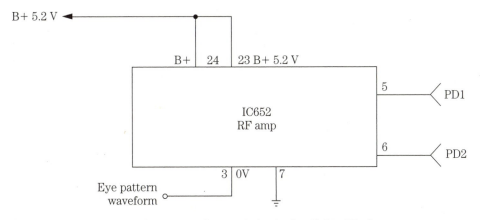

7-47 The EFM waveform (eye pattern) was missing in the Alpine CD player.

In an Onkyo DX-200, the CD player seemed to work without any digital panel numbers or operation

Isolation   Usually the front panel fluorescent display is operated by a high negative voltage from the low-voltage power supply.

Location   Locate the fluorescent tube on the front panel.

Repair   Voltage measurements on the tube indicated that the negative voltage was missing. The negative voltage regulator transistor (Q907) was located with D905 wired in backwards to the transformer blue lead. The collector terminal had –40.5 V with no voltage on the emitter terminal. Regulator transistor Q907 was open and replaced, restoring -28 V to the display tube (Fig. 7-48).

Troubleshooting charts

The block diagram shown in Fig. 7-49 for a Realistic boom-box radio, cassette, and CD player (14-529) might help in solving CD player problems.

The troubleshooting chart shown in Table 7-1 for a Realistic compact disc player might help in solving CD player problem when a schematic is not available.

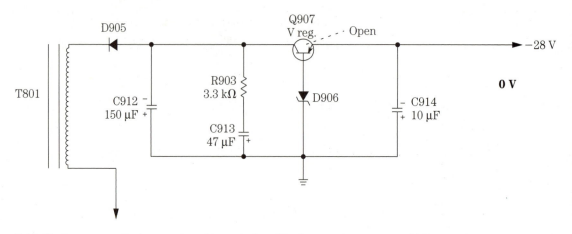

7-48 No fluorescent display was found in an Onkyo CD player with a missing –28 V caused by an open regulator transistor.

Table 7-1. Portable CD motor troubleshooting chart.

Symptom	Location	Repair
Defective slide or SLED motor operation	Does optical assembly move?	If not, check the voltage on the driver IC. Check the voltage on the motor. Take the continuity of the slide motor. Check for a defective mechanism. Check the belt and gear assembly.
	Does optical assembly continue after reaching outside edge of disc?	Take waveforms on the servo IC. Check for a defect inside the switch. Check the defective mechanism. Check the soldering of the IC terminals. Check the defective reset circuit.
	Does SLED motor reverse?	Check the motor polarity. Check the motor voltage. Check the SLED driver IC. Replace the defective IC.
Defective disc motor operation	Check the driver IC	Test for both positive and negative voltage. No voltage—Check low-voltage power supply. Check the voltage regulators. Replace the driver IC if there is voltage present.
	Defective spindle or disc motor?	Check the voltage on the motor terminals. Check driver IC voltage. Check the driver IC. Take the continuity of the disc motor. Replace the defective motor.
	Does disc motor stop in stop mode?	Check for voltage on the motor terminals. Check the driver IC voltage. The voltage should be zero. Check for a leaky or open driver IC or transistors. Test the

Table 7-1. Portable CD motor troubleshooting chart.

Symptom	Location	Repair
		or transistors. Test the voltage at the driver IC. Suspect the servo IC.
Defective loading motor operation	Does loading motor rotate?	If not, check the voltage on the motor terminals. Check the motor driver IC. Check the voltage on the motor driver IC. Check the jammed gear assembly. Check for a broken belt. Check for foreign material in the belt or gear assembly. Check the reverse switch.
	Pushed tray out—Will not load	Check the reverse switch. Clean the contacts on the switch. Check the polarity on the motor. Check the voltage on the driver IC or transistors. Suspect the driver IC. Sub another motor and not mounted.

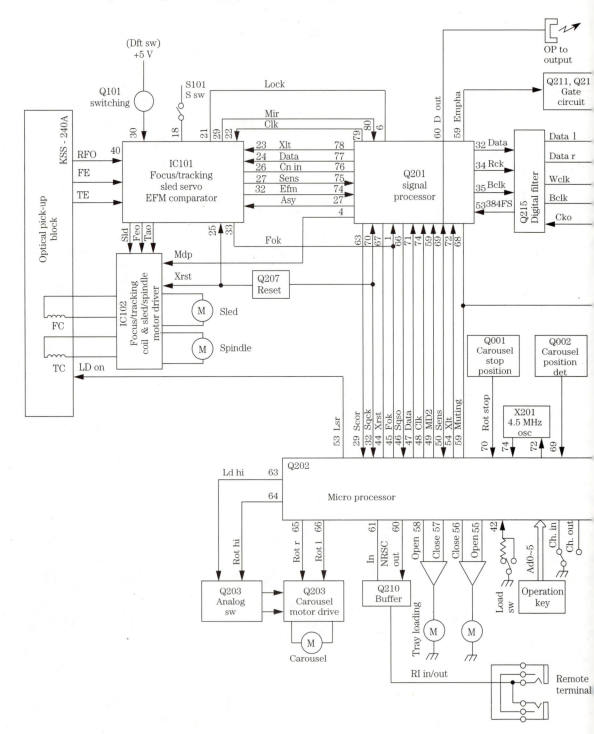

7-49 A block diagram of a Realistic CD player.

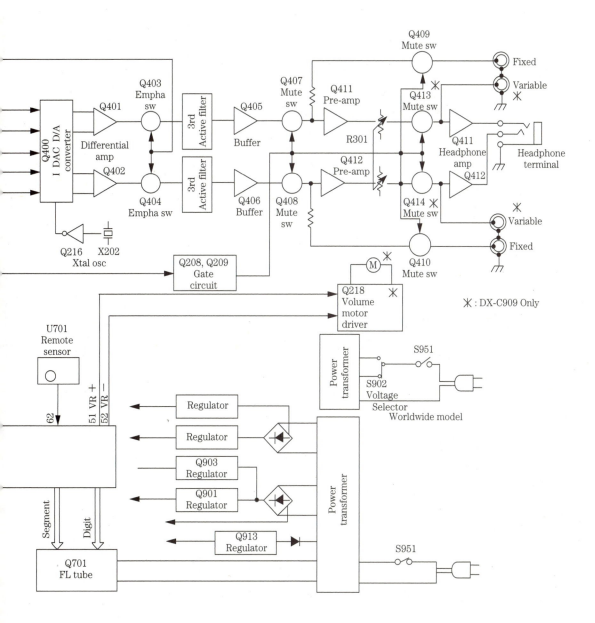

8
CHAPTER

Troubleshooting the TV chassis

About 85 percent of the troubles found in the TV chassis are located in the horizontal and vertical sweep circuits. The horizontal sweep circuits are the most difficult to service, especially with horizontal and high-voltage shutdown problems. Besides providing horizontal sweep to the yoke circuits, the horizontal output transformer supplies high voltage to the picture tube (Fig. 8-1).

The shoot out

Before tearing the chassis apart and jumping into troubleshooting the TV chassis, check the symptoms. Do you have a raster? Is the picture pulled up from the bottom (Fig. 8-2)? Do you have sound but no picture or raster? Start with the symptom shown upon the picture tube and the sound in the speaker. Use your eyes, ears, and nose to help troubleshoot the TV chassis.

No raster or high voltage might result with problems in the horizontal or high-voltage circuits. No raster, no picture, and no sound can be caused by the low-voltage power supply. Hum in the sound and hum bars in the picture might result from dried- up or open filter capacitors in the power supply or leaky voltage regulator. Only a horizontal white line indicates a defective component within the vertical circuits, and on it goes.

You can hear a frying noise in the sound, indicating a defective transistor, IC, or ceramic bypass capacitor. The tic-tic-tic of the flyback indicates problems within the horizontal output and flyback circuits. The arc-over of a spark gap upon the CRT might be caused by excessive dust and dirt in the gap area or too high a voltage. A cracking and arcing around the anode terminal of the CRT might result from excessive high voltage or poor rubber insulation of anode plug.

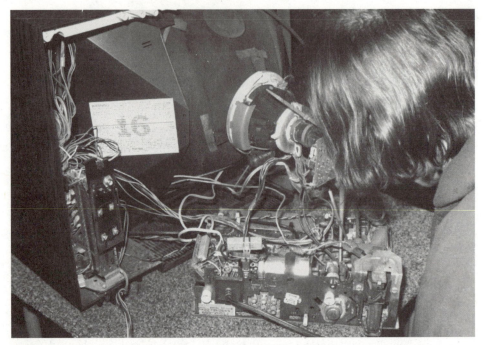

8-1 The electronic technician testing the horizontal TV circuits.

8-2 Check the symptom on the TV screen and speaker before tearing into the chassis.

On first base

After the symptoms, isolate the defective section upon the chassis. If the symptom points to the horizontal and flyback section, locate the flyback and horizontal output transistor upon a heat sink. Remember, the regulator, vertical output, and horizontal output transistor might look alike. Really high dc voltage upon the regulator and output transistor separates away from the vertical circuits. A scope waveform upon the horizontal output transistor and mounted close to a driver transformer locates the horizontal output transistor.

Likewise, the vertical output transistors are found mounted upon separate heat sinks. The vertical power output IC is found upon a heat sink, which is much fatter and long compared to the line voltage regulator (Fig. 8-3). Both the vertical and horizontal sweep signal are found in a large deflection IC with other circuits. Horizontal and vertical circuits are easily located upon the TV chassis. Check the top numbers and the letters on top of the IC, and look up the correct sweep IC in the semiconductor manual.

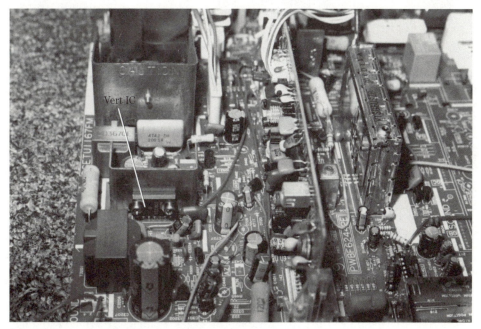

8-3 The vertical output IC is mounted on a separate heat sink.

The different TV sections

TV circuits consist of the tuner, IF, AF, sync, AGC, video, color, vertical and horizontal sweep, and low-voltage power supply. The low-voltage power supply provides voltage to most of the TV circuits. The tuner selects the broadcast stations, and signals are converted to the intermediate frequency (IF), with audio taken from the detector stage and video information fed to the video circuits. Color is added in the color and video circuits and is supplied to the three color guns of the picture tube (CRT). The horizontal and vertical circuits provide sweep to the yoke windings on the CRT (Fig. 8-4).

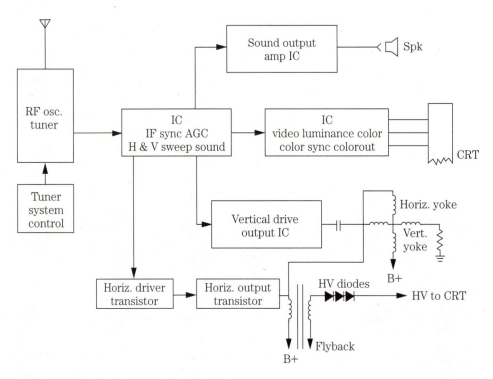

8-4 A typical block diagram of a TV chassis.

The horizontal circuits

The horizontal sweep circuits start at the horizontal oscillator or countdown circuits, providing a drive pulse to the horizontal driver transistor where the transformer couples the drive signal to the horizontal output transistor. The horizontal output circuit provides horizontal sweep to the yoke assembly and high voltage from the secondary winding of the flyback transformer. High voltage is rectified with high-voltage diodes built inside the integrated high-voltage transformer (IHVT).

Horizontal circuit problems

There are many different horizontal problems and interesting circuits found in the horizontal circuits. The horizontal oscillator circuits can cause no sweep, horizontal drifting, and off-frequency symptoms. The oscillator circuits might be contained in one IC or transistor, or all horizontal circuits might be found in one large IC.

Large defection IC

8-5 A large deflection IC might also contain other TV circuits.

The countdown circuits in the TV chassis provide both vertical and horizontal sweep circuits (Fig. 8-5).

The horizontal driver transistor takes a weak drive signal and amplifies and transformer couples the drive signal to the base of the horizontal output transistor. A leaky driver transistor can damage the primary winding of the driver transformer or voltage-dropping resistor, causing chassis shutdown. If the horizontal drive voltage is left off of the base terminal of the horizontal output transistor too long, the output transistor can also be damaged.

Horizontal output transistor

Go directly to the horizontal output transistor when the line fuse keeps blowing or the chassis will not start up. Check the resistance from the collector terminal (body) of the output transistor to the chassis ground (Fig. 8-6). A low resistance on the DMM indicates a leaky transistor or damper diode. In the latest TV chassis, you might find the damper diode built inside the horizontal output transistor. An open output transistor might have high dc voltage at the collector terminal with no high voltage or sweep. In some Sears and Sanyo TV chassis, the metal body of the output transistor might not be the collector terminal.

A leaky horizontal output transistor can cause chassis shutdown, open both fuses, overload low-voltage power supply circuits, or destroy the flyback. Likewise, a leaky or arc-over flyback or output transformer can destroy the horizontal output transistor. Always check the damper diode when replacing a leaky output transistor.

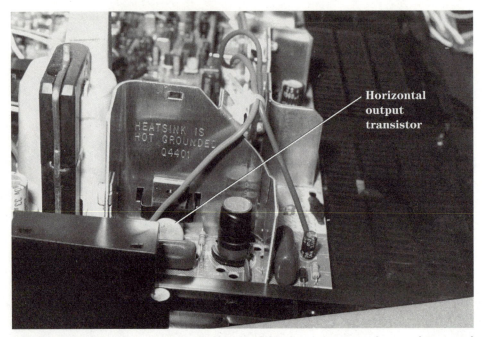

Horizontal
output
transistor

8-6 This horizontal output transistor Q4401 in an RCA chassis is mounted upon a hot ground heat sink.

Improper or no horizontal drive signal can destroy the horizontal output transistor (Fig. 8-7). When checking for a horizontal output waveform, hold the scope probe next to the flyback.

Always use a variable isolation transformer when servicing the horizontal circuits. After replacing the leaky output transistor and checking the horizontal circuits, apply about 65 V ac from the variable transformer. Check the base of the output transistor for a drive waveform and dc voltage at the collector terminal or flyback. If the transistor appears warm, shut down the chassis and check for improper line voltage, a leaky flyback, or overloaded circuits. Keep raising the line voltage if the voltage and waveforms appear normal at the base of the output transistor.

Horizontal oscillator problems

When improper or no horizontal drive signal is a problem, suspect a defective oscillator transistor, countdown IC, driver transistor, or low-voltage source (VCC). In derived secondary voltage circuits from the secondary winding of the flyback, which might supply dc voltage to the horizontal countdown circuits, the horizontal oscillator circuits will not function until all horizontal circuits are working. You must apply external voltage from a low-voltage source or batteries (Fig. 8-8).

If the horizontal circuits are operated from the regular low-voltage power supply, check the voltage source applied to the horizontal circuits. Low voltage at the countdown IC supply pin can indicate a leaky IC or a defective low-voltage source.

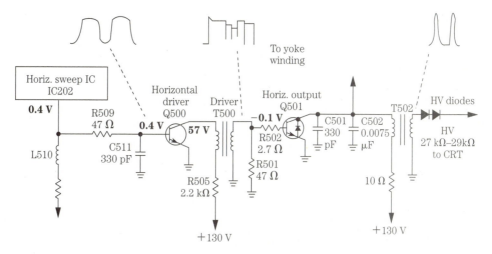

8-7 A typical horizontal IC sweep, driver, and output transistor within the horizontal circuit.

8-8 Connect the external power supply to power up the horizontal deflection IC when dc voltage is derived from the secondary of flyback.

Remove the pin terminal, and see if the voltage returns to normal. If it does, replace the leaky IC. Locate the countdown IC on the PC board. A scope test of the pin terminals will indicate the correct sweep.

When the oscillator sweep circuits are powered by a derived dc voltage from the horizontal output transformer, apply dc voltage to the supply terminal from an external power supply. Usually this voltage is from 10 to 35 V. Try a 10-, 12-, 18-, or 20-V source, and slowly bring up the voltage to get the correct waveform from the IC. You can use one or more 9-V batteries in series to reach the required voltage.

Attach the voltage source to the V_{CC} supply pin and common chassis ground. Solder in a piece of hookup wire, if necessary, to make a good connection and not short out other pin terminals. Pull the ac cord from the power plug for this test. Apply voltage from an external power supply, and scope the sweep output terminals. Locate the output terminal by tracing the wiring from the driver transistor base terminal (Fig. 8-9).

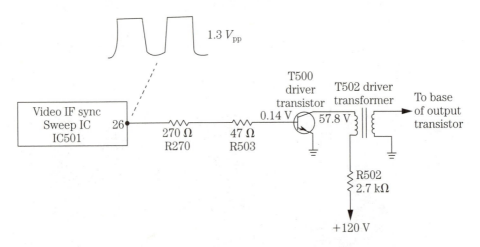

8-9 Scope the horizontal deflection IC if there is no horizontal output waveform.

If the waveform is normal at the sweep IC, you can assume the horizontal IC is normal. When the waveform is weak or is not found at the countdown or oscillator IC, suspect that the IC is bad. Suspect a leaky IC if the voltage goes below half of the external dc voltage. Remove and replace the defective IC.

Hold-down safety circuits

When there is excessive arc-over on the CRT or flyback transformer, suspect an open or defective safety capacitor. These hold-down or safety capacitors are found in the horizontal output transistor collector circuit to ground. There might be more than one in the circuit. Simply tack in another capacitor of the same rating across the suspected one, and see if the chassis shuts down or if arcing occurs. Excessive voltage applied to the horizontal output circuits can cause high-voltage arc-over. Check for a poor low-voltage regulation system.

The narrow road

Poor width can be caused by defective components in the horizontal, flyback, high-voltage, and low-voltage circuits. Check the latest high-voltage regulator circuits for poor width in the latest TV chassis. The high-voltage regulator transistor, SCRs, and zener diodes in the regulator circuits can produce insufficient width. Poorly soldered connections upon the pincushion, regulator, and driver transformers can result in poor width (Fig. 8-10). Open bypass or coupling capacitors in the horizontal and high-voltage circuit can cause poor width.

8-10 For improper width, check for poorly soldered connections of pincushion terminals on the PC board.

Low high voltage at the anode terminal of the picture tube can result in poor width. Of course, this low high voltage might result from a defective component in the horizontal output circuit. Low drive voltage upon the base of the horizontal output transistor can cause the transistor to become quite warm and produce a narrow picture. Low-voltage source applied to the horizontal output circuits can cause poor width. Check the low-voltage source for a leaky regulator transistor, zener diode, and filter capacitor.

The defective flyback

A defective horizontal output transformer can arc over internally when there are leaky high-voltage diodes and capacitors built inside the molded transformer wind-

8-11 A leaky horizontal output transformer can destroy the horizontal output transistor and produce chassis shutdown.

ings. A cracked case can indicate a defective transformer. The flyback might have an open primary winding or shorted turns between coil layers. Sometimes the winding will arc between the winding and the powered iron core (Fig. 8-11).

A leaky transformer can destroy the horizontal output transistor. Use a variable isolation transformer to vary the ac line voltage. Remove one terminal of each rectifier diode found in the secondary voltage circuits to eliminate a heavy overload. Also, remove the horizontal yoke lead from the flyback circuits in case the yoke is leaky and loading down the horizontal circuits.

A leaky horizontal output transformer or firing between windings can cause firing lines in the raster. When the core material is broken or cracked, firing lines are found in the picture. Sometimes you can hear vibration or singing noises from the flyback. Replace the transformer for all of the previously described symptoms.

Hot output transistor

Most output transistors are destroyed by a leaky flyback transformer, open damper diode, or open or leaky safety capacitors. If the output transistor becomes warm without any high voltage present, suspect either component. A leaky deflection yoke and pincushion transformer can destroy the horizontal output transistor. Insufficient drive voltage upon the base terminal can cause output transistor damage. Improper high voltage to the output transistor from low-voltage source can damage the output transistor.

When the horizontal output transistor runs red hot and sometimes ends up with a dead short, suspect improper drive voltage waveform, driver transformer, and electrolytic capacitor. Determine if the base waveform is the same amplitude as on other output transistors (Table 8-1). Solder all terminals of the driver transformer. Check the primary resistance to see if it is the same as a schematic, if available.

**Table 8-1. Random list of waveform
p-p voltage, base voltage, and collector voltage.**

Manufacturer	Waveform on base	Base voltage	Collector voltage
Goldstar CMT-2614	12.2 V p-p	–0.1 V	121 V
Emerson ECR2100	20 V	–0.1 V	123 V
Emerson MS1980R	14 V	–0.1 V	123 V
Sears 564.42071850	10 V	–0.2 V	135 V
Sharp 19SB60R	18 V	–0.5 V	118 V
RCA CTC146B	14 V	0.1 V	139 V

If the output transistor is still red hot, replace the small electrolytic capacitor (usually 1 µF to 10 µF, 250 V) that is located on the primary winding of the horizontal driver transformer. Also replace the filter capacitor connected to the primary of the flyback in the B+ source. Replace the driver transformer when all the other parts have been replaced. Do not overlook a leaky flyback transformer.

Red hot H.O.T.

The horizontal output transistor in an Emerson ECR2100 model would run red hot and keep destroying the output transistor. Right away the driver transformer connection was soldered, with the same results. The amplitude waveform upon the base terminal was around 19 V p-p with +122 V applied to the collector terminal. The driver transformer was suspected, and without a schematic, the resistance of the primary winding could not be compared. The primary resistance was around 57.1 Ω with the DMM (Fig. 8-12).

The new driver transformer was installed with the same results: a red-hot output transistor. R448 and R447 tested normal. The voltage upon the driver transistor was 32.6 V and quite normal. The only components left in the circuit were capacitors. Because electrolytics cause the most problems, C446 was shunted with a 10-µF, 250-V electrolytic. Now the 2SD1555 transistor ran warm but not red hot. There is no longer any hot output transistors in this TV chassis.

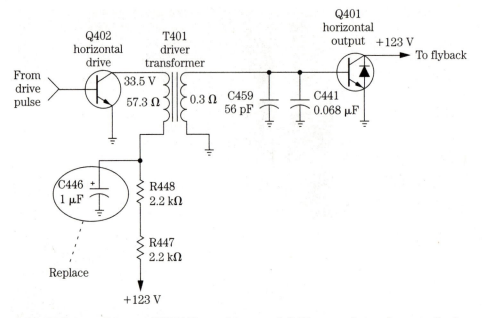

8-12 C446 in an Emerson ECR2100 portable caused Q401 to run hot and eventually destroyed it.

Intermittent raster

Monitor the dc voltage source, waveforms at the base of the horizontal output transistor, and high voltage at the anode connection on the CRT when an intermittent raster is the problem. Next scope the countdown sweep waveform to determine if the output and high-voltage circuits are intermittent. If the high voltage disappears with normal dc voltage and a sweep waveform, suspect the horizontal driver and output transistor circuits (Fig. 8-13).

Inject a drive signal at the horizontal output transistor to see if the horizontal and high-voltage circuits are normal. If they appear normal after several hours of operation, suspect the driver circuits. Spray coolant on the driver transistor to make it act up. Resolder all the terminals of the driver transformer. Monitor the drive waveform on the driver transistor and the collector voltage to determine if the transistor is intermittent. Locate the driver transistor tied to the primary winding of the horizontal driver transformer.

High-voltage shutdown

Excessive high voltage can make the x-ray or shutdown circuits close down the horizontal circuits and shut down the chassis. This prevents excess radiation from the picture tube and prevents damage to other components within the TV chassis. Defective shutdown circuits can cause premature high-voltage shutdown.

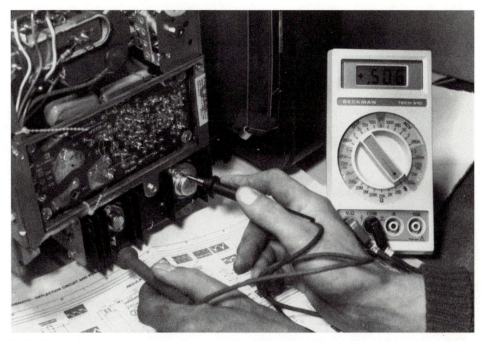

8-13 Check the horizontal output transistor for leaks when the fuse keeps blowing.

Monitor the high voltage with a high-voltage probe or meter at the CRT anode button. Connect a dc meter to the horizontal fuse or flyback primary winding to check the voltage applied to the horizontal output transistor. Slowly raise the variable isolation transformer line voltage while watching the high-voltage meter. Notice the voltage when the chassis shuts down (if the high voltage is excessive before reaching the 120-V ac power line voltage) and the high-voltage measurement.

Slowly raise the variable transformer to just under the high-voltage reading that shut down the chassis. Notice if the low-voltage source is higher than normal. If it is, check the low-voltage regulator circuits. When the high voltage is high compared to the line and dc voltage, suspect a defective hold-down, safety capacitor, or horizontal output transformer. If the chassis shuts down before reaching the normal high voltage, suspect a defective high-voltage shutdown circuit.

Disconnect the high-voltage shutdown circuit by removing the diode or resistor terminal from the PC board wiring. The high-voltage shutdown circuit takes a pulse from a winding of the flyback and rectifies it with the voltage regulator diode to the transistor or SCR circuit that shuts down the driver or horizontal oscillator stage. Some chassis have a shutdown circuit that is fed back to the countdown oscillator IC and driver circuits, shutting down the horizontal circuits.

If the high voltage is normal and does not shut down with the shutdown circuits disconnected from the flyback circuits, repair the shutdown circuits. Always replace any removed terminals after the chassis has been repaired.

Chassis shutdown

The TV chassis might shut down with a defective part in the horizontal or vertical circuits or the low-voltage power supply. Most shutdown problems occur in these stages. Chassis shutdown can be caused by poor terminal connections, defective transistors and ICs, poorly soldered terminals, or cracked PC board wiring. Monitor the horizontal circuits at the countdown sweep circuits, check low voltage at the horizontal output transistor, scope waveforms at the base of the output transistor, and check high voltage at the anode of the picture tube. Notice which circuit begins to malfunction when the chassis shuts down.

In the horizontal circuits, shutdown can be caused by the horizontal driver and output transistors. Resolder all the terminals around the PC board connections of the driver transformer. Don't overlook a defective flyback transformer. Replace the output transistor when nothing else turns up.

Defective high-voltage circuits can cause the chassis to shut down. A dirty, arced-over picture tube spark gap can shut the chassis down. Blow the dust out of all spark gap assemblies. A leaky picture tube can cause excessive brightness, arc-over, and shutdown of the chassis. Critical waveform tests at different stages in the horizontal or low-voltage circuits can locate what is causing the shutdown.

A pulled muscle

Excessive horizontal pulling of the picture can be caused by poor filtering in the low-voltage power supply. Higher than normal voltage applied to the horizontal oscillator transistor or IC can be caused by a defective zener diode in the power supply circuits. Horizontal flicking of the picture can result from poor voltage regulation in the hold down, safety, or regulator circuits. Pulling at the top of the picture can be caused by a small electrolytic bypass capacitor in the horizontal circuit.

Horizontal pulling at the top of the picture was noted in a Sanyo 91C560 portable TV. Filter capacitors in the 123-V source were shunted without any results. The drive waveforms were fairly normal on the base of Q301 and Q302. All collector voltages were fairly normal. Small bypass capacitors were shunted in the deflection and driver circuits. When C321 was shunted, the pulling at the top stopped (Fig. 8-14).

High-voltage problems

Besides high-voltage arc-over, the IHVT flyback can have arcing diodes or capacitors within the molded component. The high voltage might not come up or cause shutdown with leaky voltage diodes in the secondary winding of the flyback. Overloaded circuits in these voltage sources can be caused by leaky components within the audio, video, and vertical circuits.

Improper screen grid voltage can be caused by a burned isolation resistor or leaky diode in the derived secondary circuits (Fig. 8-15). The symptoms include not being able to turn down the brightness, excessive brightness, or chassis shutdown. Disconnect each diode from the secondary circuits to determine what section is

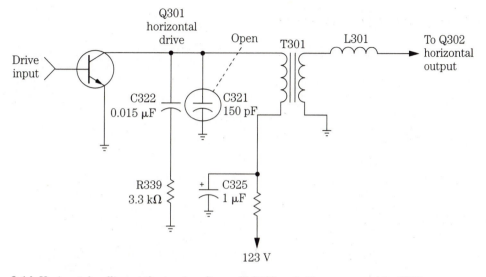

8-14 Horizontal pulling at the top in a Sanyo 91C560 portable was caused by C321.

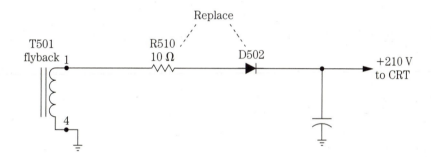

8-15 A leaky D502 silicon diode knocks out R510 when the brightness cannot be turned down.

overloading the horizontal output transformer. Check each circuit after removing each diode.

Poor brightness problems can be caused by low high voltage, defective luminance circuits, or a weak CRT. A defective picture tube might have loose particles in the gun assembly and, when lightly tapped, will flash a negative or black picture. Improper high voltage or a weak picture tube results in a dim or weak raster. Check all picture tube voltages on the pins of the CRT socket (Fig. 8-16).

Bowed legs

When the picture is bowed inward on each side of the picture, suspect a defective pincushion circuit. In large TV screens (27" and up), a pincushion correction circuit corrects distortion on the face of the large picture tube. You might have to

8-16 Take voltages on the CRT socket for a black or weak raster.

look closely with a door frame or building in the picture to really notice the bowing of the picture. The pincushion circuits are defective if the outside edge of the TV raster bows. Improper adjustment of the pincushion coils might bow the picture at the top and bottom of the raster.

Usually, the primary winding of the pincushion transformer is in series with the horizontal yoke assembly. By modulating the current through the pin transformer, most of the regulated B+ voltage appearing across the yoke winding charges, resulting in a change of horizontal width. This width will increase as more B+ appears across the yoke winding.

A correction circuit develops a parabola waveform that is applied to the pincushion transformer. Suspect a pincushion output transistor if the raster bows inward (Fig. 8-17). Suspect an intermittent power output pincushion transistor if the sides are bowed in and suddenly return to normal. First solder up all pincushion transformer terminals upon the PC board. Replace pincushion output transistor for bowed pictures in large TV screens.

The vertical circuits

Always try to use the correct schematic to quickly locate a defective part. Although vertical circuits are easy to service compared to others, try to locate a schematic of the same manufacturer or a similar circuit. Vertical circuits have not changed too much over the years, except now all the vertical circuits might be found in one large IC (Fig. 8-18).

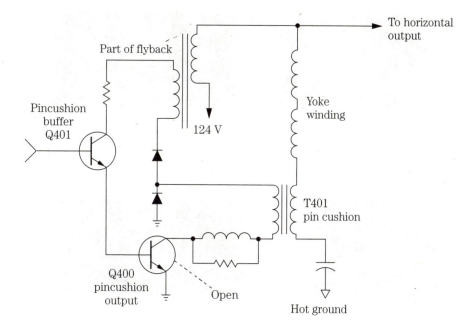

8-17 In 27" or larger screens, a defective pincushion output transistor can bow in the sides of the picture.

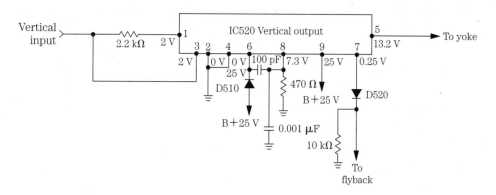

8-18 The vertical output sweep circuits might be found in one IC in a recent TV chassis.

Sometimes an import TV chassis schematic might be difficult to find. Try to locate the vertical output transistors or IC on the chassis, and compare it with other vertical circuits. The same manufacturer might make several different brands of TV chassis. A basic TV block diagram of the vertical circuits can help in determining where the defective component is mounted.

Vertical circuit problems

The most common vertical problems found in the TV chassis are the horizontal white line and no vertical sweep. Insufficient height can produce a raster of 4" or 5"

8-19 Only 4" of vertical sweep can result from defective output transistors, IC, bias resistors, and diodes.

high (Fig. 8-19). Dead vertical countdown circuits might not have a vertical output waveform but might have a normal horizontal pulse. Vertical fold-over and rolling problems go hand in hand, while retrace lines at the top can be caused by a leaky vertical output transistor or a change in resistance of the bias resistors. Dried-up filter capacitors can produce vertical crawling, while the intermittent vertical sweep problem takes a little more time and is difficult to locate.

Locating vertical circuits

Try to locate a couple of vertical output transistors on the PC board with their separate heat sinks or the output IC mounted on the metal chassis or heat sink. In earlier chassis, the two vertical output transistors were mounted close together (Fig. 8-20). You might find one large countdown IC that supplies a vertical pulse to the output circuits. Usually this same IC serves many other functions.

The early vertical output IC served in place of the two output transistors in the output circuits. Today the vertical output IC might contain all the vertical circuits on a metal heat sink. Start by locating the vertical components, and take waveforms and critical voltage tests on the suspected ICs or transistors.

Only a white line

The horizontal white line indicates no vertical sweep. Usually the no-sweep problem occurs in the vertical output, vertical yoke winding, or vertical oscillator circuits.

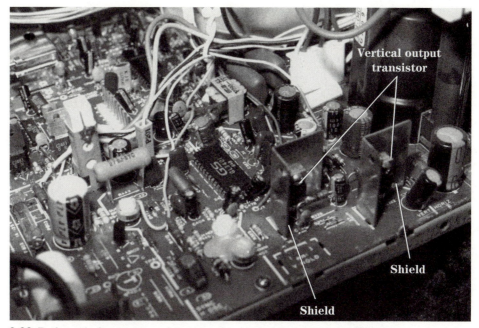

8-20 Both vertical output transistors are mounted on separate heat sinks.

A waveform test at the vertical countdown IC or oscillator will indicate if the stage is working. Check the vertical input circuits when there is no vertical waveform. ·

Check the output transistors with in-circuit beta or voltage measurements. Measure the voltage supply terminal if the output IC is suspected. In vertical stage transistors, especially directly coupled circuits, scope waveforms might not be useful. With IC countdown and output circuits, the vertical waveforms can be checked at the vertical output terminal on the countdown IC, the input of the vertical output IC, or the output terminal of the vertical output IC (Fig. 8-21).

When a signal is found on the input terminal of the vertical output IC and not at the output terminal or yoke winding, suspect a defective IC. Check all components tied to the IC before replacing them. Likewise when a vertical oscillator waveform is found with an improper waveform on the output transistors, suspect the output transistors. Replace both vertical output transistors when one is found to be defective. Check each base and emitter bias resistor while the transistors are out of the circuit.

Insufficient vertical height

Not enough height is noted when the picture has only 2" to 4" of raster or cannot fill out the entire raster on the picture tube. Sometimes the raster will fill out when the unit is turned on and then collapse to a couple of inches. Monitor the voltage source on both transistors.

Suspect one or both transistors when insufficient vertical height is a problem. An in-circuit transistor test might find one or both transistors to be leaky. Sometimes

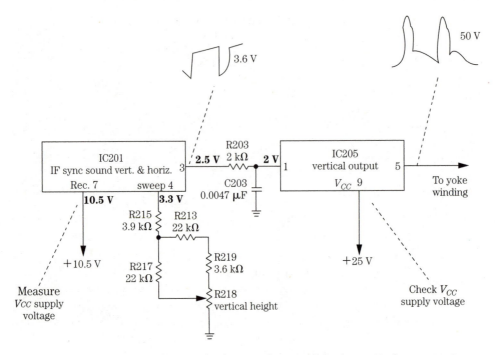

8-21 Check the various waveforms and voltages to locate a defective vertical component.

when they are removed from the circuit, they might test good. Replace both vertical output transistors. Don't overlook burned bias resistors.

Intermittent vertical circuits

Intermittent vertical circuits are more difficult to locate and repair. Attach a scope probe to the vertical oscillator or countdown circuits. If the countdown or oscillator circuits are normal, suspect the output transistors or ICs. Scope the input and output of the vertical output IC to determine if the output IC is intermittent. Take a waveform found on the electrolytic coupling capacitor from the vertical circuits to the yoke winding.

The yoke coupling capacitor might be open, producing a horizontal white line and insufficient sweep, or it might be dried up, producing a bunching of white lines at the top of the raster. Always check the vertical return resistor (under 50 Ω) or the capacitor found at the ground end of the yoke winding (Fig. 8-22).

Vertical crawling

When lines bunch together and slowly crawl up the raster, suspect the large filter capacitors or poor low-voltage regulator circuits. The raster might have dark horizontal lines along with the crawling lines. Shunt each filter capacitor with one of the same capacity or higher and with the same or higher working voltage. Shut down the chas-

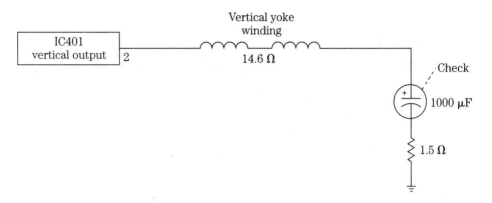

8-22 Only a horizontal white line can result from an open yoke winding, coupling capacitor, or return resistor.

sis, and clip the capacitor across the suspected ones, so as not to damage solid-state devices. Shunt the vertical output coupling capacitor when line bunching occurs.

If more than one capacitor is found inside a component and only one section is causing the problem, replace the whole capacitor component. Look for the tallest electrolytic capacitor on the chassis. Sometimes a large capacitor (650 μF or more) might be mounted on the metal TV chassis.

Lines at the top

In directly coupled transistor circuits, check for leaky or open transistors or a change in resistance within the base or emitter circuits. First check the resistors for correct resistance. Remove one lead from the circuit for an accurate measurement. Change one transistor before removing both. Replace the vertical output IC when there are white lines at the top of the raster.

At the top

Within the vertical transistor output circuits, retrace lines might be caused by the two output transistors, bias resistors, and diodes. Retrace lines in the power vertical IC output can result from a defective vertical IC, electrolytic capacitor, and drive from the deflective IC. Do not overlook a defective filter capacitor within the output IC voltage source.

Retrace lines at the top of the picture in a Goldstar CMS4841N portable were caused by C312 (Fig. 8-23). When the 1000-μF electrolytic was shunted with a normal capacitor, the lines disappeared. Trace the 24-V source back to the low-voltage sources in the flyback circuits. This 24-V source is fed from a winding (pin 5) on flyback (T403) and rectified by D302 and filtered with C312. In another chassis, IC201 caused retrace lines at the top.

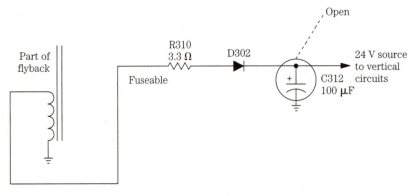

8-23 Lines at the top of the picture were caused by a defective C312 in a Goldstar CMS4841N portable.

Vertical oscillator or countdown circuits

Today vertical oscillator or countdown circuits might be inside one large IC. The deflection countdown circuits provide sweep for both the horizontal and vertical circuits (Fig. 8-24). A scope waveform at the vertical drive output indicates a drive pulse. If this vertical drive waveform is missing, suspect improper dc voltage at the countdown IC or a defective IC. Scope all the terminals of the IC to locate the correct vertical drive pulse.

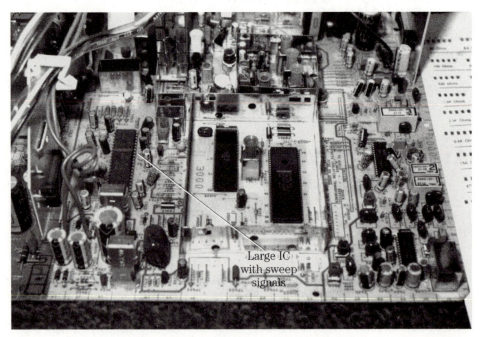

8-24 The countdown vertical and horizontal sweep circuit might be found in one large IC.

Vertical fold-over

Vertical fold-over appears at the top or bottom of the raster. Most fold-over problems occur in the vertical output circuits. Replace the vertical output transistors or ICs when fold-over occurs. Check each bias diode or resistor by unsoldering one terminal lead and testing for the correct resistance or an open resistor. Don't overlook feedback resistors or capacitors. Make sure the supply voltage has not increased or decreased.

Vertical rolling

Improper vertical sync applied to the oscillator or countdown IC can cause vertical rolling. The vertical raster will not stay in one place. Sometimes vertical rolling and fold-over problems are caused by the same defective part. Scope the vertical sync applied to the vertical circuits. A change in the resistance of the base or emitter circuits can cause improper vertical lock. Remove one end of the resistor or diode for accurate measurement.

Improper dc voltage fed to the vertical circuits can produce vertical rolling. Poor filtering in the low-voltage power supply that feeds the vertical circuits can make the picture roll. Suspect a defective vertical hold control if the raster collapses at once when the control is rotated. Replace the hold control.

Identifying vertical components

Look for vertical output transistors on a separate heat sink mounted on the PC board. These are flat transistors with the collector terminal screw-mounted to a metal heat sink (Fig. 8-25). Remember, the heat sink might have a high dc voltage. Sometimes they are hidden under other parts on the metal TV chassis.

In later TV chassis, one large IC might contain all of the vertical sweep circuits and be found mounted on a large heat sink. Usually heat sink paste or compound is applied to the back of the transistor or IC for cool operation. Unsolder all the terminal leads from the underside of the chassis before removing the mounting screws. Make sure each pin is free from the PC board wiring. Remove the screws, and pull out the suspected IC.

The numbers and letters found on the body of the transistors or ICs might be the part number. Look up these numbers in the universal solid-state replacement manual. Both output transistors might be NPN types, or one might be an NPN and the other a PNP transistor.

If no numbers or letters are found on the body of the defective transistors, check the working voltage and compare it with those found in the universal replacement manual. Choose the correct working voltage, type, and mounting for the transistors. Universal transistors and IC components work well in vertical TV circuits. Check Howard Sams Photofacts for the correct vertical replacement.

Try to find the part number on the body of the vertical IC for universal or direct part number replacement (Fig. 8-26). Check with the manufacturer's service center

8-25 Locate the vertical output IC or transistor upon a separate heat sink.

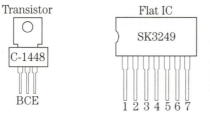

Transistor

C-1448

BCE

Flat IC

SK3249

1 2 3 4 5 6 7

8-26
The numbers and letters found on the body of ICs and transistors can help you order the correct replacement.

for the correct vertical output replacements. Vertical circuits in TV chassis are much easier to locate and service than horizontal sweep circuits.

Tuner/IF circuits

Often the tuner and IF stages are located on the left side of the TV chassis, looking from the back, and close to the front (Fig. 8-27). The old tuner might be a turret or mechanical wafer type with manual tuning. The varactor tuner can be controlled with a system control, control processor, memory, or frequency synthesizer control units. The tuner control can have a soft-touch, keyboard, push-button, selector board, auto board, or remote-control tuning.

The wafer-type tuner must be cleaned at least once a year to prevent erratic or intermittent tuning. These are still found in low-price color and black-and-white portables. Remove the metal cover, spray cleaning fluid into the wafer switch contacts, and rotate the tuning knob back and forth (Fig. 8-28).

8-27 The varactor tuner and IF components are mounted on the left side of this RCA chassis.

8-28 Remove the bottom cover, spray the contacts, and rotate the tuner knob back and forth to clean manual tuners.

The varactor tuner can be mounted on edge right on the PC board with the bottom contacts soldered into the PC board wiring (Fig. 8-29). Tuners under warranty should be removed and returned to the manufacturer's service center. Older tuners out of warranty can be sent into a tuner repair center such as PTS Electronics, P.O. Box 272, Bloomington, IN 47402.

8-29 The varactor tuner is mounted flat on the PC board with pin terminals through PCB chassis.

A defective tuner can cause a snowy, weak, dim, intermittent, or flashing picture, erratic or drifting channels, or a white raster. Check for a defective tuner by plugging a tuner subber into the IF socket or connection. Voltage measurements, tuner subber injection, tuner substitution, and correct AGC voltage can help you discover a defective tuner. A defective tuner module should be removed and sent in for repair (Fig. 8-30).

IF/video circuit problems

In addition to the DMM, the scope with demodulator probe, sweep marker, and color-bar generator are useful test instruments when servicing IF circuits. A dead or weak IF stage can be located with a strong local station tuned in and scope waveforms taken at each IF stage. The oscilloscope must have a demodulator or detector probe attached to it. Most IF stages are found under a large shielded area (Fig. 8-31).

The no-picture, no-sound symptom can occur if the tuner or IF video components are defective. Isolate the tuner by connecting the tuner subber to the IF cable. Check

8-30 When damaged, a tuner module in an RCA FM27227 can be sent in for repair or can be exchanged.

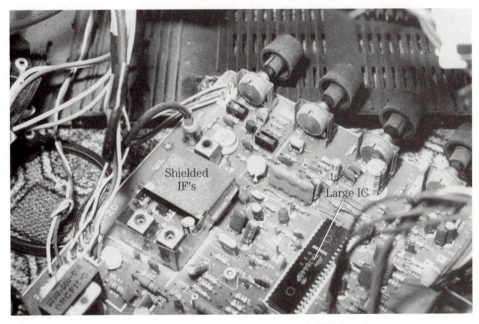

8-31 The IF circuits are found in a shielded area in this early TV chassis.

the first and second IF stages for a snowy picture. Improper AGC can cause a snowy picture. Intermediate frequency alignment is not required with a SAW filter network.

A weak or washed-out picture, little or no brightness, heavy scanning lines, or a snowy picture can be caused by the video circuits. Usually the video circuits are found near the IF circuits in one large IC. Take critical voltages on, and scope the terminals for loss of signal. Start at the first video amp with the scope probe, and check the signal at the base and collector of each video transistor until the signal is lost.

An intermittent picture or no picture can be caused by a defective video delay line. Scope the signal in and out of the unit. A leaky or open brightness limiter can cause no brightness.

Color circuit symptoms

With a normal picture and good sound, suspect the color circuits when there is poor or no color in the picture. Locate the chroma-luminance IC on the chassis. This IC contains the IF video, chroma, and deflection circuits. The chroma IC might be located on a separate video-luminance-color PC board (Fig. 8-32).

Scope the color IC for missing waveforms when there is no color. Check for the color oscillator waveform (3.58 MHz) taken from the crystal pin on the IC. Check the terminals for a missing horizontal pulse from the flyback circuits. Check the three-color output waveforms that feed the color output transistors on the CRT board (Fig. 8-33).

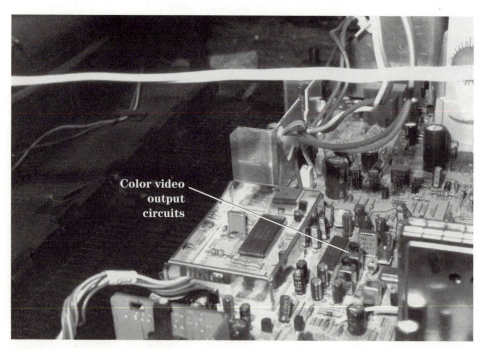

8-32 Locate the color and video circuits on the TV chassis with a color 3.58-MHz crystal.

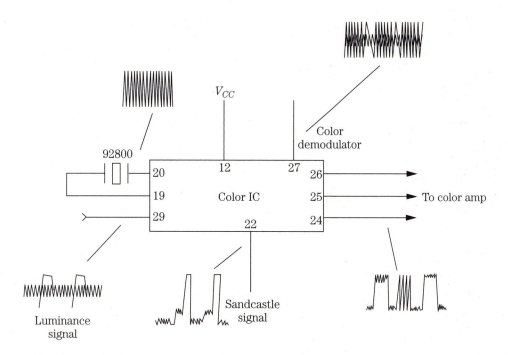

8-33 Check the various input and output waveforms on the color IC terminals.

Look for open coils and poor component terminal connections. Check for leaky capacitors to ground. Intermittent color can be caused by leaky capacitors in the color killer circuits. Take a resistance test from each terminal pin to the chassis ground.

Check the color output transistors when one color is missing, and test the guns in the picture tube. Weak or intermittent color problems can occur in the color output amps. Locate the color amps on the CRT socket board. Check the voltages on each collector terminal. Test each transistor in-circuit. Universal transistors can be used in the color output circuits.

Sound problems

Suspect the sound circuits when the CRT picture is normal but there is poor or no audio. In earlier chassis, the audio circuits were made up of transistors. Today all of the sound circuits might be included in one large IC. Locate the audio IC next to the shielded discriminator coil, or trace the speaker wires back to the IC or transistors (Fig. 8-34).

Check the sound circuits with an external audio amp from the speaker terminals to the output terminal of the IC. The speaker coupling capacitor can dry up, go open, or cause intermittent sound. Take critical voltage measurements on the audio IC. Scope the sound coming into the audio IC.

Weak sound can be caused by an open audio take-off coil, transistor, IC, or coupling capacitor. Garbled sound can be caused by a maladjustment of the discrimina-

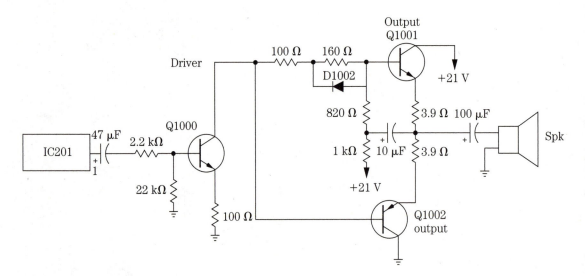

8-34 You might find transistors in the output sound circuits on the latest TV chassis.

tor coil. Turn up the sound and rotate the tuned slug to either side (not over one-quarter turn) until the audio is clear. Don't overlook a defective speaker when there is mushy or distorted sound.

Surface-mounted components

Besides being used in camcorders and CD players, surface-mounted devices (SMDs) are now found in TV chassis and varactor tuners (Fig. 8-35). Surface-mounted transistors have three end connections, while resistors and capacitors have only two soldered end connections.

Surface-mounted ICs or processors might have gull-wing terminals (Fig. 8-36). Usually a white dot or number indicates terminal one. Look up the part numbers stamped on the body of the ICs and processors to locate the correct circuits.

The resistors and capacitors might have the correct value stamped right on the body of the component. You might find the part number corresponding with the schematic stamped along side. Surface-mounted transistors, capacitors, and resistors are very difficult to locate without a schematic and magnifying glass.

Take voltage and resistance measurements on the resistors and capacitors of suspected SMD parts. Test each transistor in-circuit with a beta tester. Be very careful not to short the terminals during critical voltage measurements. Take care when removing and replacing SMD parts. Always replace an SMD part when it tests normal after removing it from the PC board wiring. Don't try to use it again. Surface-mounted resistors and capacitors can be replaced with universal SMD parts. Replace transistors, ICs, and processors with exact replacements.

8-35 SMD circuits are found on the bottom side of an RCA CTC140 chassis.

8-36 Check for correct pin numbers stamped on the IC or processor with gull-wing terminals.

Grapes come in bunches

Besides finding the original problem within the TV chassis, you might think the set is fixed, except that there are still problems in the TV chassis. For example, the horizontal output transformer can become shorted, knock out an isolation resistor in the B+ line, damage the line voltage regulator, knock a couple of fuses, and take a couple of diodes in the bridge circuits.

When lightning or a power outage occurs, look for possible damage to the tuner input cables, power cord, silicon diodes, line voltage regulator, fuse, and stripped wiring (Fig. 8-37). On a power outage condition, the power lines sway in the air and short-circuit one another, increasing the voltage at the fuse box. Besides replacing the fuses within the TV chassis, several power line fuses at the house must be replaced. When the neutral wire on a power transformer has a poor connection or goes open, extra voltage is applied to every electronic and electrical device in the house, causing extensive damage.

8-37 After lightning damage, check for a defective line voltage regulator with a dead chassis.

Look around the chassis for possible burned or broken components that might be tied to another defective component. Take critical voltage and resistance measurements to uncover a leaky or damaged component. Trace each circuit out to determine if additional damage can be found. If the chassis does not perform after replacing a defective component, trace out corresponding PC wiring to other possible defective components.

Dogs everywhere

A tough dog might take hours, weeks, or several months to repair. Sometimes a repair might be called tough by one technician while the next technician might locate it at once. Tough dogs seem to come all in one week or one month. Always set the chassis aside after working on it an hour or so. Time can be lost if you work half a day on it. Give it up for now, and tackle it the first thing in the morning when your mind is fresh.

Ask for help if you cannot locate the defective component after several attempts. Go to the TV distributor or parts depot for help. If there aren't any service personnel there, contact the factory. A simple telephone call and 10 minutes of your time might save several hours of frustrations. Check with factory repair depots that repair boards or the entire chassis.

Down on the farm

Have you ever received or picked up a TV set with mouse tracks all over the chassis? Some parts are chewed upon. A nest might be built next to a warm component. The TV chassis failed because a dead mouse was shorting out high-voltage components. Several parts might be found burned on the board due to those outside pests who like to come inside during the winter. Some TV sets have small cracks that allow rodents to get inside a console model. They can scamper straight up the cabinet wall and sneak in a crack to the TV chassis.

To prevent mouse damage, plug all holes in the bottom of the chassis. Place wire screen over cracks and angle holes where the chassis sets upon. Of course, you can place a trap inside along the chassis. Then again, children might get their fingers in the trap. Just plug the holes with a wire screen. Finally, have the owner call the exterminator.

TV symptoms

The following symptoms are actual case histories occurring in many different TV chassis. These symptoms might help you locate and service other TV brands and models with similar problems.

Smeary picture, vertical flipping

Isolation   A voltage test on the 24-V source of a Goldstar NC-07X1 chassis indicated problems in the low-voltage supply.

Location   Check the components in the low-voltage supply.

Repair Replace overheated R313 in the 24-V supply and C307 and C308 in vertical circuits (Fig. 8-38).

Dead, nothing

Isolation   A voltage test at R801 and C802 indicated no dc voltage in a Samsung CT505-XD portable.

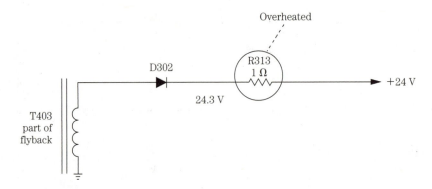

8-38 For a smeary picture in the Goldstar NC-07X1 chassis, replace the overheated R313 in the 24-V source.

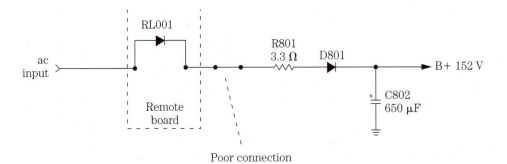

8-39 The dead symptom in a Samsung CT505-XD was caused by a broken PC wiring connection.

Location The ac line voltage was found at the remote board with no resistance measurements between RL001 and R801 (Fig. 8-39).

Repair A poor wiring junction connection on the PC board wiring had opened up.

Intermittent diodes

Isolation Normal voltage was found on the VIPUR output transistor (Q4100) in an RCA CTC140CH chassis.

Location Q4100 is located under a plastic and metal cover on a heat sink of the main chassis.

Repair In the early models, intermittent sound and picture were caused by cracked diode terminal soldered connections of the SIP board (Fig. 8-40).

An RCA CTC108C chassis blows fuses

Isolation Check for a leaky horizontal output or damper diode.

Location Locate the output transistor.

Repair Both the damper diode and the transistor were normal. C417 was found to be leaky (Fig. 8-41).

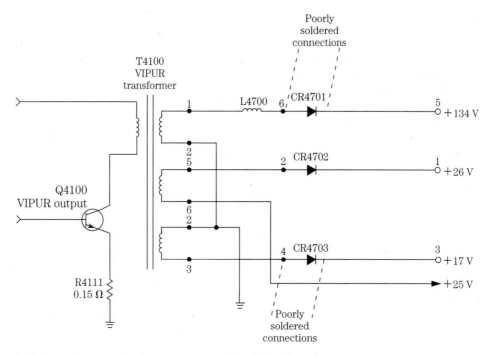

8-40 Intermittent or dead operation in an RCA CTC140 chassis can be caused by poor terminal connections of silicon diodes on the SIP board.

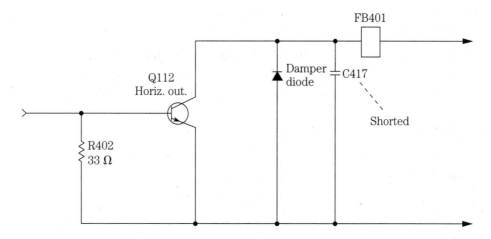

8-41 A leaky C417 in the output circuit of Q112 caused the fuse to blow in the RCA chassis.

No picture, no sound in an RCA G27251

Isolation   The green stereo light would come on when the push button was pressed and sometimes would not shut off.

Location   Go directly to the VIPUR power supply circuits at the rear of the chassis.

Repair   Found CR4201 diode open in the leg of the winding on pin 11 of the VIPUR transformer (Fig. 8-42).

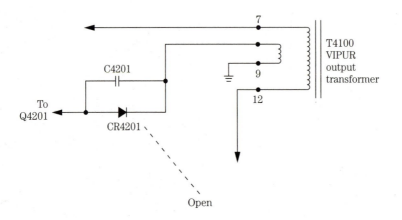

8-42 An open silicon diode CR4201 in the RCA G27251 caused the light to stay on.

No tuner control in a Goldstar CMS4841 TV

Isolation   Check voltages upon the MICOM tuner control IC1.

Location   IC1 is located toward the back of the TV in the middle of the chassis. It is a large IC with 42-pin terminals.

Repair   Found no supply voltage at terminal 42. Found D20 leaky with a charge on resistance of R814 and R813 (Fig. 8-43).

Green power light was on all the time in an RCA CTC140 chassis with no sound or picture

Isolation   Check the circuits that turn on the green power light in the VIPUR circuits.

Location   Locate the VIPUR transformer (T4100), and trace out pin 11 to the part.

Repair   Replace the defective CR4104 diode off of pin 11 of the VIPUR transformer (Fig. 8-44).

Vertical fold-over in a Sharp 19SB60R

Isolation   Go directly to the vertical output circuits.

Location   Locate IC531.

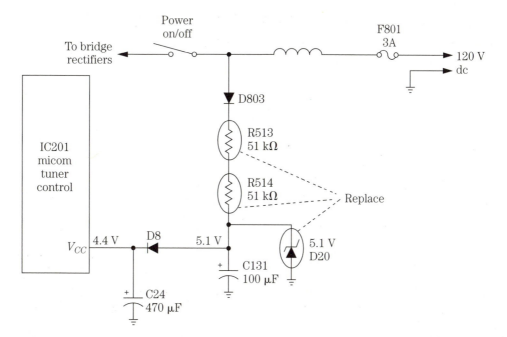

8-43 A leaky D20 and a change in resistance of R513 and R514 caused a Goldstar CMS4841 TV to have no tuner control.

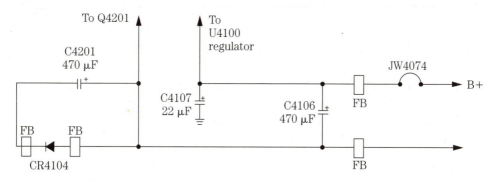

8-44 CR4104 caused the power light to stay on in the RCA CTC140 chassis with chassis shutdown.

Repair   A good waveform came into pin 4 of IC531. The pulse was weak. Ordered out IC531 (1X0238CE) from the service center (Fig. 8-45). The same problem was found with a poorly soldered connection in the input module in a similar chassis.

In an RCA GJR2038P, the picture tube brightness was very high with retrace lines at the top and sometimes the chassis would shut down

Isolation   Check the boost voltage to the picture tube, and check the brightness circuits.

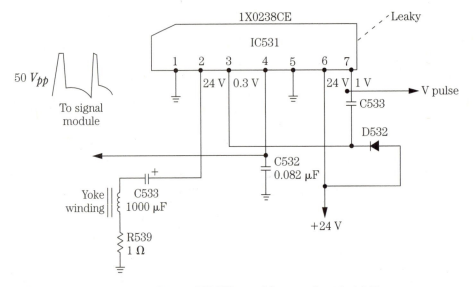

8-45 A defective IC531 in a Sharp 19SB60R portable caused vertical foldover.

Location   Locate the brightness transistor (Q703).

Repair   Replace the brightness transistor (Q703) with the exact part.

High brightness then shutdown in an RCA CTC107

Isolation   The brightness would come up very high, then the chassis would shut down.

Location   Check the CRT and voltages on the CRT board.

Repair   When the brightness came up, there was an arcing noise in the CRT socket. Removed the cover, and blew the dust out of the spark gaps on the elements of the picture tube. This cured the shutdown problem.

No color with a loss of picture and some noise in an RCA FJR2020T

Isolation   Check the color circuits.

Location   Locate the large IC processor and comb filter processor.

Repair   Found a leaky capacitor C611 (0.24 µF) between U600 and the video buffer transistor (Fig. 8-46).

An RCA CTC146B chassis would not turn on

Isolation   The TV chassis cannot be turned on manually or by remote.

Location   The standby power supply should be 4.9 V, but it was only 1.7 V. The 12-V source measured 11.2 V.

Repair   A resistance measurement from 4.9-V source at emitter of Q3101 and C3121 (15µF) showed a 19 Ω measurement to ground. The source was traced to pin terminal 29 and 4 of U3300 (Fig. 8-47). The voltage source was removed from IC pins, and resistance was still 19 Ω. Replace U3300.

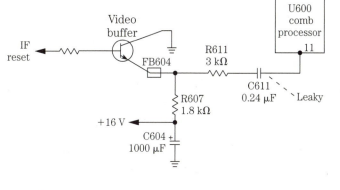

8-46 C611 caused a loss of picture and caused noise in an RCA FJR2020T in the comb processor circuits.

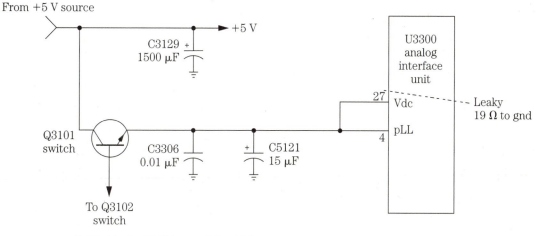

8-47 A leaky U3300 in an RCA CTC146 chassis caused a no turn on symptom.

One-half the screen was dark in a Zenith K2592

Isolation   (See Fig. 8-48.) Try to locate the possible defective capacitors in the AGC, video, and horizontal sweep circuits.

Location   Locate and check each section.

Repair   Found a leaky C3377 on a 9-153 module.

Remote will not operate after the chassis warms up in an RCA CTC140 chassis

Isolation   Checked the remote transistor, and it tested okay.

Location   Locate the system control IC (U3300).

Repair   Found a higher voltage on TP3303 and pin 36 of the IC. CR3302 was found to be leaky (Fig. 8-49).

Channels move upward in a General Electric EXR345ER

Isolation   Sometimes the channel would change upwards by itself.

Location   Locate the IC processor (U1001).

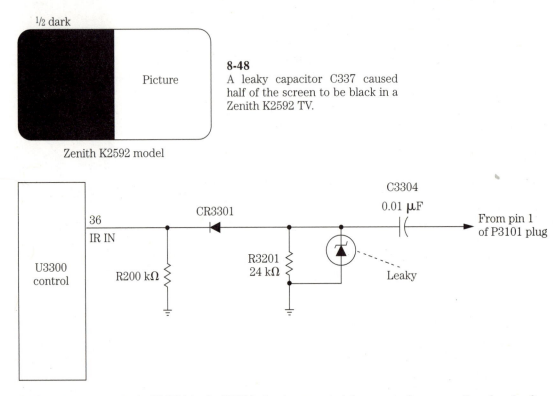

¹/₂ dark

Picture

Zenith K2592 model

8-48
A leaky capacitor C337 caused half of the screen to be black in a Zenith K2592 TV.

8-49 A leaky zener diode CR3302 in the U3300 circuit prevented the remote from operating chassis after warm up.

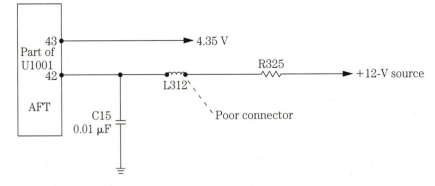

8-50 Intermittent connection of L312 in the U1001 circuits caused the channels to change in a General Electric EXR345ER.

Repair Although L312 looked good, the voltage was very low on pin 42 of U1001 (1 to 2 V). Resoldered both terminals of L312 to restore 5.95 V on pin 42 (Fig. 8-50).

9
CHAPTER

Repairing power supplies

The power supply is the most important section of any electronic device. If improper or no voltage is measured at the power-supply source, you might find a weak or dead unit. Always check the power supply voltage before checking any other component. Electronic devices will not operate without a proper-supply voltage.

The different power supplies can have half-wave, full-wave, or bridge rectifiers in the transformer or have transformerless power circuits (Fig. 9-1). In early TV chassis, large power transformers provided voltages to the various circuits. Today TV chassis have no power transformer, operating with direct power line circuits.

You will find small ac step-down power transformers in CD players, cassette decks, and table and alarm-clock radios. Sometimes the small ac/battery radios operate from the ac adapter that plugs into the power line and radio.

Only half way

The half-wave rectifier delivers a half-cycle of dc voltage for every half-cycle of applied ac voltage. The half-wave cycles are 180° apart with the same polarity. The half-wave rectifier is found in the transfomerless TV chassis, radios, and industrial applications. The half-wave rectifier consists of a silicon diode of 1 to 3 A. Often, high-capacity filtering is noted in half-wave rectifier circuits (Fig. 9-2).

The full-wave rectifier circuit has two or four silicon diodes. Usually, the full-wave circuit with two silicon diodes has a center-tapped step-down transformer. Each diode rectifies an alternate half-cycle of secondary voltage. The ripple in the dc output voltage is equal to twice the supply frequency and is much easier to filter out the hum in the full-wave circuits. Often, smaller filter capacity is used in full-wave circuits compared to half-wave rectification.

Bridge over water

In bridge rectifier circuits, the secondary winding of the transformer is not center-tapped. Four silicon diodes are found in the bridge circuit (Fig. 9-3). The nega-

9-1 Large power transformers are found in high-powered amplifiers.

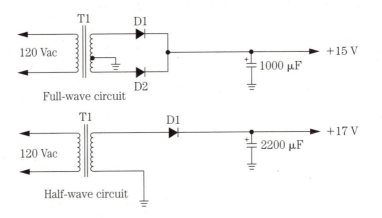

9-2 The half-wave rectifier rectifies one half of each cycle of applied ac voltage.

9-3 The bridge circuit consists of four separate components, or an all-in-one component, of full-wave rectification.

tive (anode) end of the two diodes are grounded, while the two diodes connected together with positive (cathode) ends provide positive dc supply voltage. Bridge diodes might be included in one component and can be replaced with single diodes when the exact part is not available.

Low-voltage regulators

You might not find any voltage regulators in the small portable radios or cassette decks, while in other circuits, zener diodes and transistors are used as voltage regulators. TV power-supply circuits use transistors, zener diodes, and large IC voltage regulators.

Low-voltage transistors and zener diode regulators in the combination AM-FM-MPX and cassette decks might have regulated voltage applied to the dc tape motor, critical FM, audio pre-amp, and output audio circuits (Fig. 9-4). You might find several different regulators in the deluxe radios and tape players.

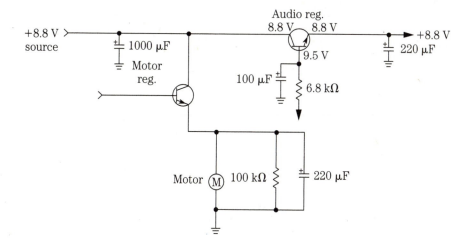

9-4 Low-voltage transistors and zener diodes provide regulated voltages in the different consumer products.

Open regulator transistors and burned zener diodes cause most of the voltage regulator problems in radios and tape decks. A defective diode can have a burned or charred body. An open regulator transistor has no voltage at the emitter terminal. Replace an open transistor regulator when voltage is applied to the collector terminal and there is no voltage at the emitter in an NPN-type transistor.

Voltage-regulator circuits found in TV chassis can have IC, transistor, or zener diode regulation. In many TV chassis, one large voltage regulator IC provides voltage to the horizontal output transistor circuits. Double-check the large voltage regulator when a leaky horizontal transistor or flyback is found in the TV chassis. Transistor and diode regulators can be found in the derived secondary voltages of the horizontal output transformer.

The buck stops here

With a defective product, always take a voltage measurement across the main filter capacitor. Most electronic technicians check the supply voltage the first thing. If the low-voltage circuits are not operating, nothing operates. The dead power supply might result from leaky diodes, open primary winding on the transformer, leaky filter capacitors, and voltage regulators. The voltage regulators might consist of a transistor, a zener diode, or a combination of both (Fig. 9-5).

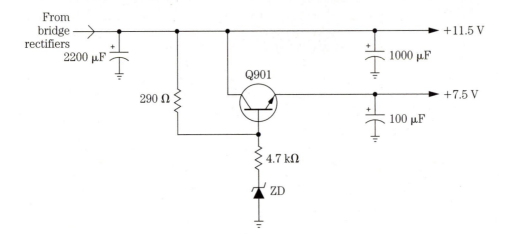

9-5 An open voltage regulator (Q101) cancels the +7.5 V source.

A leaky transistor or IC regulator might have low output dc voltage. The open transistor or IC regulator has no output voltage. The leaky zener regulator can have lower output dc voltage. The transistor regulator becomes leaky or open, while the zener diode regulator becomes leaky and overheated with burned marks. You can easily spot a leaky or overheated zener diode in the low-voltage sources. Take voltage measurements in and out of the suspected regulator transistor or IC to determine if the transistor is open or leaky.

Servicing the radio power supply circuits

Most radio power supplies consist of batteries, ac power circuits, or ac adapters. When the adapter is inserted, the batteries are disconnected from the circuit. Likewise, in the large battery and ac power supply circuits, the batteries are switched in and out of the circuit (Fig. 9-6).

In some models, an external battery source can be plugged into the radio, disconnecting the batteries within the radio. The ac power supply is not in use when the batteries are switched into the circuit (Fig. 9-7). Poor switch and dirty jack contacts can prevent the radio from operating. Check the switch contacts and plug with the low-ohm scale of the DMM. Clean the switch and jack contacts with cleaning fluid.

9-6 Batteries are switched out of a boom-box player when the ac adapter is used.

Check the batteries. Weak batteries can cause weak and noisy reception or slow tape motor movement. Poor battery contact can be caused by leaky batteries and corroded terminals.

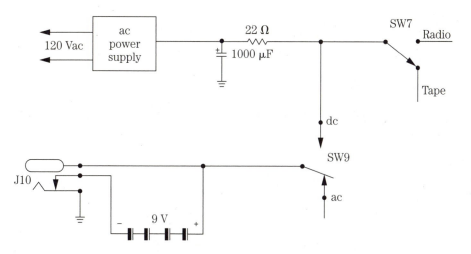

9-7 SW9 switches in external battery voltage when operating on dc.

Alarm-clock radio ac power circuits

The tabletop or alarm-clock radio contains a simple full-wave rectifier circuit. A step-down power transformer reduces the line voltage from 10 to 20 Vac. The step-down transformer has two silicon diodes in a full-wave circuit. A simple capacity and resistance network provides hum-free reception (Fig. 9-8).

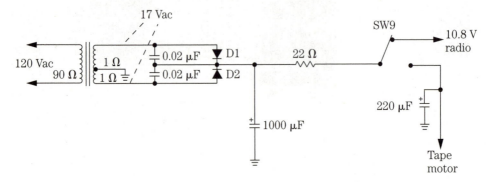

9-8 Full-wave rectification results with D1 and D2 with voltage from step-down transformer.

Larger models use diode and voltage regulation. If the table radio is also battery-operated (9 V), the highest supply voltage will be around 9 to 10.5 V. A 10.5-V source feeds the power IC and transistors, while a 6.1-V source is tied to the cassette pre-amp tape circuits. The lower voltage is tied to the AM-FM radio circuits.

Large table model radios with an LED clock display have several secondary windings on the power transformer. High ac voltage (25 to 50 V) is rectified with two silicon diodes, and the resulting voltage is applied to the clock display. Another ac winding (3 to 5 Vac) goes directly to the clock display. A bridge rectifier circuit attached to another winding provides dc voltage to the tape motor and audio circuits.

Check the clock-radio power-transformer circuits by checking the resistance across the ac line cord plug (Fig. 9-9). The resistance measurement should show the resistance of the primary winding of the step-down transformer with the power switch on. If you don't get a measurement, suspect an open primary winding or power switch. Usually, one or two silicon diodes become leaky or shorted and place a load across the secondary winding. Because the primary winding is wound with small wire, the winding opens up.

Troubleshooting clock power-supply circuits

Check the primary winding and ac plug with the low-ohm scale of the DMM (50 to 150 Ω) if the clock is dead. Often, shorted diodes and leaky filter capacitors will open the primary winding of the transformer. The transformer produces voltage when plugged into the wall outlet. The dc voltage is switched to the radio or tape player.

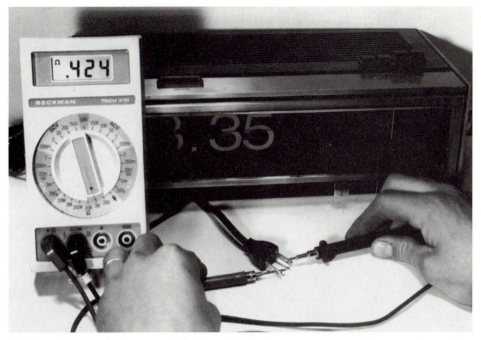

9-9 Check the ac transformer winding by taking resistance measurements across the ac plug.

Check the different voltage sources by measuring the voltage across the large filter capacitor. If a voltage is not found at this point, check each silicon diode in the circuit. Measure the ac voltage across the different transformer wire leads. Check for dc voltage at the emitter terminal of the suspected voltage regulator.

In today's table, clock, and portable radios, only one power transformer is found. The transformer can be mounted on or off the large PC board. The large filter capacitor is found close to the silicon diodes. Look for larger decoupling capacitors in the lower-voltage sources (220 to 470 µF). You might find the power-supply circuits close to the audio output transistors.

Weak or no audio can be caused by leaky voltage regulator diodes and transistors. Open regulator transistors and burned isolation resistors can result in no sound. Don't overlook a dried-up decoupling capacitor if there is weak or distorted audio.

ac adapter repair

The ac adapter consists of a small power transformer, diodes, electrolytic capacitor, flat lead, and male plug. Check for voltage at the metal plug. If there is no dc voltage, measure the resistance across the male ac plug to determine if the primary winding is open. Replace the entire adapter if the transformer is defective (Fig. 9-10).

Often the flat rubber cord will break at the male plug or where it enters the adapter case. Lightly pull on the cord at the male end. Cut off the male plug, and in-

ac
adapter

9-10 The ac adapter is used here to power the portable CD player.

stall a new one if the wire stretches or appears broken where it enters the plastic plug.

Repairing cassette player power supplies

Like the small phono amp, the typical cassette player circuits consist of a step-down power transformer, two diodes, and a large filter capacitor (Fig. 9-11). In earlier chassis, separate ac winding supplied voltage to the pilot lights. Most problems resulted in excessive hum caused by open primary windings of the transformer or shorted diodes. The small power circuits can be found on the main PC board or a separate power supply board.

In small portable cassette players, the power supply circuits are battery- or ac-adapter-operated. An external jack provides power from the ac adapter. When the male plug was not inserted, the batteries are switched into the circuit. A small 220- to 470-µF electrolytic capacitor is used as a filter after the small leaf on/off switch (Fig. 9-12). Clean dirty jack and plug connections with cleaning fluid.

Humming along

Go directly to the large filter capacitor or voltage regulator for a low or excessive hum in the audio. Pull the ac plug, and discharge the large filter capacitor (1000 to 3300 µF) with a test clip or screwdriver. Solder or tack in a good electrolytic capacitor of the same capacity or higher. Observe the correct polarity. If the hum disappears, replace the dried-up filter capacitor.

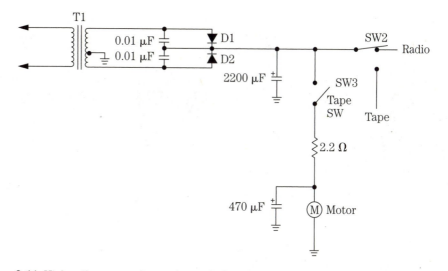

9-11 Higher than normal capacitance is found in the low-voltage power supply of the cassette player.

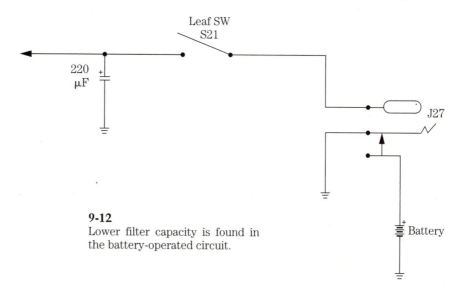

9-12
Lower filter capacity is found in the battery-operated circuit.

Hum can be caused by a leaky voltage regulator transistor or diode. Defective decoupling capacitors in the voltage-dropping circuits can cause a low hum in the sound. Check both the regulator transistors and diodes in the circuit for leakage. Shunt the suspected decoupling electrolytic capacitor with one of a known value.

Locating low-voltage parts

The power-supply circuits will be located around the small power transformer, diodes, and large filter capacitor. Often the power transformer is found off the main

9-13 The small power transformer might be mounted off to one side of the chassis.

board (Fig. 9-13). Look for the main filter and decoupling capacitors on a separate power board. When the transformer is found off the main board, trace the secondary wires to the main board to locate the diodes and filter capacitors. You might find the voltage regulator or zener diode close by in lower voltage circuits (Fig. 9-14).

Troubleshooting car radio power supplies

Because the car radio is operated from the car battery (13.8 V), you would think that the power-supply circuits would be quite simple. This is not true in the deluxe or high-powered car receivers. In the early models, filter parts were mounted on the audio board (Fig. 9-15). You might find voltage transistors and IC regulators with zener diodes in the deluxe chassis.

Car supply circuits consist of a fuse, on/off switch, choke coil, filter capacitors, decoupling capacitors, zener diodes, and several different voltage sources. Because most car receivers have a cassette player, another function switch and motor circuits are included. The different voltage sources are fed to the audio output ICs or transistors (13.8 V), cassette pre-amp IC or transistor (11.6 V), RF AM and IF (8.4 V), FM RF and converter (8.25 V), and FM-MPX circuits (8 V) (Fig. 9-16).

Voltage regulators

You will find transistors and zener diode regulators within the deluxe car radio. In fact, you will find several transistors, zener diodes, and IC regulators in circuits with eight or nine different voltage sources. Some car power sources have dc voltage

Capacitors, diodes, and
regulation board

9-14 The low-voltage transistor and zener diode regulators with filter capacitors are found together in the amp.

Decoupling
electrolytic
capacitors

Main filter
capacitor

Audio board

9-15 Check for hum in car radio caused by open filter capacitors.

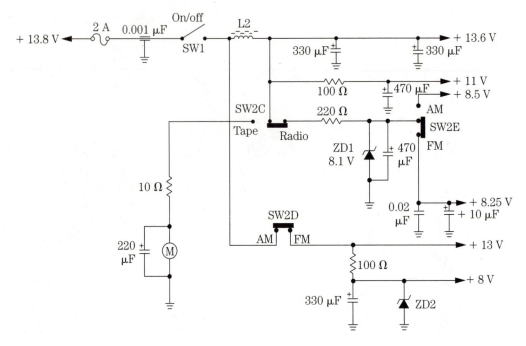

9-16 The low-voltage source in the auto receiver is regulated with zener diodes.

sources coming from the IC output components. A combination CB and car receiver might have separate power sources from the car radio to power an external CB radio (Fig. 9-17).

Check the various power-supply sources when one section of the car radio does not function. Usually the zener diode will run hot and sometimes appear leaky and burned. A defective transistor or IC regulator might not have any visual signs of failure, except improper voltage found at the emitter or output terminals. Leaky regulators and ICs provide a higher voltage source. Don't overlook burned isolation and voltage dropping resistors in the various voltage sources.

Power supply location

Look for a separate power supply board in the deluxe car receiver. The voltage regulators, filters, and decoupling capacitors are found close by (Fig. 9-18). You might find two low-voltage electrolytic capacitors across the same voltage source. The smaller electrolytic capacitors are the decoupling types. In some car receivers, the filters and decoupling capacitors are found on the audio board.

Troubleshooting power supply sources

First measure the dc voltage at the fuse and power output transistors or ICs. Suspect a defective on/off switch, fuse, or stripped PC board wiring with no voltage at the large filter capacitor. Check each voltage source at the zener diodes, transistors, or IC regulators when one section will not function (Fig. 9-19).

9-17 The high-powered 250-W amplifier with connecting harness drives several speakers in the auto.

Power voltage source board

9-18 Locate the large filter capacitors, transistor, and zener diode regulators on the PC board.

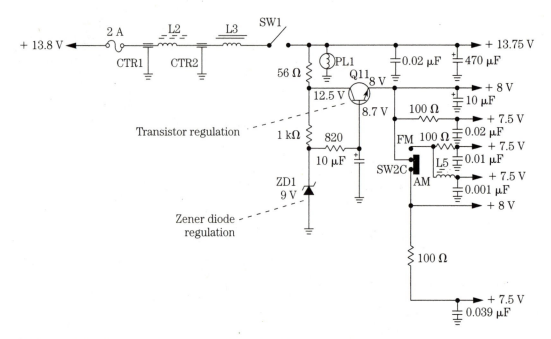

9-19 Check the different voltage sources in the AM, FM, and tape player sources.

Excessive hum in the power-supply source can be caused by the large filter capacitor or voltage regulator. Low hum can be caused by a leaky zener diode or dried-up decoupling capacitors. Sometimes these decoupling capacitors shake loose or break one of their wires.

Check for a leaky polarity diode, filter capacitor, or audio output transistors and IC components when the fuse keeps blowing. If the fuse opens before the on/off switch is rotated, suspect a polarity diode or burned wiring. Leaky zener diodes and burned isolation resistors can cause weak audio and distortion.

No tape sound

The car radio had excellent AM-FM-MPX reception but no tape sound, indicating that the audio amp section was good. The trouble must exist in the pre-amp stages of the head circuits. Voltage measurements on pin 7 of IC801 was very low. Either the IC pre-amp (IC801) was leaky or the voltage was incorrect. The voltage source feeding pin 7 was traced back to a decoupling capacitor and resistor (Fig. 9-20). Voltage was found on one side of the voltage dropping resistor (R803) but was low on the other side. R803 was replaced.

Servicing large stereo amp power supplies

Several different full-wave or bridge rectifier circuits are found in high-powered amps. Large power transformers are found with higher dc voltages to operate high-

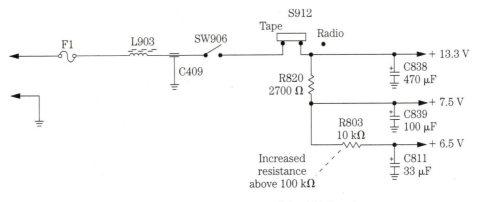

9-20 No cassette sound was found with an increase of the 10 kΩ resistor.

powered transistors and IC components. Transistor- and IC-voltage regulators supply several voltage sources.

The power-supply units are easy to locate in the larger amplifiers (Fig. 9-21). Just look for a large power transformer and electrolytic capacitors. It's possible to see very high capacitance in these low-voltage circuits. The voltage can range from 25 to 90 Vdc.

9-21 The large power transformer, filter capacitors, and power supply components are easily located in the stereo amp.

There she blows

Electrolytic capacitors have a tendency to blow their top, so to speak, with higher-than-normal voltage applied and gas building up inside the capacitors (Fig. 9-22). The electrolytic can become warm and blow the top off if the polarity is reversed or installed backwards. Small electrolytic capacitor can run too warm when a lower-voltage capacitor replaces one of higher working voltage.

9-22 You might find electrolytic capacitors with blown tops in large stereo amps.

When replacing electrolytic capacitors, make sure the working voltage is the same or higher. Likewise, replace defective electrolytics with the same capacity or higher. Check for correct polarity before soldering up. These electrolytic capacitors can blow up in your face. Excess tin foil and insulating material might be found over the entire chassis when the top blows off.

Dead, blows fuses

Suspect leaky diodes, bridge rectifiers, filter capacitors, power-output transistors, or ICs when the fuse blows. Go directly to the large filter capacitor, and take resistance measurements across the main filter capacitors. The meter hand should charge up and discharge with a very high measurement. Low ohm readings below 1 kΩ can indicate a leaky or shorted capacitor. Here the electrolytic capacitor measures no voltage across the terminals (Fig. 9-23).

Check each diode within the bridge rectifier or full-wave rectifier circuit. A low ohm measurement is noted in one direction. If the reading is low with reversed test

9-23 The technician taking voltage measurements in the large stereo receiver.

leads, remove the suspected diode and test it out of the circuit. Replace the whole bridge unit when one diode is found to be leaky or open.

Sometimes the power transformer can be damaged, run warm, and blow the main ac line fuse. If the transformer is overheating or extremely warm, disconnect all leads going to the diode circuits (Fig. 9-24). Mark each color-coded wire so that they can be correctly replaced. Then power up the transformer. If it runs warm or hot without a load, replace it. Remember, these transformers are quite expensive and sometimes difficult to obtain.

Very loud buzz in all speakers

A very loud buzz was heard in all speakers of a Sylvania R73-3 amp. Shut down the chassis, and discharge all large filter capacitors. Clip another electrolytic capacitor of the same or higher capacity across each capacitor with alligator clips. Don't shunt outside capacitors across others while the chassis is operating. You could damage critical processors, ICs, transistors, and diodes in the process. Capacitor C512 (5000 μF, 35 V) was the culprit and was replaced (Fig. 9-25).

Different voltage sources

A low hum can be caused by defective electrolytic capacitors and burned resistors in the decoupling stages. Locate these capacitors on a separate board with the bridge rectifiers. Weak and distorted audio can result from improper positive and/or negative voltage applied to the power output ICs or transistors. In large amps, a high positive (+40 V) and negative (−40 V) can be applied to the power output semicon-

9-24 Trace the secondary wire of the power transformer to the PC board.

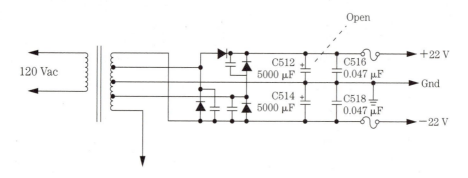

9-25 An open C512 caused a loud buzz in a Sylvania R733 power supply.

ductors (Fig. 9-26). Leaky zener diodes and transistor voltage regulators can cause weak and distorted sound.

No power shutoff

The power could not be turned off with the regular switch (S504) in a Soundesign 5959 cassette-stereo amplifier. The switch tested normal. In this model, the power turn-off circuits are found in the low-voltage power supply. The same voltage was found on all three terminals of Q604. Replacing transistor Q604 with an ECG374 universal replacement solved the unusual power turn-off problem (Fig. 9-27).

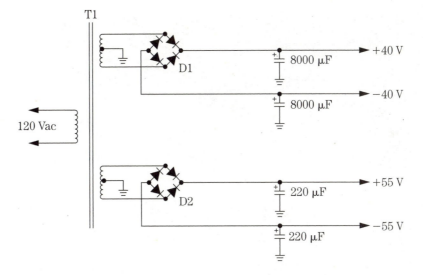

9-26 Two separate low-voltage sources in the high-wattage amp.

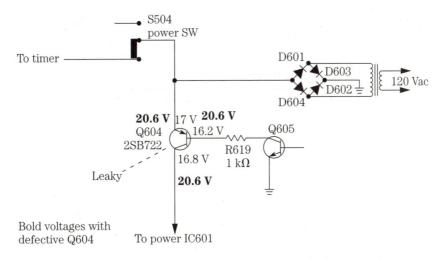

9-27 A leaky Q604 shut down the sound in a Soundesign 5959 cassette deck.

Troubleshooting CD player power supplies

The small portable CD player can be operated from batteries or an ac adapter, while the larger player has a power transformer. Some power transformers are quite small in size with several different ac windings. The power transformer is mounted on the chassis, while other power supply components are mounted on the main PC board (Fig. 9-28).

You might find three or more secondary ac windings on the transformer with full-wave, half-wave, diode, and bridge rectification. Often, one diode is used as a

9-28 The power transformer is mounted off of the PC board in a Magnavox CD changer.

half-wave rectifier to furnish higher voltage to the control and display circuits. Different voltage sources might be needed for servo, decoder, control, and display circuits. A positive and negative voltage are found with each bridge rectifier circuit.

Transistor, zener diode, and IC regulation

The zener diode can be used by itself or with a transistor for voltage regulation of the low-voltage power supply. In positive and negative voltage sources, the zener diode regulates the voltage applied to the transistor regulators.

Zener-diode regulation

Look for overheated zener diodes, producing leakage or an open circuit. When the diode becomes leaky, the voltage applied to the circuit is lower. Check the body of the overheated diode for burned, charred, or white marks. Check the diode like any low-voltage diode with the diode test of the DMM. A leaky diode shows a low resistance measurement in both directions. Remove one end, and test for accurate measurement (Fig. 9-29).

Transistor-regulation circuits

Power-transistor regulation can be found with a zener diode or by itself. You might find two or more regulator transistors in a CD player power supply. These reg-

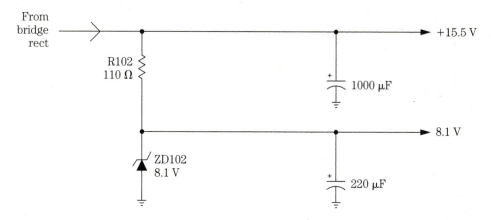

9-29 The 8.1 V is regulated by zener diode ZD102.

ulators provide a positive and negative voltage source. The positive voltage source is taken from the emitter terminal of an NPN transistor, and the negative voltage is taken from the emitter terminal of a PNP transistor.

Regulator transistors have a tendency to open or leak. When the regulator opens, there is no output voltage at the emitter terminal. The output voltage might be lower or higher with a leaky transistor. Sometimes, in a combination zener diode/transistor regulation circuit, the leaky transistor can damage the zener diode. Replace both parts when one is defective.

IC-voltage regulation

Some units use a combination of transistors, zener diodes, and ICs for the different voltage sources. Look for the transistor, IC, and zener diode regulators next to the large filter and decoupling capacitors. Usually all power-supply components are found in one corner bunched together, including separate diodes and bridge rectifiers.

An IC regulator can become leaky and produce a lower or higher output voltage. If leaky, the IC regulator might run warm. If it is open, no voltage will be found at the output terminal. Check for a higher voltage at the input terminal and a lower voltage at the output terminal. If the output voltage source is low after replacing the IC or transistor regulator, suspect a leaky component tied to the voltage source. When one of the positive or negative voltages is low or missing, suspect a defective IC regulator (Fig. 9-30).

Repairing black-and-white TV power circuits

Low-voltage power supplies in black-and-white TV chassis might be transformer- or ac-line-operated. Usually transformer power circuits operate at a low voltage, with a combination of battery and ac operation. The ac power line low-voltage power supply operates at a much higher voltage with some derived voltages from the flyback secondary circuits.

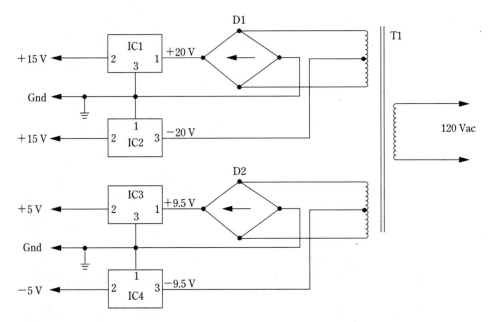

9-30 A defective regulated IC can cause low or no voltage in the low-voltage supply.

The step-down power transformer secondary voltage is rectified by full-wave and half-wave rectification. Very large electrolytic capacitors are used (2000 to 3300 μF) for filtering out ac ripple. A line-voltage fuse (0.5 to 1.5 A) protects the low- voltage power supply circuits (Fig. 9-31). Several transistor and zener diode regulators might be found in lower voltage power supply sources feeding the IF, sound, and sync circuits.

When the dc plug or batteries are inserted, the ac circuits are switched out. Often the dc cord is found with a fuse (2 to 3 A). Suspect a different car battery polarity when the fuse keeps blowing (Fig. 9-32).

Besides several dc voltage sources, you might find two or more voltage sources from the secondary winding of the flyback transformer. These circuits are rectified by a half-wave diode and filter capacitor. Lower-voltage sources are developed with resistance and electrolytic decoupling capacitors (Fig. 9-33). The horizontal and flyback circuits must operate before dc voltages are developed in the horizontal output transformer.

Two different views

Power supplies that have two different polarities (+/–) can be found in TVs, CD players, camcorders, VCRs, and amplifiers (Fig. 9-34). The positive supply originates at the positive source of a bridge rectifier, and the negative voltage from the center top winding. You might find transistor, IC, and zener diode voltage regulators in these circuits.

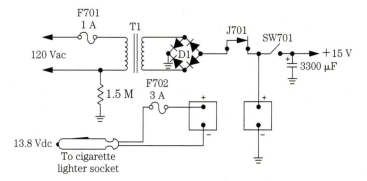

9-31 F701 protects the low-voltage circuits in a black-and-white TV.

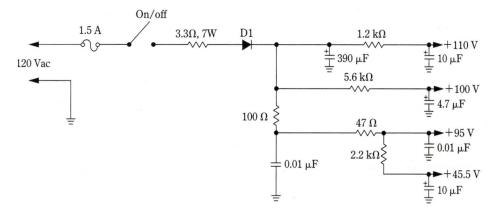

9-32 Voltage-dropping resistors and filter capacitors provide many different voltage sources.

When either one of these power sources is missing in the audio amplifier, you will hear weak volume and extreme distortion. CD or camcorder motors that operate from transistor or IC drivers will not respond to the applied signal. Look for the ungrounded center-tap wire from the power transformer to determine if two separate power supplies are used. Check the negative terminal of several electrolytics to see if it is grounded. The ungrounded negative terminal of the electrolytic capacitor indicates a negative voltage source.

Locating low-voltage components

Always look for a large filter capacitor, diodes, and an ac line fuse. Sometimes the line fuse might be off to the side for easy replacement. The zener diode and transistor-voltage regulator components are nearby. In the transformer-operated chassis, the fuse and transformer might be found on a separate chassis. On bridge circuits, single diodes, half-wave, or full-wave rectifier units are located near the fuse and low-voltage filter capacitors. Low-voltage components are easily located in black-and-white TV chassis.

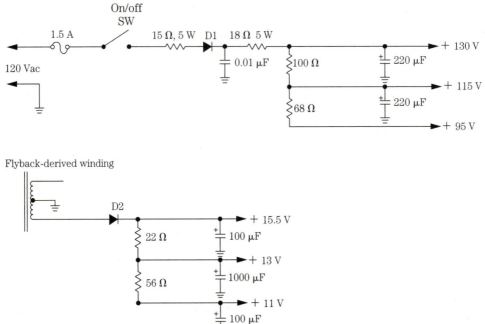

9-33 Low-voltage sources are found in the power supply and flyback circuits of the TV chassis.

9-34 The positive and negative voltage sources are required in high-wattage amplifiers.

Keeps blowing the line fuse

Suspect a leaky diode, filter, or horizontal output transistor when the fuse keeps blowing. Locate and check for leakage across the main filter capacitor. Low resistance can indicate a leaky filter capacitor or output transistor. Check the resistance between the collector (metal body) of the horizontal transistor and common ground. If the resistance is below 400 Ω, suspect a leaky transistor or damper diode. Remove the transistor, and test it out of the circuit. Remove the filter capacitor if leakage is still found with the output transistor out of the circuit.

Don't overlook a leaky diode in the low-voltage power supply when the main filter resistance is above 1 kΩ. Place the test probes across each diode, and test for leakage. A low resistance measurement in both directions indicates a leaky diode. A normal silicon diode should have a measurement above 200 Ω.

Hum with pulled-in sides

Suspect a defective filter capacitor, voltage regulator, or horizontal output transistor when the sides of the raster pull in. One or more heavy dark horizontal lines can indicate a defective filter capacitor. A wavy raster can be caused by defective filter capacitor or leaky regulator.

Measure the voltage at the metal body of the horizontal output transistor or flyback. If it is lower than normal, check the voltage across the filter capacitor. Shunt another electrolytic capacitor across the suspected one if you hear a hum with low voltage. Shut down the chassis. Discharge the main filter capacitor. Clip a known good capacitor with the same or higher capacity across the suspected capacitor. Make sure the working voltage is the same or higher.

Remember, in transformer-operated chassis, the voltage is lower with high-capacity filters, while in the power line, the capacity is lower with very high operating voltage. Leaky zener diodes and voltage-regulator transistors can cause the same problems. Burned voltage-dropping resistors and dried-up decoupling capacitors can cause a low hum and lines in the picture.

Open fuse: Sanyo 61T64 portable

A Sanyo black-and-white TV was found dead with a blown 1.5-A line fuse. The fuse would open when the switch was turned on. A resistance measurement across the filter capacitor was normal. It was found that the half-wave silicon diode (D701) was leaky. D701 was replaced with a 2.5-A silicon diode.

Servicing color TV power supplies

Today the low-voltage power supply connects directly to the ac power line, while in earlier TV chassis, power transformers were used. Always use a variable-isolation transformer between the power outlet and TV chassis when servicing the ac chassis. If you don't, you can damage components within the TV chassis or test instruments. Also, a shock hazard exists between the test instrument and a hot TV chassis.

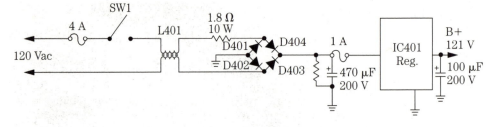

9-35 A defective line-voltage regulator can cause chassis shutdown in a TV chassis.

The typical ac power supply consists of a bridge rectifier, filter capacitor, voltage regulators, and fuse (Fig. 9-35). Some chassis have one diode with half-wave rectification. Look for the ac and dc line fuse and bridge rectifier circuit on the PC board main chassis. A large high-voltage electrolytic capacitor (330 to 870 µF) is found nearby. Voltage-regulator transistors and zener diodes can be found around the electrolytic decoupling capacitors. In the latest TV chassis, look for a large voltage regulator IC supplying high dc voltage to the horizontal output transistor.

Troubleshooting typical power-supply circuits

Look at both ac and dc fuses. Check for B+ voltage at the collector terminal (metal body) of the horizontal output transistor or across the large filter capacitor terminals. Low or no voltage at the output transistor can indicate an open dc fuse or defective voltage regulator. If you don't find dc voltage at the large filter capacitor, check for an open or leaky diode or high-wattage isolation resistor or for poor ac wiring connections. Measure the ac voltage at the large resistor and common ground. One or more leaky diodes in the bridge circuit can cause the power line fuse to open. Be careful; this is power line voltage that you are measuring.

If the main line fuse keeps blowing, suspect a leaky or shorted diode in the bridge rectifier circuit. When both fuses open up, suspect a defective horizontal output transistor or voltage regulator IC. Don't overlook a burned isolation resistor in the flyback primary winding.

Heavy dark lines across the raster can indicate a defective main filter capacitor or regulator (Fig. 9-36). Clip another electrolytic capacitor across the suspected one with the power off. Watch for correct voltage, capacity, and polarity before clipping in the capacitor. Replace a defective regulator when lower dc voltage and hum bars appear in the raster. These IC regulators have a tendency to open or appear leaky.

Lower output voltage from the voltage regulator might be caused by overloading in the horizontal output and flyback circuits. Remove the horizontal output transistor, and test the voltage regulator source. If normal, this voltage will be a little higher with the horizontal load removed. Replace the voltage regulator if the voltage is still low after removing the horizontal output transistor and the input voltage to the regulator is fairly normal.

Power line regulators

The fixed power line IC regulator operates at a higher voltage regulator than those found with step-down transformers. Most of these regulators are tied into the

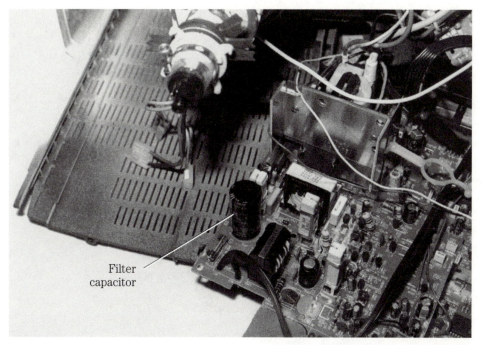

9-36 The main filter capacitor in the TV chassis can cause hum in the sound and hum bars in the raster.

power line circuit after the bridge rectifiers. These regulators can have an output voltage from 110 to 135 Vdc. The voltage output is a fixed voltage with regulated output up to +1 or –1 V (Fig. 9-37). The early fixed line-voltage regulators looked somewhat like a transistor. Later on, they are placed in a larger container.

Today, the new line-voltage regulator represents a transistor with five terminals. You can spot a line-IC regulator with the numbers stamped upon the body. The last three numbers are the output voltage (Fig. 9-38). The dc input voltage can vary from 145 to 175 V, depending upon the filter capacitor and surrounding components. An STR30125 fixed-line regulator has an output voltage of 125 V. Test the dc voltage going into the regulator and the fixed output voltage. If proper voltage is going in and improper voltage is coming out, suspect a defective regulator.

Check all components that are tied to the regulator terminals.

Troubleshooting SCR-switching regulator circuits

In early ac chassis, the SCR regulator was used to control the dc voltage applied to the horizontal driver and output transistor. The ac input circuits are the same as the typical low-voltage circuits, except an SCR with driver, phase detector, and error amp provide adjustable B+ voltage. In other SCR circuits, you can find the IC regulator with B+ adjust in the derived voltage circuits of the horizontal output transformer.

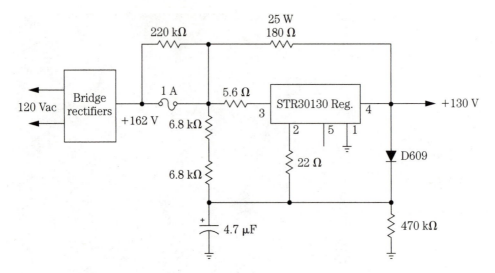

9-37 The STR30130 line-voltage fixed regulator supplies 130 V output.

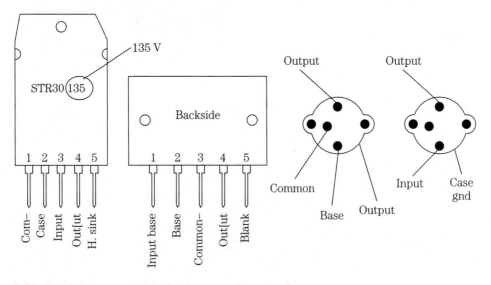

9-38 The different pin terminals of power voltage regulators.

Often the horizontal circuits must perform before B+ regulation can be obtained at the B+ voltage source. A primary winding of the flyback is connected after the main filter capacitor and bridge rectifier, in series with the SCR-switching regulator. B+ can be regulated with the B+ adjust control to set the correct voltage to the horizontal output circuits. Suspect a defective voltage-regulator circuit or horizontal-output component when the B+ control has no effect in changing the B+ voltage.

Servicing SCR regulator circuits

Check for an open ac fuse (F501). If it is open, check the diodes in the bridge rectifier circuits. Measure the dc voltage across the main filter capacitor (C505). Suspect SCR501, ZD701, and associated components if there is no voltage at the output source (Fig. 9-39). Measure the dc voltage at the anode terminal of SCR501. No voltage here can indicate an open winding of the flyback or bad connections at the PC board. Measure the resistance of the flyback winding. An open winding or poor connections can prevent 165 V from reaching the anode terminal.

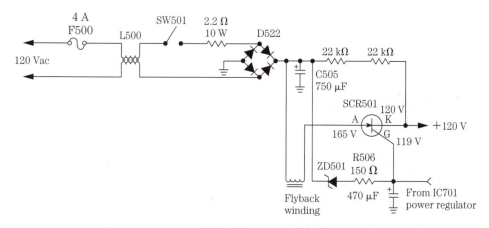

9-39 SCR regulators are found in the early TV power-supply circuits.

If you find the 165 V at the anode terminal of SCR501 and there is no or low gate voltage (119 V), check the IC regulator, zener diode, and associated circuits. Check waveforms and voltages on the SCR driver, phase detector, and error amp when attached to the SCR regulator circuits. Check voltage sources from the horizontal output transformer and horizontal circuits when you cannot find B+ at the cathode terminal of the SCR regulator. Don't overlook the IC power regulator in the horizontal flyback circuits.

SCR power-supply location

Locate the large filter capacitor, ac and dc fuses, and silicon diodes on the main PC board. Sometimes these components are found close to the flyback transformer area. The SCR-switching regulator might be found on a separate metal heat sink nearby. Usually the B+ control, SCR, transistors, and zener diodes are found close to the SCR regulator. Don't confuse the SCR regulator with the horizontal output transistor, if both are on separate metal heat sinks. The switching regulator is always smaller than the output transistor.

Night and day: power circuits

The standby power supply is on all the time and provides voltage to the control circuits so that the remote control can turn on the TV. Usually a small standby power supply and bridge circuits provide the required standby voltage. The small power transformer might be fused but is wired directly to the ac power line. The standby voltage provides 5 to 20 Vdc to the control circuits.

Suspect the standby circuits when the TV can be turned on manually but not with the remote. Locate the standby transformer and the corresponding parts nearby (Fig. 9-40). Locate the standby transformer fuse and bridge rectifiers. Visually trace the transformer leads to the bridge rectifiers. You might find a combination transistor and zener diode regulation circuit in the output voltage source. Sometimes this transformer and fuse are located on a separate metal chassis. The protection fuse might be in the primary or secondary winding of the transformer.

Standby
transformer

9-40 The standby transformer is located with the low-voltage power supply in this RCA TV chassis.

Check for dc voltage at the output of the bridge rectifier and filter capacitor (Fig. 9-41). Suspect the fuse (1 A), transformer, or silicon diodes when there is no voltage. Check the continuity of the primary winding of the power transformer. Replace the transformer if it is open. Double-check the power transformer's soldered connections on the primary and secondary leads. Suspect an open voltage regulator transistor or zener diode when there is no voltage at the emitter terminal.

See chapter 8 for information on the switched-mode power supply (SMPS) and chopper low-voltage circuits.

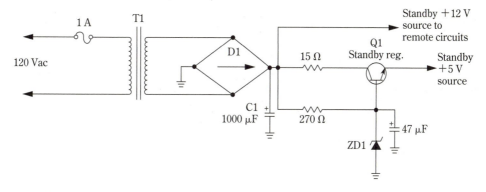

9-41 Check the standby low-voltage circuit when there is no remote operation.

No high voltage, no raster (J.C. Penney 685-2012)

The fuse was inspected and tested good. A voltage measurement across the main filter capacitor (C814) (800 µF) was zero. The low voltage was traced to the anode terminal of the SCR regulator (115 V). The SCR voltage regulator tested normal. A resistance check across filter capacitor C814 did not show leakage. In fact, the DMM did not charge up or down. C814 was found to be open after an 850-µF, 200-V capacitor was clipped across the filter terminals.

Intermittent turn on (Sharp 19A61)

Sometimes the chassis would turn on and, at other times, shut off. Usually after operating for a long period of time, the chassis would shut down and not come on right away. Low voltage (120 V) was missing when the chassis was in shutdown. Voltage measurements were made on the SCR driver, phase detector, error amp, and SCR701 (Fig. 9-42).

SCR701 and all transistors checked normal. Next the silicon and zener diodes were checked. Resistance measurements were fairly close, leaving the possible breakdown of a transistor under load or a defective electrolytic capacitor. When C707 was clipped in, the 10-µF electrolytic capacitor brought the 120-V power source back for good.

TV power supply symptoms

A dead chassis in the Goldstar CM54841 TV

Isolation Check the low-voltage power source at regulator or horizontal output transistor.

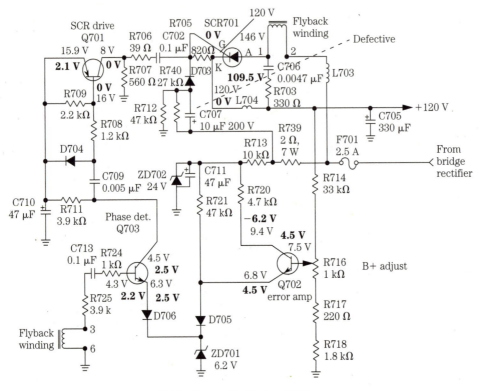

Bold voltages with dried-up C707

9-42 A defective C707 in a Sharp 19A61 chassis resulted in intermittent turn on.

Location The 130-V fixed regulator (IC801) was located on a large heat sink with part number STR30130.

Repair After replacing the 3-A fuse, 164 volts was found at the blown 1-A fuse. F802 kept blowing with no voltage out of the regulator. All connections to the regulator were checked. The leaky 130-V regulator was replaced to restore the 130- V source (Fig. 9-43).

No voltage source to the horizontal output transistor with a dead chassis in a General Electric 19PM-C chassis

Isolation Because a power transformer with many leads was mounted next to the flyback, the low-voltage circuit was suspected of switching power supply.

Location The switching transformer was mounted close to the flyback, and the switching transistor was located upon a small heat sink.

Repair The 2.5-A fuse would blow without any output voltage at the large 330-μF, 200-V filter capacitor. The switching transistor was located and tested leaky in the circuit. Q901 was replaced with ECG375 universal replacement (Fig. 9-44).

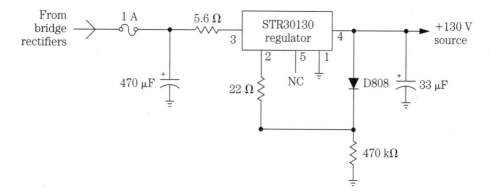

9-43 A STR30130 regulator in a Goldstar CMS4841 portable produced a dead chassis.

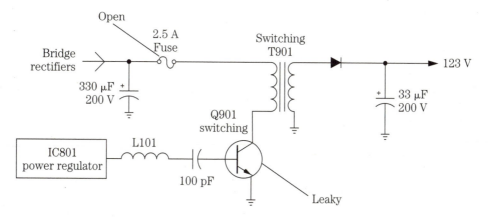

9-44 A leaky switching transistor (Q901) caused a dead chassis in GE 19PM-C.

Intermittent shut down was noted in an Emerson MS1920K TV

Isolation Monitor the 155-, 125-, and 6.6-V sources within the low-voltage power supply.

Location The 155-V source is fed to the sound circuits, while the 125-V source feeds the horizontal output circuits.

Repair When the 125-V source went low in the horizontal output transistor, very little voltage can be adjusted and was noted with VR501 (20K). R506 had increased in resistance and was replaced with a 100-kW, 1-W resistor (Fig. 9-45).

Power-supply troubleshooting chart

Check the troubleshooting chart (Table 9-1) for additional power-supply problems.

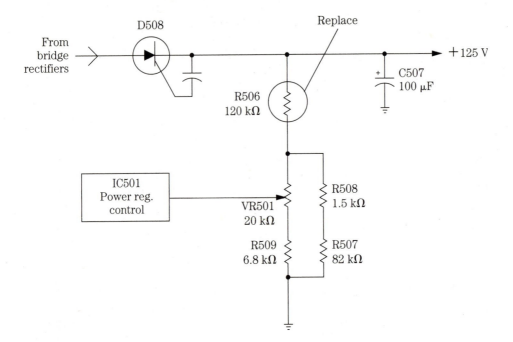

9-45 Replace the increased resistance of R506 with a 100-kΩ resistor for intermittent shut down in an Emerson MS1980R TV.

Table 9-1. Low voltage troubleshooting chart without a schematic.

Symptom	Location	Repair
Dead chassis	Check low-voltage power supply.	Measure the dc voltage across the large electrolytic capacitor. No voltage—Check fuses.
	Diodes	No voltage—Check silicon diode with the diode test of the DMM.
	Locate bridge rectifier	Check each diode inside the bridge component. Test for leakage. Most diodes become shorted or leaky.
	Locate transformer	Measure the ac voltage across the transformer wire terminals. Check the primary winding for an open winding.
Loud hum in sound	Locate filter capacitors	Shunt the large filter capacitors. Clip the leads across the terminals with the power off.
Low hum	Check decoupling capacitors	Shunt each small electrolytic capacitor in the circuit.

Table 9-1. Continued.

Symptom	Location	Repair
Intermittent voltage	Check regulators	Monitor the output source with the DMM. Check for an open or intermittent regulator transistor. It often goes open. Check for an open regulator IC. Check all the wiring connections. Check the zener diodes.
Keeps blowing the fuse	Check diodes	Check for shorted or leaky silicon diodes. Suspect a leaky filter capacitor. Check for shorted turns in the power transformer.
Transformer overheating	Diodes	Check for one or more shorted rectifiers. Check for a leaky filter capacitor. Test for shorted turns in the transformer. Remove the transformer wire leads and see if the transformer still runs warm. Replace the defective transformer.
Flyback voltage sources		
Dead	Horizontal circuits	The horizontal circuits must operate before any voltage sources.
No low-voltage source	Low-voltage circuits	Check for an open isolation resistor. Check for a shorted or leaky silicon diode. Shunt the electrolytic capacitor. Check the winding of the flyback to the voltage source.
Really low source	Diodes	Test the diode for openness or leakage. Check the regulated transistor. Check for an overloaded circuit in that particular voltage source.
Hum, smeary, or odd symptoms	Filter capacitors	Shunt the small electrolytic capacitors in the voltage circuits. Check for overload in that voltage source.

10
CHAPTER

Servicing stereo sound circuits

The stereo sound circuits are found in AM-FM-MPX radios, large portables, boom-box players, cassette decks, car radios, compact disc players, and TV chassis (Fig. 10-1). Servicing stereo sound circuits is fairly easy because comparison of signal, voltage, resistance, and components can be made in each channel without a schematic. Simply isolate each stage with an external amp, signal trace with a scope, and make in-circuit tests of the transistors and ICs.

Radio stereo circuits

The AM-FM-MPX radio circuits break off at the FM-MPX circuits and divide into left and right audio channels. The FM stereo music is switched into the audio stereo circuits with a function switch. A dirty function switch can cause erratic or intermittent FM stereo reception (Fig. 10-2).

When the FM stereo indicator will not light or one channel is dead, there is a good chance that the FM-MPX section is defective. Check the stereo audio section by switching to the cassette or phono player. Turn the switch to the FM position to see if both speakers function. If they do, the FM-MPX section is defective. See chapter 11 for information on servicing the FM-MPX section (Fig. 10-3).

Car radio stereo circuits

The car radio AM-FM-MPX circuits are similar to those found in table or portable radios. Besides AM-FM-MPX stereo circuits, most car receivers have a cassette player. Some recent car radios have stereo compact disc control.

If the unit's stereo function is okay but the stereo bulb or LED will light, suspect a defective bulb, improper bias adjustment, defective indicator light circuit, or improper voltage source. The audio signal can be signal traced right up to the stereo in-

10-1 The Montgomery Ward deluxe AM-FM MPX portable with double cassette and stereo sound.

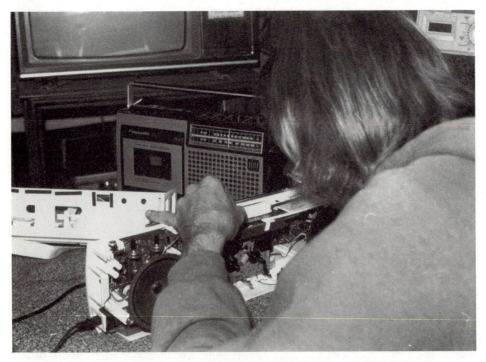

10-2 The electronic technician checking the sound circuits in the AM-FM-MPX radio.

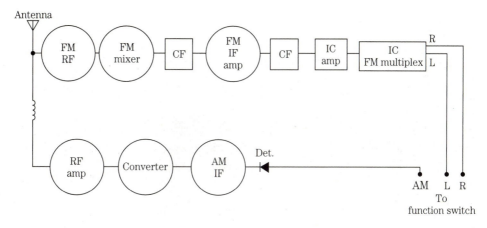

10-3 A block diagram of an AM-FM-MPX front-end stereo receiver.

dicator circuits. Signal trace the audio to the input of the indicator IC. Take critical voltage tests on the IC and LEDs. Remember, FM-MPX stereo channels can function with a defective indicator circuit.

When indicator lights stay on all the time, it is either the indicator light circuitry or an improper threshold adjustment. Usually the indicator light or bank of LEDs is powered by a transistor or IC. Suspect a defective transistor, IC, or improper voltage source when the stereo light stays on very bright (Fig. 10-4).

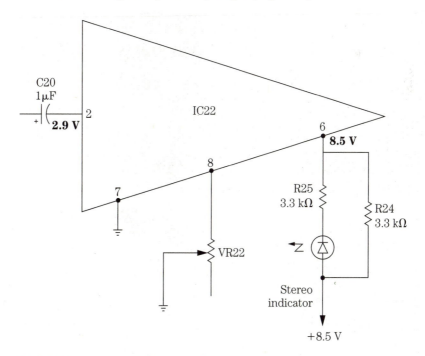

10-4 IC22 controls the LED stereo FM indicator.

Auto sound circuits

The auto tape sound circuits might consist of a single IC that includes both tape head input signals. The audio signal is picked off of the tape by the stereo tape heads and fed directly into a dual-IC (Fig. 10-5). This cassette audio is coupled to a function switch or has fixed diodes in the line to prevent radio audio from entering the tape circuits.

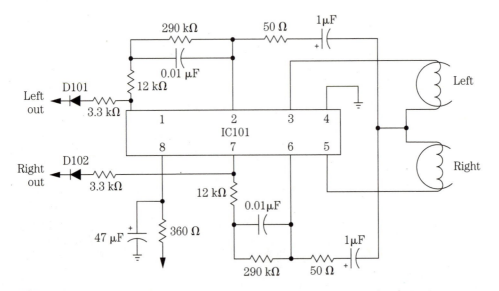

10-5 D101 and D102 serve in place of a function switch in an auto tape head circuit.

D101 and D102 isolate the tape head stereo sound from the other audio circuits. When the cassette is played in the auto radio player, the picked up sound is amplified and allowed to pass through the two small diodes to the input circuits of the stereo output amplifier. Likewise, when the audio stereo signal is played, the stereo sound cannot enter the tape circuits because D101 and D102 will not let the signal enter the tape circuits. Fixed diodes are used to switch the two signals, eliminating a radio-tape function switch (Fig. 10-6).

Big amps

In today's auto or truck, a high-wattage amplifier system might include AM-FM-MPX radio, cassette and CD player, equalizers, auto control, audio control, and an amplifier for each set of speakers. These amplifiers might start from 100 W up to 1000 W or more (Fig. 10-7). The high-powered amp can drive 10" to 15" woofers, several mid-range, and five to seven tweeters. A couple of auto batteries can be added to keep the system operating.

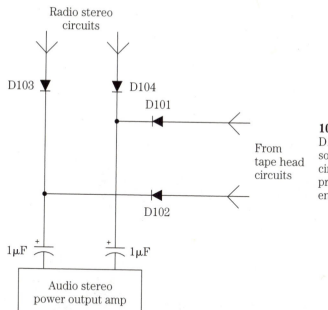

Radio stereo
circuits

D103   D104

D101

From
tape head
circuits

D102

1μF   1μF

Audio stereo
power output amp

10-6
D103 and D104 prevent tape
sound from entering the radio
circuits, and D101 and D102
prevent radio signal from
entering tape channels.

10-7 A Craig 250-W auto receiver amplifier.

The high-powered amp might consist of high-voltage ICs or transistors. In this 250-W amplifier, five power-output transistors are found in each stereo channel. All power-output transistors and IC components are mounted upon a large heat sink, usually the metal-fin body of the amplifier (Fig. 10-8). Each power-output transistor

Power output
transistors

10-8 All 10 power output transistors are bolted to the large metal body heat sink.

can be tested in or out of the circuit. Remove the suspected transistor for accurate leakage test. Always, replace each audio component with exact replacement part.

If a universal transistor replacement is used, make sure the terminal leads are the same length. Mount the transistor in the same spot. Apply silicon grease between insulator and transistor and between insulator, transistor, and heat sink. Before removing the output transistor, check for a piece of mica insulation. You might find one or two transistors mounted directly upon the heat sink. Arrange the piece of insulation to match mounting holes.

Portable stereo circuits

The boom-box can have AM-FM-MPX radio circuits, cassette, or compact disc player. Although the boom-box might have more audio power and several speakers, the front-end circuits are similar to the table model radio. Some large players can have up to three speakers in each channel. The function switch provides audio from the AM-FM-MPX, cassette, and compact disc player.

In boom-box stereo circuits, the audio components are easiest to locate (Fig.10-9). You will find power transistors and IC components bolted to a separate heat sink or the metal chassis. Locate the output components on separate heat sinks with the AF audio section nearby. Wherever you find the large electrolytic capacitors and power output capacitors, the audio section is close by.

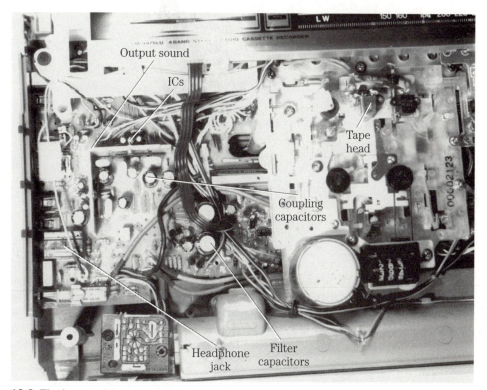

10-9 The layout of the sound circuits in a boom-box cassette player.

The portable audio stereo circuits might consist of transistor, ICs, or both. Muting transistors can be found at the line output or audio output to the volume control. A right and left meter system might be amplified by one or two transistors from the muting system. The radio, tape, or line output signals are switched into the circuit and amplified by a driver transistor or ICs. Often, a dual audio IC is found for driver of both audio channels.

In low-powered portable stereo units, a power IC component can be found between the volume control and speakers. A tone and balance control might be found before the volume control. The dual-volume control is found with a separate balance control. Separate volume controls can be used in some audio circuits. Capacity coupling is found between volume control and power IC (Fig. 10-10).

The audio signal is amplified by IC103 in each channel and coupled through a 1000-μF capacitor to the speakers. Most problems found in the stereo channels are

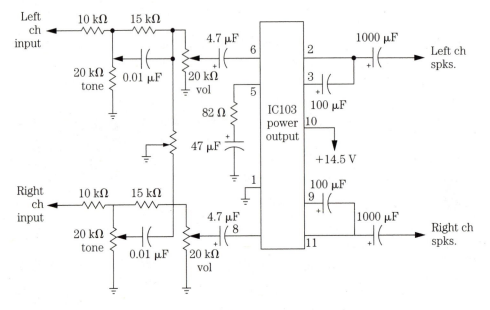

10-10 IC103 provides power amplification for both stereo channels.

a defective IC, worn volume control, and open coupling capacitors. Weak, intermittent, and no sound can be caused by a defective electrolytic coupling capacitor. Suspect IC103 when either channel is distorted, weak, or dead. Both channels might be affected by a leaky output IC103. Sometimes only one channel is defective and the other is normal. The suspected IC must be replaced in either case. Do not overlook improper voltage supply source. Low-supply voltage can result from leaky IC103. Replace IC103 if it is running red hot.

Compact disc stereo circuits

The typical portable CD audio section consists of a D/A converter, sample hold, low-power filter network, audio IC amplifiers, and headphone circuits. The audio output circuits are switched into the regular audio stereo circuits within the boombox player, car CD receiver, and compact disc player, with or without headphone reception (Fig. 10-11). By rotating or pushing the function switch to CD, the audio portion is switched to the internal audio circuits.

The audio signal can be traced from the D/A converter IC to the function switch or headphones. Use an external audio amp, and signal trace each audio channel at the D/A converter. The audio signal can be traced with the scope with a disc playing. Rotate the function switch to the AM-FM-MPX radio or cassette player to determine if the CD circuits are defective and that the audio circuits are normal.

Check the signal at the output of the D/A converter, input S/H IC, and the pre-amp stages through the low-pass filter network and pre-amp audio circuit (Fig. 10-12). When the music stops, check the IC voltage on each terminal. Leaky audio ICs

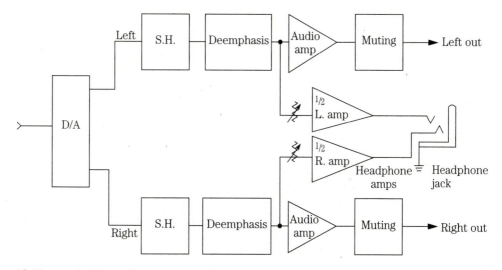

10-11 A typical block diagram of auto CD stereo circuits at the D/A converter.

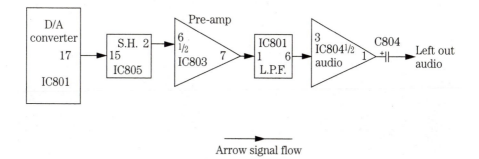

Arrow signal flow

10-12 Follow the arrows for signal flow from the D/A converter to the audio output.

can run warm and have a low supply voltage. Remember, the CD signal is separated at the D/A converter into left and right audio channels.

Stereo cassette decks

The stereo circuits found in cassette decks or players function around two tape heads with audio in the left and right channels. The two tape head windings are found in one tape head to pick up recorded audio to both channels. These magnetic tape heads are fed to identical pre-amp circuits that amplify the weak signal. This audio signal is amplified and switched to high-power audio circuits (Fig. 10-13).

A dirty tape head can cause weak or distorted sound or a dead channel. Sometimes the head gap becomes packed with excess oxide dust, creating poor pickup to the transistor or IC pre-amp circuits. Look for broken head wires when one channel is intermittent or dead. Loss of high frequency can be caused by a worn tape head. Cross-talk is caused by poor adjustment of the tape head (Fig. 10-14).

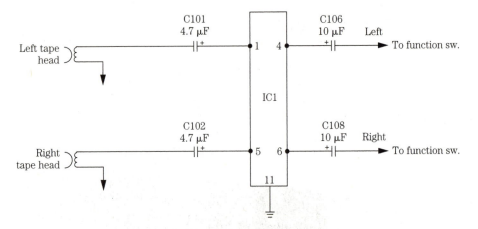

10-13 IC1 provides tape head amplification for both stereo channels.

Stereo tape
head

10-14 Improper alignment of the tape head can cause cross-talk.

Isolate the pre-amp circuits from the tape head to the function switch. Signal trace the audio with an external amp from the tape head winding to the base of the transistor or the input terminal of the IC. Remember, the signal directly off of the tape head is very weak. The audio should be amplified after the pre-amp and AF circuits. With a 1-kHz test cassette, signal trace with a scope through the entire audio circuit.

Stereo audio circuits

The stereo audio circuits consist of a left and right audio channel. The audio signal might be amplified several times before being heard in the speakers. Today's small cassette players and radios have only one or two IC audio components. Medium-power stereo amplifiers have transistors or ICs and transistors in the sound circuits (Fig. 10-15).

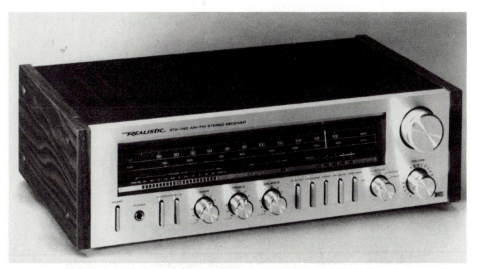

10-15 The Realistic STA-740 AM-FM-MPX stereo receiver can drive several different speakers.

The high-power amps found in car receivers, CDs, and amplifiers have several stages of audio with high-power transistors and ICs. Separate high-power amps found in cars might contain 600 W per channel, feeding into a 4-Ω power output speaker load, with a price of over $2000. Regular power amplifiers range from 50 to 300 W with Darlington, MOSFET, and bipolar transistors and ICs.

The vacuum tube is still found in some high-power amplifiers. Some have tubes with MOSFET power output circuits. High-power components should be replaced with high-power parts. These special components should be replaced with original part numbers.

Too hot to touch

Transistors and IC components that run red hot should be replaced. Check the body of the suspected IC for overheated white and brown marks. These power-output components can run warm but not too hot to touch. Usually, the transistor or IC might be leaky, or a driver transistor might be open or leaky (Fig. 10-16). Before replacing the hot transistor or IC, check the bias resistors, diodes, and driver transistors. Measure for higher-than-normal voltage upon the output component. A missing

Power output transistors

10-16 If the output transistor is hot, check the bias resistors and voltage.

negative or positive power source can cause weak sound with distortion and hot output transistor or IC.

Besides hot ICs and transistors, a balanced audio circuit might end up damaging the speakers with a dc voltage found upon the voice coil. The balanced output circuit should not have any voltage upon the speaker terminals (Fig. 10-17). If a speaker fuse is found between the speaker and amplifier, the fuse should open with dc voltage present. When a large electrolytic coupling capacitor is found between the amp circuits and speakers, the speakers are protected from voltage damage.

In high-powered audio circuits, five or more high-powered transistors are found in each stereo output channel. When a transistor goes open or becomes leaky in directly coupled circuits, the power output transistors can be damaged. A change in resistance can upset the bias on a driver transistor. The leaky bias diode can cause transistors to overheat and be destroyed (Fig. 10-18). Check for burned or leaky zener diodes in the circuit to upset the whole output amplifier. Suspect a change in resistance, leaky diode, or open or shorted driver transistor when the output transistors run red hot and repeatedly become damaged.

Dead to the left

The dead channel can be caused by a leaky or open transistor, IC, coupling capacitor, power resistor, speaker, or low-voltage power supply. Signal trace the audio circuits with an external amp or scope tests. Slip in a 1-kHz test cassette, and check

10-17 If the voice coil is damaged, check the speaker terminals for dc voltage.

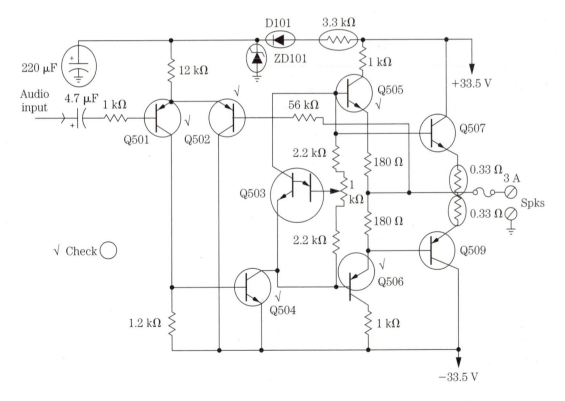

10-18 Open driver transistors, electrolytic capacitors, burned bias resistors, and zener diodes can cause output transistors to run hot.

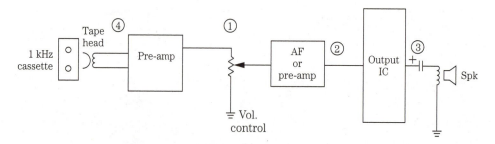

10-19 Check by the numbers with external audio amp to locate a dead or intermittent circuit.

the audio circuits with the scope to the main speaker. To check out a dead stage, inject a 1-kHz sine waveform to the input and use the scope as a monitor (Fig. 10-19).

When the signal quits, take critical voltage measurements and inspect the parts found around the high-power transistors and ICs. Test the components in-circuit, and when one is found to be leaky or open, remove it and test again. Look for shorted or open capacitors in the IC power output circuits. Leaky or open electrolytic capacitors in power output circuits can produce a dead channel.

Look for open fuses and burned or open bias resistors in the audio circuits (Fig. 10-20). Check all the transistors in-circuit when burned resistors are found. Remove

AF and driver sound circuits

10-20 Look for burned bias resistors, fuses, and diodes in the overheated output transistors.

the suspected transistors, and check the bias resistors while they are out of the circuit. Replace resistors with the exact resistor found in the good channel. Look for power output transistors on heavy heat sinks.

Weak to the right

Weak reception can be caused by open or dried-up coupling capacitors, open emitter capacitors, AF and driver transistors, and dirty tape heads. Signal trace the weak audio, stage by stage, and compare it with the normal channel until the defective part is located. If both channels are weak, suspect a poor voltage source or common IC.

Weak signal symptoms that are found with some distortion occur in the AF, driver, and output circuits. Check the audio output transistors for leaky or open conditions. Make sure all bias resistors are normal. Compare them with the good channel. Open driver transistors and a change in value of bias and emitter resistors can produce a weak and distorted condition. Locate the output transistor and IC parts on the large heat sink (Fig. 10-21).

Intermittent left channel

Defective transistors, ICs, coupling capacitors, electrolytic capacitors; poor element terminals; poor board wiring; and improper plug connections can cause a chan-

10-21 Locate the power output transistors and ICs in the stereo receiver.

nel to be intermittent. Check the intermittent channel at the volume control with an external speaker to determine if the intermittent occurs in the front-end or output circuits. Intermittent power supply can cause intermittent sound.

If the intermittent is in the output circuits, use a regular speaker or an external amp as a monitor. Sometimes, when a test instrument is touched on one element of a transistor, the sound will return. Check the voltages on the transistors and ICs when there is an intermittent. Applying coolant and heat on transistors and capacitors can make the part act up (Fig. 10-22).

Fuses and filter capacitors
power supply

BIAS   AF-driver
circuits   circuits

10-22 Apply coolant and heat on resistors, diodes, and transistors for parts that can cause distortion.

Probe around on small components with an insulated tool to find poorly soldered terminals or parts. Pushing up and down on sections of the PC board can cause the chassis to become intermittent. Resolder the intermittent PC board wiring section to locate a badly soldered joint. Don't overlook large blobs of solder on wiring junctions. Check for a worn or broken volume control.

Now and then

Intermittent sounds that only occur once in a while with long periods in between make it very difficult to locate the defective component. If the intermittent becomes active after 4 or 5 hours of operation, heat might become a factor, pointing to transistors, ICs, and resistors that run warm. Monitor the intermittent at the volume control to determine if the defective component is in the power-output circuits.

Then monitor the sound at the speaker terminals with an 8-Ω 100-W load resistor on large amplifiers. Connect a known speaker if no voltage is found at the speaker terminals. Monitor the sound at the driver stage with external amplifier. Connect a voltmeter to both the positive and negative power sources. When the intermittent occurs, notice if there is a voltage change in both voltage sources. By dividing the circuits, the slow but sure intermittent can be located.

Look for defective coupling capacitors, transistors, ICs, and bias pots and a change in a resistor (Fig. 10-23). Replace both electrolytic coupling capacitors. Clean

Stereo
sound circuits

10-23 Look for a change in bias resistors, electrolytic coupling capacitors, transistors, and ICs for intermittent problems.

up bias controls with cleaning fluid sprayed inside the control. Readjust the bias voltage. Check for overheated resistors, and replace them. Remove one end of the resistors or bias diodes, and make a correct resistance test. Compare the same components to the other stereo channel when color codes or numbers are not visible.

Holy cow

Sometimes a loud squeal or warbling sounds are heard in the speakers. Notice if the disturbing noise is heard in the left or right channel, or both. Suspect electrolytics in the power supply if the noise is heard in both channels. Shunt electrolytic capacitors in both the positive and negative power supplies (Fig. 10-24). Check for dried-up decoupling electrolytic capacitors in the different voltage sources. Always, shunt each capacitor with the amplifier turned off. Discharge each shunted electrolytic before applying into another capacitor circuit.

10-24 Shunt electrolytic capacitors within the power supply to locate noisy sound.

Signal tracing audio circuits

Often, power output transistors within the power-output circuits are directly coupled to one another directly or with a resistor in between. Notice that the signal path is found at the base of Q101 and that one half of the Darlington transistor amplifies the audio signal and directly couples to the base terminal of Q103 (Fig. 10-

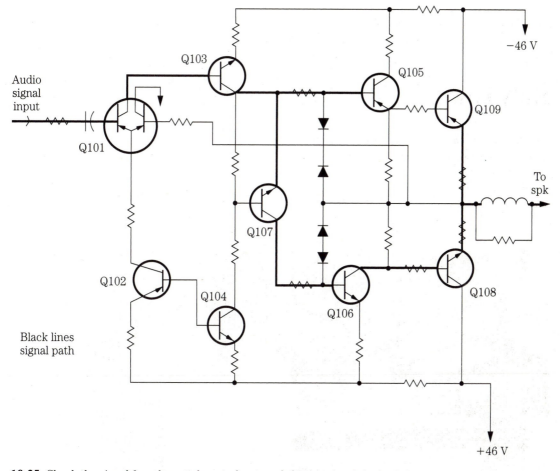

10-25 Check the signal from base to base to locate a defective circuit in the stereo amp.

25). The output audio from the collector of Q103 is fed through a resistor to the base of Q105. Also, the audio signal is fed to the emitter of Q107 with the collector fed through a resistor to the base of Q106.

Q105, Q106, Q108, and Q109 are in a series push-pull amplifier circuit. Q105 and Q106 operate in a push-pull audio circuit and feed the output signal from the emitter terminals to the base of Q108 and Q109. Again, Q108 and Q109 are in push-pull operation. The high-powered audio output signal is taken from the emitter resistors of both Q108 and Q109, through a coil-resistor network and speaker relay switch.

Check the audio signal at the base of each transistor for audio gain with the external audio amp or scope probe. Notice the gain or loss of each stage. The signal at each base can be compared with the same circuit of the normal channel. A loss of signal or distortion can be signal traced at a defective transistor circuit. When the signal or distortion is noted on the base of a transistor, check the preceding circuits. You might find an open driver and leaky AF transistor producing weak and distorted sound.

Distorted audio

Go directly to the audio output circuits when either channel or both channels are distorted. Locate the audio output transistors or ICs. Use a speaker or external amp to determine which channel is distorted. Keep the external amp volume low at all times when checking for weak and low distortion. Check the distortion at the volume control or output stages.

Leaky or open output transistors and IC components can produce a distorted channel. Check for burned or open bias resistors in the output stages. Compare these burned resistors with those found in the normal channel for correct resistance. You might find one transistor leaky and the other open in a push-pull output circuit. Always replace both transistors when one is found to be leaky (Fig. 10-26).

10-26 Check the sound output circuits for distortion.

Low distortion can be caused by a leaky or open driver or AF transistor. Make sure that the bias resistors are good. Check the collector or emitter resistors of the driver for burn marks or a change in resistance. Poor bias adjustment can cause distortion.

Look for problems within the low-voltage circuits when distortion is found in both channels. Check the negative and power-supply voltages to see if they measure the same at the output transistors and ICs. Improper or low voltage can produce distortion.

In a Craig AM-FM-MPX stereo player, the normal power-supply voltage was +24 V. Only 9.8 V was found at the output transistors. When the large filter capacitor was tested, the voltage was found to be only 9.8 V. All four silicon diodes were checked. Two of the diodes had leakages of 1200 and 1500 Ω (Fig. 10-27). Replacing both diodes solved the distortion problem.

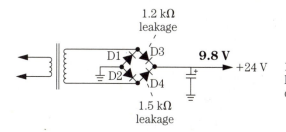

10-27
Low voltage caused extreme distortion with a leaky D3 and D4.

Noisy sound

Noisy transistors, ICs, and coupling and bypass capacitors can cause frying or crackling noises in the audio. Poor contacts on the on/off switch can cause a buzzing noise in the amplifiers. Locate noise in the output, driver, and AF transistors with an external amp. Another method is to short the base terminal to the emitter terminal of each transistor and, when noise is present, proceed toward the front end of the amplifier. Intermittent noise from a loosely grounded board can cause noise in both speakers.

A loud hum in both speakers can be caused by open filter capacitors or broken foil around PC board leads. Check the large filter capacitors if they appear leaky. Suspect a decoupling capacitor when a hum appears after a few minutes. Shunt these capacitors with the power off so as not to destroy solid-state components. Turn the power on, and listen for a hum in either channel.

Solid cone

The sound is weak and muffled out of a large 12" PM woofer speaker. The speaker was dismounted and found to have a frozen voice coil. The voice coil was frozen against the metal core. Sometimes this happens when too much power or volume is applied to the speakers. The voice coil can collapse around the metal area when dc voltage is applied to the voice coil (Fig. 10-28). A foot or object pushed through the speaker cloth can cause the voice coil to sag. Always, check for a dc voltage upon the speaker terminals when a frozen cone is found.

Different sound symptoms

The following sound symptoms might help you isolate, locate, and repair the next sound chassis. These case histories can be applied to just about any audio channel.

10-28 The frozen voice coil can be caused by cone damage, too much volume, and dc voltage on the speaker terminals.

The fuse keeps blowing in a J.C. Penney MCS 683-3207 stereo receiver

Isolation   Check the audio output transistor.

Location   Locate Q210 and Q212 upon the heat sinks.

Repair   Found Q210 leaky. Replaced both Q210 and Q212 output transistors.

Blows fuses

Isolation   Replace the fuse. If it still opens, check the low-voltage power supply and audio output transistors.

Location   Locate the output transistors, and check the voltage at the large electrolytic capacitors. Check ICs on the large heat sinks.

Repair   Test the driver and output transistors in-circuit for leaky conditions. Check the input and output terminals of the output ICs with an audio amp. Check each silicon diode in the low-voltage power source for leakage.

Right channel dead in a J.C. Penney 3845

Isolation   The right channel fuse is open.

Location   Locate the right channel output IC with an external amp.

Repair   While checking voltages on the right channel output IC (STK-0050), a burned resistor was found. Replaced a leaky TR210 and also the IC power output

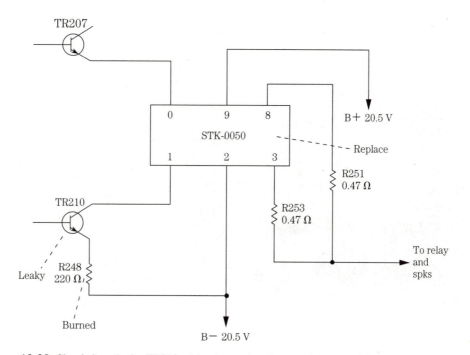

10-29 Check for a leaky TR210 with a burned emitter resistor in a J.C. Penney 3845.

(Fig. 10-29). Checked R248 within the left channel to acquire correct resistance (220 Ω).

FM stereo drifting in a J.C. Penney 3233 solid-state stereo receiver

Isolation   Try to locate the FM front end and MPX circuits.

Location   Because the drifting was related to the FM stereo circuit, locate the FM MPX IC.

Repair   Voltage measurements were taken upon all terminals of IC2. Although some of the voltages were only off 0.05 to 1 V, IC2 was replaced with an RCA SK3147 universal FM MPX replacement (Fig. 10-30).

No left channel in a J.C. Penney 683-1991

Isolation   Signal trace the left channel starting at the speaker. Check the low-voltage source.

Location   Located signal on the input terminal of IC101A (pin 8) and none on output terminal 5.

Repair   After locating the loss of signal, checked the voltages on IC101A and found high voltage at pin 5. Should be 8.5 V, according to the right channel. Replaced IC101A with universal ECG-824 replacement (Fig. 10-31).

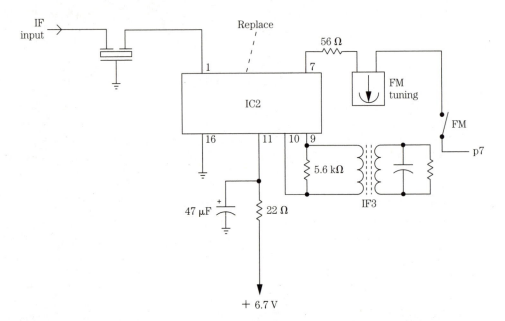

10-30 FM drifting was caused by IC2, which was replaced with an RCA SK3147 universal replacement.

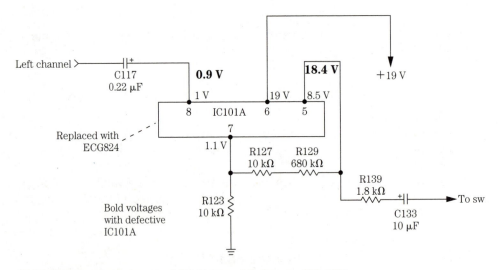

10-31 Higher voltage destroyed IC101A in a J.C. Penney 683-1991.

Pioneer SK-1000TW blows left channel fuse

Isolation   Go directly to the output transistor or IC.

Location   After locating a leaky output transistor with an external amp, found a burned bias resistor (0.7 Ω).

Repair   Replaced the leaky power output (2SC897) with SK3535 universal replacement. Before replacing the transistor, the bias and predriver transistors were tested. Replaced burned resistor with 5-W type.

Distorted left channel

Isolation   Check for an open fuse. Go directly to the output transistors or ICs, and check the signal with an external amp. Keep the volume on the amp very low.

Location   Locate the left channel with signal picked up with an external amp.

Repair   Found a leaky output transistor (2SC789), and replaced it with an SK3054. Checked all the bias resistors.

Left channel distorted and weak

Isolation   Check for an open fuse in the left channel. Signal trace with an external amp. Take critical voltage measurements, and compare them with the normal right channel.

Location   The left channel AF amp (121-272) was operating very warm.

Repair   Checked AF amp in-circuit, found it to be open, and replaced it. Also replaced driver transistor 121-271 for leakage (Fig. 10-32).

Excessive distortion on both channels

Isolation   Check common IC or power-supply circuit.

Location   Look for large filter capacitors within the low-voltage power source when voltage is low or improper.

Repair   Found only –7 V in a Magnavox AM-FM-MPX receiver instead of –18 V. When the large filter capacitor (1000 μF) was shunted, both channels returned to normal. Broken wiring foil around the positive terminal produced the distortion (Fig. 10-33).

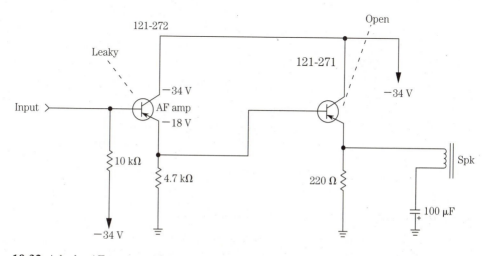

10-32 A leaky AF amp transistor caused a weak and distorted symptom in a Zenith output stage.

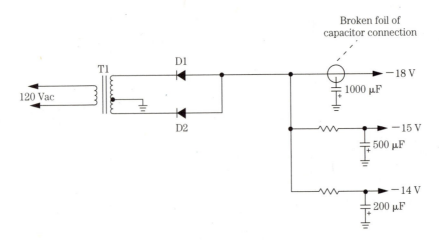

10-33 Broken foil around the terminal of a 1000-μF capacitor caused extreme distortion in a Magnavox unit.

Noisy left channel after set was in a couple of hours; could be shocked in with the noise by rapidly turning set off and on in an Onkyo TX-V940 receiver

Isolation Isolate the frying noise to the audio output circuits by turning the volume control down with no effect on noise.

Location The left channel audio output transistor can be located on a heat sink and PCB.

Repair Signal traced the noise starting at the speaker and going from stage to stage until noisy Q501 transistor was found in amplifier input stages (Fig. 10-34).

Weak right channel

Isolation Signal trace with an external amp or scope starting at the volume control.

Location Locate the weak right channel with a speaker or external amp.

Repair Started at the volume control, and went from base to base and found an open 10-μF coupling capacitor between the two AF stages.

In a Soundesign 4485, the left channel was very weak

Isolation The stereo lights that flash when music is played would not light up.

Location Start at the volume control on the left channel, and compare with the right channel audio.

Repair The volume was normal at both center terminals of the volume controls. The weak part must be ahead of the left channel volume control. Using external audio signal tracing, C213 was found to be open (Fig. 10-35).

Noisy left channel in an RCA RZC-800W

Isolation Signal trace with an external audio amp. The noise would start very low.

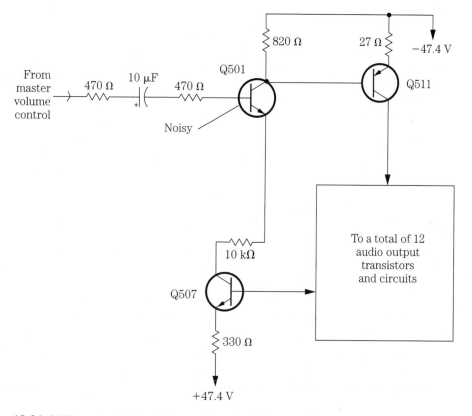

10-34 Q501 produced a low frying noise in the Onkyo TX-V940 receiver.

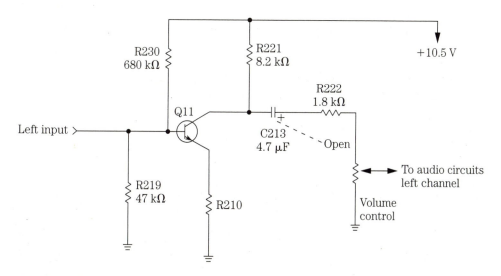

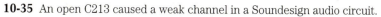

10-35 An open C213 caused a weak channel in a Soundesign audio circuit.

Location   Locate the noisy channel with a speaker or external amp.

Repair   Let the amplifier warm up for at least 30 minutes. Sprayed each transistor with coolant. Replaced the leaky output transistor. There was still noise. Replaced the noisy driver transistor.

Right channel noisy

Isolation With the volume down, you could hear popping and cracking noises in the speaker.

Location   Signal traced with an audio amp to locate the noisy channel.

Repair   The noise was signal traced up to the collector. Shorted the base to the emitter terminals, and the noise disappeared. Replaced Q106 with SK3020 universal replacement (Fig. 10-36).

Motorboating in both channels

Isolation   Because both channels were motorboating, suspected the power-supply circuits.

Location   Locate large filter capacitor in low-voltage power supply.

Repair   When the 47-μF capacitor was shunted in a Morse Electrophonic AM-FM-MPX, receiver the motorboating ceased (Fig. 10-37).

Amplifier troubleshooting chart

For further amplifier problems and solutions, check troubleshooting chart (Table 10-1).

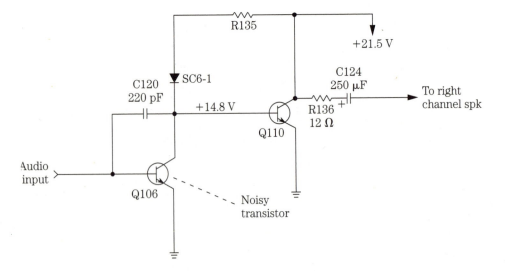

10-36 A noisy Q106 was found in a Sylvania table model radio.

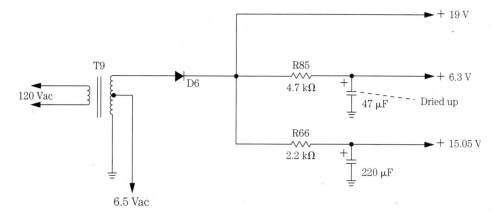

10-37 An open 47-µF capacitor caused motorboating in a Morse-Electrophonic chassis.

Table 10-1. Servicing stereo sound circuits.

Symptom	Location	Repair
Dead—No sound	Check at volume control	Signal trace the audio at the volume control to determine if the problem is in the output circuits.
	Output circuits	Measure the voltage across the main filter capacitor. Check the voltage on the output transistor or IC. If the fuse is blown, suspect a leaky output transistor or IC. Signal trace the volume control to the base of each transistor with a scope or external amp.
Intermittent sound	Input or output	Monitor intermittent audio at the volume control to determine if the problem is in the input or output stages. Monitor at the speakers and on the base of the second AF or driver transistor. Monitor at the input and output of the suspected stereo output IC. Look for an intermittent coupling capacitor, burned bias resistors, diodes, and transistors.
Weak sound	Input or output	If the cassette player sound is weak, check the tape heads. Clean with alcohol and a cloth. Check the weak channel against the normal stereo channel to see what stage the problem is in. Go from one side of the coupling electrolytic capacitor to the other for weak audio. Check the signal from base to base of each transistor. Check the input signal and output of the suspected IC. Weak sound is

Table 10-1. Continued.

Symptom	Location	Repair
		caused by open coupling capacitors, bias resistors, diodes, and transistors.
Distorted sound	Output stage	Clean the tape heads. Check the input and output of power IC for distortion with an external amp. Check each transistor for openness or leakage with in-circuit tests. Remove the transistors and diodes, and check them out of circuit. Check the audio output bias resistors for correct values. Check the bias diodes for leakage. Make sure the voltage source is proper and both the positive and negative sources are equal. Check the distortion at a given point in the circuit and compare it to the normal channel. If both channels are distorted, suspect a common IC or improper supply voltage.
		Distortion is caused by open transistors, ICs, bias resistors, bias diodes, and improper voltage. Remember, in directly coupled circuits, an open or leaky transistor down the line can affect all transistors in the stereo line-up.
Hum in sound	Filter capacitors	Shunt each filter capacitor with a good one. Shut down the chassis to shunt the capacitor. Discharge each capacitor after shunting each one in circuit.
	Low-level hum	Check and shunt each decoupling capacitor. Check for an increase in resistance at the base terminal. Check the input cables and wires.
Motorboating in the sound	Audio output	Replace the audio output or ICs.
	Low-voltage power supply	Shunt all electrolytic capacitors in the power supply.
Output transistors run red hot	Output stage	Check for open or leaky driver transistors. Check the bias diodes and resistors. Check for excessively high voltage source. Check for either a positive or negative voltage source that is missing.

11
CHAPTER

Troubleshooting
AM-FM-MPX circuits

With the chassis removed from the case or the back cover removed, you can locate each section and narrow down the hunt for the defective component. Take a peek at the whole receiver layout (Fig. 11-1). Often the AM-FM front-end transistors and IC components are located near the tuning section. AM RF and converter coils have a lot of wire, while the FM RF, oscillator, and mixer stages have a few turns of bare copper wire.

Look me over

When a schematic of a given receiver is not handy, a typical block diagram is the next best thing (Fig. 11-2). Locate the separate AM and FM front-end section. The FM coils are made of solid wires and mounted upon the PC board. Locate the AM ferrite antenna rod with a coil, and trace the wires to the tuning section and RF AM transistor. The FM-MPX stereo section should be mounted close to the FM stages. The early AM-FM radios had IF transformers sticking up from PCB. The audio section might have power transistors or ICs on heat sinks, while the low-voltage power supply can be located by the power transformer and large filter capacitors (Fig. 11-3). A quick voltage test across the filter capacitor can indicate if the power supply is working.

Which end is up

Take the receiver symptoms, and apply them to the components located upon the chassis. If the AM reception is weak or intermittent or there is no reception, go directly to the front-end circuits. Locate the antenna ferrite coil and tuning mechanism (Fig. 11-4). In the early chassis, two or three ganged tuning capacitors tuned the RF and oscillator stages. Today, the varactor diodes tune the RF and oscillator

11-1 The AM/FM/MPX circuits are found in the cassette and CD player.

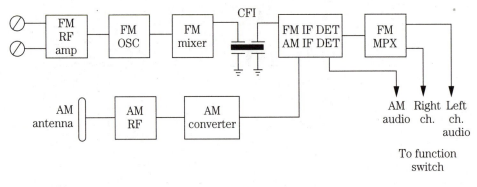

11-2 A typical block diagram of an AM/FM/MPX receiver.

sections. In digitally tuned receivers, a tuning controller IC provides control and tuning data applied to the RF, mixer, and oscillator circuits.

Common circuits

Look for circuits common to both AM and FM when stations are not received. Check the low-voltage power supply and regulator source feeding the AM and FM circuits. A leaky regulator transistor or zener diode can produce a dead AM and normal FM section. Critical voltage measurements on the AM or FM transistors and IC components can help locate a defective low-voltage source.

Check the IF stages or a combination of the FM IF and the AM converter circuit. Sometimes the first FM IF transistor serves as the AM converter. Notice if both the AM and FM IF stages are common to one another (Fig. 11-5). For example, an open second IF transformer connection can destroy both AM and FM reception. A dirty AM-FM slide or function switch can cause dead AM or FM reception.

No AM reception

Dead AM reception can be caused by a leaky or open AM transistor or IC component. Check each transistor with in-circuit tests. If in doubt, remove the transistor from

Large
filter
capacitors

11-3 Locate the low-voltage power supply by locating the large filter capacitors.

11-4 Locate the antenna coil, which is sometimes wound on a ferrite form.

IF transformers

11-5 Locate the small IF transformer to locate the AM and FM front-end circuits.

the circuit. You will find that the suspected transistor is open or leaky with in-circuit tests but is normal when removed. Replace the transistor if in doubt. Sometimes the AM-FM front-end IC might be defective in AM and normal in FM circuits (Fig. 11-6).

Clean the function switch assembly with cleaning spray when the AM or FM is erratic or intermittent. Spray cleaner into the switch contacts, and work the switch back and forth to clean the contacts. Replace a broken function switch with an original part number, if one is available.

Suspect a defective AM RF transistor when a local station can be heard with a voltage probe on the collector terminal. Test the transistor in-circuit for open or leaky conditions. Critical voltage tests will indicate if the transistor is open or leaky. Check the antenna coil for open or broken leads when there is no AM.

Cracked ferrite antenna

Radio receivers that use a ferrite antenna coil can have the weak radio symptom if the ferrite core is cracked or broken. Some of these ferrite iron cores are round, and others are flat. Sometimes the core will break inside the coil winding, resulting in weak AM reception.

Weak AM reception

Determine if the AM signal is weak and the FM reception is normal. Try to locate the AM components near the tuning assembly, antenna coil, and common AM-FM IC.

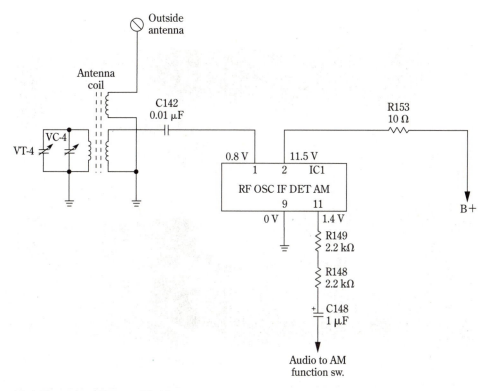

11-6 Trace the AM to an IC chip.

Weak AM reception can be caused by an open or leaky RF transistor or IC. If there is no RF transistor, look for a damaged RF coil or antenna lead or improper voltage applied to the AM circuits.

In some low-voltage power supply circuits, a separate AM voltage-regulator transistor or zener diode provides voltage to the AM section. A dirty voltage switch contact can prevent voltage from getting to the AM stages. Take critical voltage measurements, and trace the low voltage to the low-voltage power-supply circuits. All transistors and IC parts found in the AM-FM system can be replaced with universal replacements.

Intermittent AM reception

Determine if both the AM and FM stages are intermittent or just the AM circuits. A cracked or broken antenna core can weaken AM reception. The crack or break might be inside the coil winding, so take a close look (Fig. 11-7). Some cores can be replaced. The exact replacement might be available from the manufacturer.

The small Litz wire used in antenna coils is easily torn loose or broken off, and poorly soldered connections can produce intermittent AM reception. Check the windings for continuity. Use the lowest scale on the ohmmeter, and check each winding. You should get a direct short or very low measurement.

FM coils   AM-FM RF   AM-FM
Var. capacitors   OSC mixer

11-7 In early receivers, look for the AM and FM tuning capacitors to locate the AM and FM circuits.

Push up and down on the PC board within the AM section. Probe around on small coils, capacitors, and resistors with an insulated tool. Sometimes, by moving a small part, you can turn up a poorly soldered connection or terminal. An intermittent terminal lead or junction within the transistor can be located with this method.

The trouble might be between the AM converter stage and the antenna coil. Grasp the body of the antenna coil, and see if the stations become louder. If they do, there is continuity between the converter and at least part of the antenna. Look for a break or poorly soldered connection under a layer of coil dope or wax.

Weak and intermittent AM reception can be caused by a defective RF transistor. You will find a separate RF transistor in the better receivers, but many receivers use a single transistor AM converter that functions as both the RF and oscillator stages (Fig. 11-8). When a unit has only two variable capacitor sections or varactor diodes, you can be fairly sure that the unit uses a converter stage with no separate RF. If several stations can be tuned across the dial, you can assume that the converter transistor is normal.

No FM reception

If the AM section is normal but there is no FM, suspect a leaky or open FM RF, FM oscillator, or FM mixer transistor or IC. A leaky FM RF FET transistor can re-

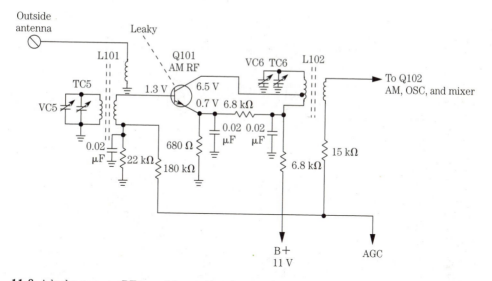

11-8 A leaky or open RF transistor can produce weak or intermittent reception.

sult in no FM reception. Take critical voltage measurements on the FM transistors and ICs. Check all transistors in-circuit. When a defective transistor is found, replace it.

A loud rushing noise on FM can indicate a defective mixer or oscillator transistor or IC. No FM reception can be caused by improper voltage from the low-voltage circuits. Check for leaky or open regulator transistors and zener diodes. Sometimes a dirty switch contact supplying voltage to the FM circuit can prevent FM reception.

Weak FM

If the AM section is normal but the FM is weak, the prime suspects are the FM RF section and the FM antenna. The FM RF transistor is typically mounted on the PC board next to self-supporting coils (Fig. 11-9). If a chassis layout is handy, use it. If it's not, signal trace the wiring from the FM RF variable capacitor section, varicap, or varactor diodes to the RF transistor.

The FM RF transistor can be tested with in-circuit transistor tests or diode transistor tests of the DMM, which will indicate if the transistor is open or has a beta reading. Don't adjust any FM screw trimmers, variable capacitors, or FM RF transformers.

Using a plastic rod or wooden pencil, tap the RF coils. Sometimes these self-supporting coils are not soldered properly and make a bad board connection. Check the FM lead from the RF coil to the antenna terminals. Your TV antenna can serve as an alternate FM antenna because the FM band is between channels 5 and 6. Be sure to use an FM coupler to avoid degrading TV reception when connecting the FM receiver to the TV antenna.

11-9 Look for the FM self-supporting coils.

No AM or FM, defective IF circuits

The power supply, IF transistors, IC components, and RF transformers and ceramic filters are common to the AM and FM sections; therefore, if both sections are dead, chances are that the cause is one of these four. Rotate the function switch to the tape or phono position to check the audio stages. If they work, the problem is located in the IF stages (Fig. 11-10).

Take voltage measurements on each IF transistor and IC. Compare the voltage with the supply source. Improper readings can indicate a defective transistor or IC. Test all transistors in the circuit, or remove suspected ones for testing. Normal voltage readings indicate that the IF transistors and power supply are functioning, but will not always identify a defective IF transformer.

A defective IF transformer can be located with signal injection or continuity tests. Use the ohmmeter to check each winding. The AM IF transformer will measure a few ohms, and the FM IF transformer less than 1 Ω. Erratic or intermittent AM and FM reception can be caused by a poorly soldered connection of the transformers or a broken IF coil winding. Ceramic IF filter networks have no resistance measurements and seldom cause problems.

You can locate a defective IF transformer by signal tracing. Turn the function switch to the AM or FM position. Inject a 455-kHz signal to the base of each consecutive IF stage, starting at the last IF stage and working towards the converter stage. Likewise inject the FM IF stages in the same manner with a 10.7-MHz signal. When

11-10 Check for IF transformers with AM and FM windings in series.

you don't hear any tone, you have located the defective stage. Signal from a white noise generator can signal trace the RF and IF stages.

Intermittent AM and FM reception

If both AM and FM sections operate intermittently, see if the phono or cassette player units are working. If both are functioning, you can assume that the intermittent condition must be in the IF or other stages common to the AM and FM circuits. Locate the AM and FM IF transformers and coils. Look for the IF transformers in a row (Fig. 11-11).

First, try to move the IF coils or transformers around with your fingers. You might have a poorly soldered connection. Move the IF transistor or IC with an insulated tool or pencil. Sometimes a transistor lead will have a poor internal connection or soldered terminal connection. Because IF transistors and ICs frequently cause intermittent conditions, spray each one with three coats of coolant. Let each coat disappear before applying the next one. Take your time with each transistor or IC before going to the next.

Monitor the low-voltage source feeding the AM and FM IF stages. Intermittent voltage can produce intermittent reception. Clean the AM-FM slide and function switches (Fig. 11-12).

If the intermittent trouble still exists and it seems to be around the IF transformer, remove the transformer from the board. You might be able to repair the IF

IF transformer
terminals

11-11 The IF transformer terminals on the PC wiring side.

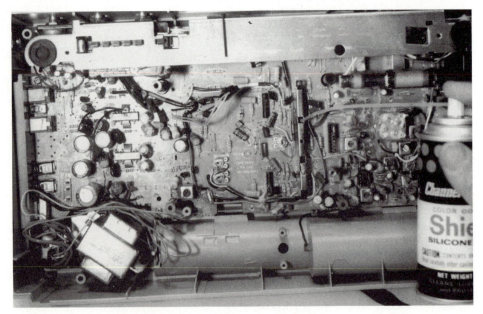

11-12 Spray and clean the function switch with cleaning spray.

transformer by removing the insides. Pull back the tabs or indented area, and pull out the coil from the outside shield. Check each winding for continuity. Place the coil under a magnifying light, and solder the suspected terminal. These transformers should be replaced with factory originals. After replacement, tap and twist the transformer slightly to check the success of the repair.

Control circuits

In deluxe receivers that are operated by remote control, the volume control is controlled by a small dc motor. The small motor is controlled by a motor driver IC901 with infrared signal amplified by Q901 and a volume indicator (Fig. 11-13). The infrared volume indicator is controlled by a signal from the remote transmitter. The infrared transmitter turns the volume up or down. The A output at terminal 8 of IC901 rotates the volume control up in both stereo channels, while the B out at pin 7 turns the volume down. The rotation of the motor is reversed by changing the voltage polarity out of IC901.

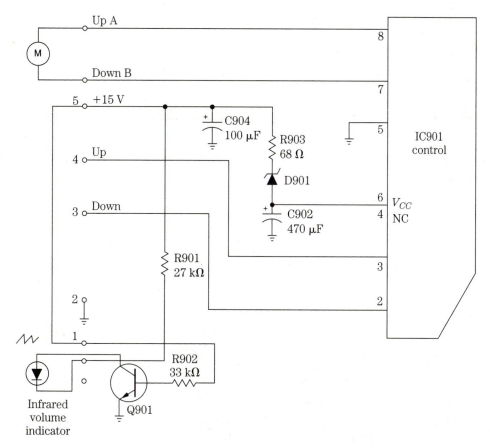

11-13 The volume control motor is operated from the IC901 driver with an infrared remote control.

When volume cannot be turned up or down by the remote control, check for +15 V at pin 6 of IC901. Measure the voltage output across the motor terminals. Suspect Q901 with no voltage output. Make sure infrared indicator is normal with a diode test of DMM. Check Q901 in-circuit for leakage or open. Make sure that the control knob is not jammed or frozen.

Weak AM and FM reception

Check the IF and power supply sections for both weak AM and FM reception. Suspect a leaky IF transistor or IC component when both sections are weak. Take critical voltage measurements with in-circuit transistor tests. Broken IF transformer leads or poorly soldered PC board connections can produce weak reception.

A low voltage fed to the AM and FM circuits can produce a weak signal. Trace the low-voltage source back to the power supply. Check the voltage regulator transistor for open or leaky conditions. Suspect an overheated or charred zener diode for low regulator voltage.

Both AM and FM were weak in a Sharp SC210 receiver. At first, IF trouble was expected, except that the low voltage in the IF section was down to 3.5 V (it should have been 8.2 V). The suspected voltage regulator diode ZD201 turned out to be normal (Fig. 11-14). The low voltage was traced to pin 14 of IC202. Replacing IC202 with an ECG1243 universal replacement solved the weak AM-FM reception problem.

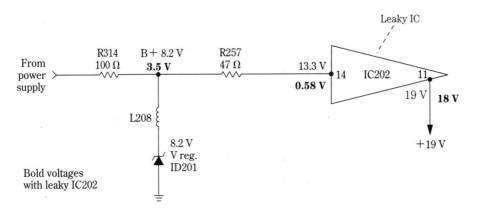

11-14 A leaky IC202 caused a weak AM and FM reception in a Sharp receiver.

Checking transistors in-circuit

Open and leaky transistor tests can be made with a beta tester or diode junction test of the DMM. The defective transistor might be open, leaky, or shorted (Fig. 11-15). Sometimes in-circuit leakage tests are inaccurate with coils, diodes, and low-ohm resistors in the base circuit. Because FM front-end transistors are difficult to remove and replace, check them with in-circuit tests.

11-15 Check each transistor in the circuit with the diode test of DMM.

If you don't know whether the transistor is an NPN or PNP type, you can find out with either tester. Remember, with the DMM diode transistor tests, the base is common to both emitter and collector terminals. Check the PC board for correct terminal connection. Sometimes the letters E, B, and C are found near or on the correct terminal on PC board wiring side. Test the transistors after they are removed from the circuit. Always test the new transistor before installing it (Fig. 11-16).

Replacing front-end transistors

After locating a defective FM transistor, check the location, lead length, and how the transistor is mounted. Obtain a new universal replacement, and cut the leads the same length as the old one. Mount the transistor back in the same spot. Redress leads around the transistor. Test the new transistor in the circuit. Sometimes, when replacing FM mixer and oscillator transistors in older receiver, FM alignment should be performed. With the latest receivers, especially with ceramic IF stages, alignment might not be needed.

Meter movements

The signal and recording meters might have analog meter movement or a bank of LEDs. The audio signal can be traced right up to the meter or LED circuits with an external amp. These indicator LEDs can be tested with the diode test of the

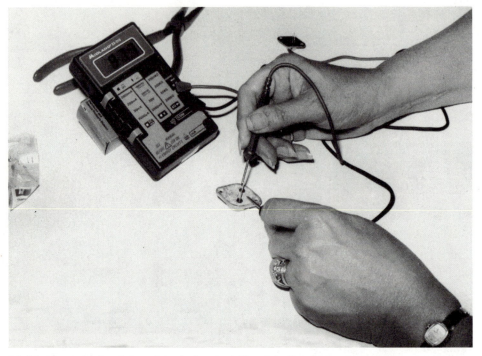

11-16 Always test each transistor before installing it in the electronic chassis.

DMM. If one becomes dead or weak, all have to be replaced in one large component. Determine if signal is applied to the indicator circuits before checking the LED indicator circuits (Fig. 11-17).

Signal-meter movement is found in large AM-FM receivers. The station is tuned in when the meter is at the highest reading. Check the signal up to the meter circuits. Test the diodes with a DMM. Take continuity tests of meter movement (Fig. 11-18). The meter hand will move when the ohmmeter is attached to the meter terminals. Sometimes the meter hand will stick in one position if the meter scale is warped. Simply remove the meter assembly, and reglue the meter dial assembly.

LCD display

The LCD display indicates the station tuning, operations, and numbers. The LCD display is driven with an IC101 driver circuit from data microprocessor IC202 (Fig. 11-19). The key matrix, analog switch, PLL, video selector switch, and motor driver tie into the data system. Check the 6-MHz crystal waveform at pin 26 and the +5-V supply voltage. A –20 V is supplied to the LCD.

The varactor tuner

Today, varactor tuner diodes are used to tune in AM-FM stations in the RF, oscillator, and mixer stages. Varactor diodes take the place of manually tuned variable

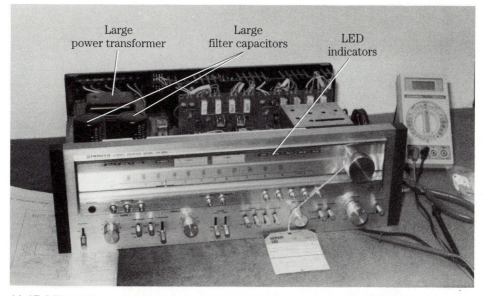

11-17 LED indicators are used for both receiving and recording within the receiver.

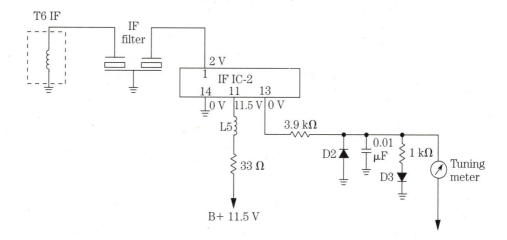

11-18 When there is no meter movement, check the IC or transistor meter amplifier.

capacitors. A varactor diode will change capacitance when a different voltage is applied to it. By varying the voltage across the diode, a variable capacitor tunes the AM and FM bands like a TV tuner. A simple tuner can be made up of a variable resistance, varying the voltage on the varactor diodes.

Besides the FM varactor tuner, you might find varactor diodes in the RF, mixer, and AM oscillator circuits. The same signal voltage is supplied by the controller. In some chassis, there is a test point to measure the controlled voltage (Fig. 11-20). The voltage applied to the varactor diode is a different voltage for each station.

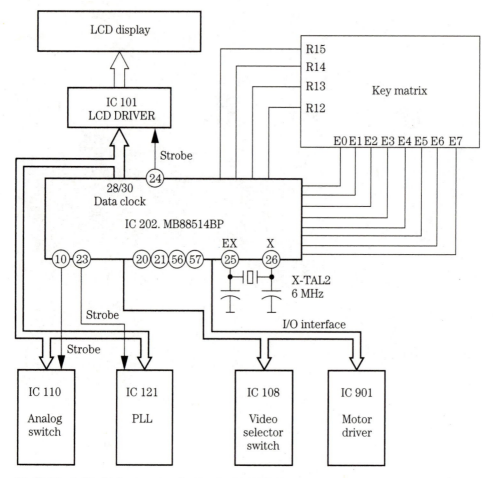

11-19 The LCD display is controlled by the IC101 LCD driver.

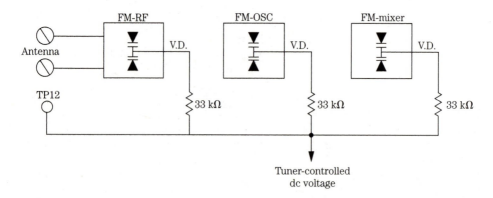

11-20 Varactor diodes tune the FM, RF, oscillator, and mixer circuits with dc voltage from the controller.

Determine if the trouble lies in the front-end tuning, control system, or IF stages of the receiver. Inject a 455-kHz IF signal at the first AM IF. If signal gets through, check the voltage applied to the varactor diodes. This voltage will change as the tuning knob is rotated. Suspect a defective tuning controller when no or improper voltages are found at the varactor diodes (Fig. 11-21).

Front LCD board

11-21 The tuning controller and radio circuits are found in the front panel LCD board.

Digital tuning controller

The digital tuning system provides control over the supply voltage fed to the front-end (AM-FM tuning), fluorescent display, signal indicator, and switching in some digitally controlled receivers. A microcomputer chip provides digital tuning control and remote data to several stages of the digitally controlled system. The synthesizer processor (IC) provides controlled voltages to the AM-FM RF and IF circuits (Fig. 11-22).

Check the supply voltage at the FM tuner. This voltage will vary as the tuning knob is rotated. Suspect a transistor regulator or improper voltage at the processor chip with no voltage at the FM tuner. Take critical voltage tests on IC603, TR603, and TR604. If voltage is found at the front-end tuner, suspect a defective tuning system. Check each stage as those found in the conventional front-end tuning circuits. Take a scope test of the crystal-controlled PLL circuit of IC603.

Other new receiver circuits

Auto stop and auto tuning

When receiving an FM station, the IF signal from CF3 is applied to pin 1 of IC122. Pin 12 of IC122 becomes low (from 3 to 0 V) and Q128 is turned off. Under

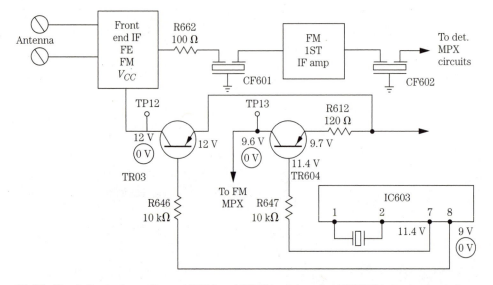

11-22 Check the tuning voltage at TP12 and TP13 to determine if IC603 is tuning in stations.

this condition, the high level (from 0 to 5 V) is provided to pin 34 of IC202. This stops the auto tuning mode in the microprocessor (Fig. 11-23).

When receiving an AM station, the IF signal from CF4 is applied to pin 9 of IC120. Pin 16 potential of IC120 becomes high and Q127 is turned on. Then Q128 is turned off. Under this condition, a high level (from 0 to 5 V) is provided to pin 34 of IC202. This stops the auto tuning mode in the microprocessor.

Check the signal in at CF3 for FM reception and CF4 for AM reception. If auto stop tuning works in FM mode but not AM, check for AM signal at pin 9 of IC120. Measure the voltage supply source at IC120. Test transistor Q127 and Q128. Check for +15 V supply voltage at IC202.

Overload thermal protection

Some of the deluxe receivers have built-in overload thermal protection circuits in case of abnormal operation. When the temperature of a posistor changes, the circuit shuts down. These posistors are found in the audio amplifier and transformer circuits (Fig. 11-24).

When the temperature of the posistor (one is installed with the heat sink [TH601] and the other with the power transformer [TH602]) rises abnormally, the resistance of the posistor becomes larger and pin 1 potential of IC602 will increase and cause transistor TR606 to turn off, causing pin 2 of IC611 to be cut off from the power supply voltage (V_{CC}) line and the power transistors to be protected.

Check D603 and D604 with a diode test of DMM when output transistors have overheated and are damaged. Make sure posistors are mounted against the heat sink. Take a resistance measurement of each posistor, and compare it with a new component. Replace IC602 if all the other components are normal. Measure correct +15-V feeding posistors and diodes.

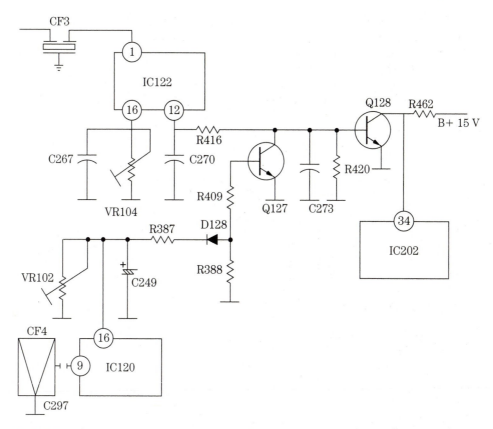

11-23 The auto stop and tuning circuits for AM and FM tuning. (Radio Shack)

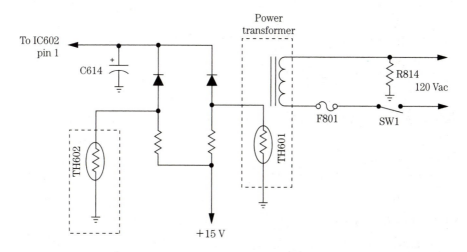

11-24 One or more posistor components can protect the transformer and output transistors from overheating. (Radio Shack)

Transistor TR606 protects the power transistors when abnormally high current flows in TR602a, b and TR603a, b due to excess input drive or when the load impedance connected across the output is too low (Fig. 11-25). If current increase is excessive, the voltage across R610a, c will turn on TR604a, b. Then the pin 1 potential of IC602 will increase and cause transistor TR606 to turn off. Now resulting at pin 2, if IC601 is cut off from the power supply voltage Vcc line.

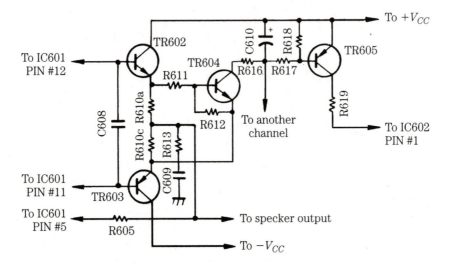

11-25 Transistor TR606 protects the output power transistors when high current flows in TR602 and TR603. (Radio Shack)

To suppress power turn-on noise, a time delay is provided by a time constant of R622 and C313. This time delay is set to activate the transistor TR606 to connect +Vcc line and pin 2 of IC601 after enough time has elapsed for the tone control amplifier and the pre-amplifier to reach a stable operating condition (Fig. 11-26).

Take critical voltage measurements on each transistor and IC. Test each transistor in the circuit. Check the supply voltage on IC601 and IC602. Measure each bias resistor for a change in resistance after the other tests are made to service the special circuits in the latest receivers.

AM-FM-MPX symptoms

These AM-FM-MPX symptoms can occur in car radios, portables, cassette decks, boom-boxes, or component stacked entertainment centers. These front-end symptoms occur only in the AM-FM-MPX stereo section. These case histories might help you repair the next radio circuit found on your service bench.

No AM, FM normal

Isolation   Locate all the circuits common to AM and FM, and check them.

Location   Check the RF, oscillator or converter, and AM mixer circuits. Suspect poorly soldered connections of the IF transistors and coils.

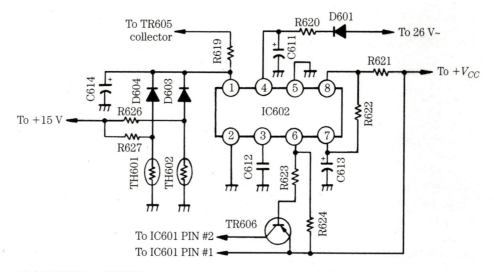

11-26 TH601 and TH602 are connected to terminal pin 1 of IC602.

Repair   No AM signal was found at the AM-detector stage. When the probe touched the diode, the AM band came alive. There was a poor board connection of D7 (Fig. 11-27) (Radio Shack.)

AM dead, FM intermittent

Isolation   Locate stages that are common to AM and FM circuits. Don't overlook the power supply.

Location   Locate the IF transistors.

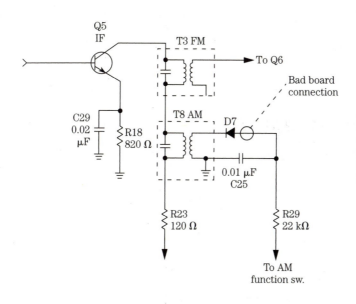

11-27
A poor board connection at the AM output secondary winding produced no AM reception.

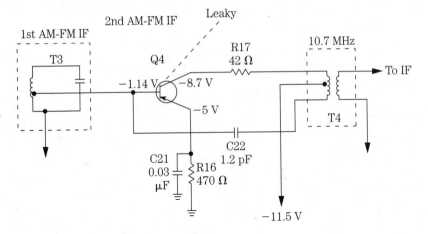

11-28 A leaky Q4 caused intermittent AM and FM reception in the IF circuits.

Repair Took critical voltage and in-circuit tests. Found a leaky IF transistor, and replaced with a universal replacement (Fig. 11-28).

No FM, AM normal in a Soundesign 5154

Isolation Check RF, oscillator, and mixer circuits because AM is normal.

Location Locate the front-end transistors with solid FM coils.

Repair When voltage probe touched the collector terminal of the FM RF FET, a local FM station could be heard. Replaced 2SK41 transistor with SK9164 universal replacement (Fig. 11-29).

Weak FM

Isolation Suspect the FM RF transistor or IC.

Location Locate the FM antenna lead-in, and trace it to the solid FM coils.

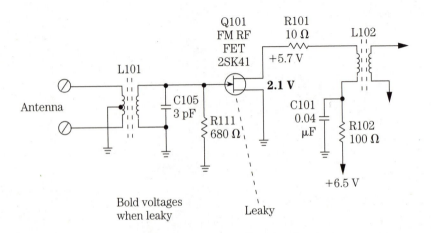

11-29 A leaky Q101 caused no FM reception in a Soundesign receiver.

Repair   Check the FM RF transistor in-circuit, and take critical voltages. Replace the transistors.

In a Fisher CA271, the receiver was dead with only a relay click on

Isolation   Try to locate the audio output transistors.

Location   The audio output transistors tested normal, and all voltage seemed normal.

Repair   Finally, ended up soldering the driver transistor lead connections on PCB.

No FM, only local AM station in a Sharp ST-3535

Isolation   It is something in common with both stages: IF or power supply.

Location   Locate the FM oscillator and mixer transistors. Locate the common IF stages.

Repair   Take in-circuit tests. Found the first AM and second FM IF amp transistor to be leaky. Replaced the transistor, but the FM still faded. Replaced the FM oscillator transistor (Fig. 11-30).

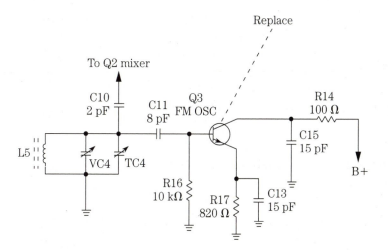

11-30 A leaky Q3 produced fading FM in a Sharp ST-3535 receiver.

Rush in FM, AM normal in a Sanyo ATR10

Isolation   Check the FM front-end transistors or ICs.

Location   Locate the FM transistors near solid turns of copper wire.

Repair   Take critical voltages on the FM transistors. Very low voltage. Found a leaky capacitor (C24) and a hot resistor (R23) (Fig. 11-31).

Intermittent and noisy FM

Isolation   Suspect the FM oscillator and mixer or IF stages.

Location   Check the IF stages between the IF transformers and filter networks.

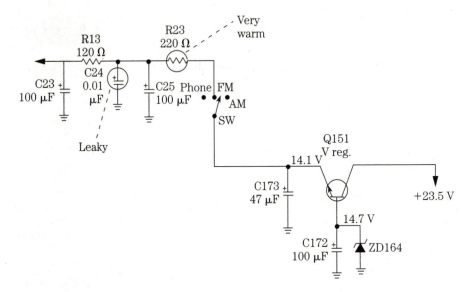

11-31 A leaky C24 caused R23 to run warm with low-output voltage.

Repair Found a poor base terminal in the first AM-FM IF stage. Resoldered the base resistor (47 kΩ) (Fig. 11-32).

No FM, AM okay

Isolation Check the FM transistors or ICs and the IF stages.
Location Locate the front-end FM transistors near the tuner.
Repair When pushing on T102, the FM reception returned. Resoldered poor IF transformer connection (Fig. 11-33).

The FM stations would not lock in with a Fisher FM586 receiver

Isolation Locate the FM stages.
Location Check to see if the AM stations locked on.
Repair Simply connect a DMM with the voltmeter range across TP6 and TP7. Adjust FM detector (T101) for 0 V or –50 mV.

FM indicator light stays on all the time in a Soundesign 5737

Isolation Locate the indicator transistor or IC.
Location Found IC301 on the PC board, and looked up replacement in a manual.
Repair Checked voltages, with no change of voltage while an FM station was tuned in. Replaced a leaky IC indicator (Fig. 11-34).

Weak FM and AM in a Sanyo MX720K

Isolation Check the IF and power supply common to both circuits.
Location Take critical voltage tests to find if there is voltage in both stages.

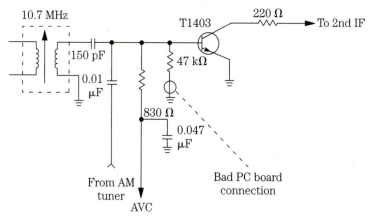

11-32 A badly soldered joint on a 47-kΩ resistor caused intermittent FM.

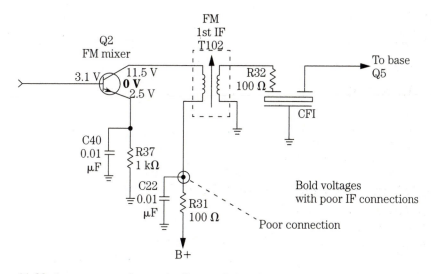

11-33 A poor connection on the B+ supply line left no voltage on the collector of the Q2 FM mixer.

Repair   Found a leaky voltage regulator transistor supplying voltage to the AM and FM circuits (Fig. 11-35).

No AM or FM in a J.C. Penney 3222 receiver

Isolation   Determine if the voltage source is normal.

Location   Locate the large filter capacitors, and check the voltage feeding the AM-FM circuits.

Repair Found a zener diode shorted in the AM-FM voltage source (Fig. 11-36).

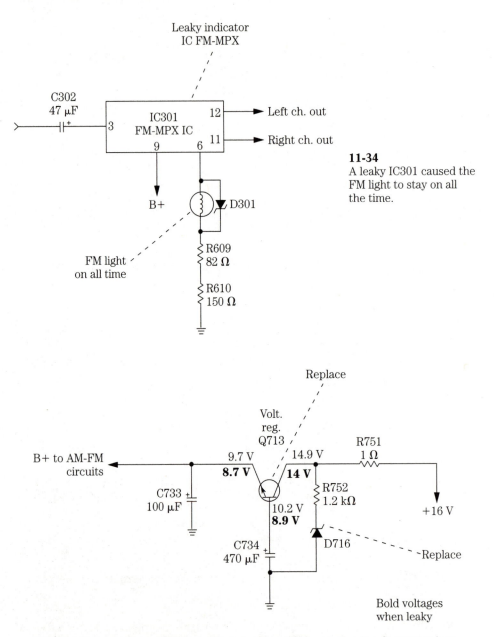

11-34
A leaky IC301 caused the
FM light to stay on all
the time.

11-35 A leaky regulator Q713 caused weak AM and FM reception in a Sanyo receiver.

No movement of the FM meter

Isolation   Determine if there is signal at the meter or in the circuit.
Location   Check the back of the meter terminals.
Repair   Found the meter needle stuck on a warped meter dial assembly.

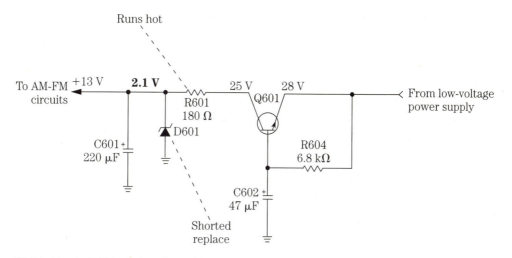

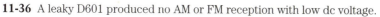

11-36 A leaky D601 produced no AM or FM reception with low dc voltage.

No sound was found in a CA862 Fisher receiver

Isolation The main fuse was blown and replaced. The fuse opened at once when the receiver was turned on.

Location Locate the audio output transistor.

11-37 Leaky output transistors or a leaky IC part can cause no or weak audio in the AM/FM/MPX receiver.

Repair Replaced outputs 2SA1302 and 2SC3281. Also replaced leaky driver 2SA1249. Check bias resistors and diodes. Finally, replaced a leaky 2.2-W zener diode (Fig. 11-37).

No AM or FM, cassette normal

Isolation Check the power supply source or IFs.
Location Cleaned the rotary switch.
Repair The rotary function switch was jammed. No possible repair. Replace the switch with an original part.

AM-FM-MPX circuit troubleshooting chart

Check Table 11-1 for additional troubleshooting data within the AM/FM/MPX receiver.

Table 11-1. AM/FM/MPX receiver troubleshooting chart.

Symptom	Cause and remedy
Receiver inoperative (LCD indicator does not light)	A) Faulty AC power cord. Replace.
	B) Defect in the power switch. Replace.
	C) Broken wire in the power transformer. Replace with power transformer.
	D) Blown power (fuse 1). Replace the fuse.
Fuse blows when power is turned on	A) Defective power transformer. Replace.
	B) Short in the primary or secondary of the transformer circuitry.
	C) Damaged rectifier D7, or D17. Replace the defective component(s).
	D) Short circuit in the rectifier circuit. Repair the short.
Power indicator lights but no sound from both channels	A) Speaker switches A or B defective. Replace the defective switch(es).
	B) Defect in transistor Q516L/R, Q517L/R on the main amp board. Replace the defective componen(s).
	C) Transistor Q509L/R, or Relay 501 defective. Replace the defective component(s).
Speaker A inopertaive	A) Speaker switch A defective. Replace.
Speaker B inoperative	A) Speaker switch B defective. Replace.
Speaker SURROUND inoperative	A) Speaker switch for surround defective. Replace.

Table 11-1. Continued.

Symptom	**Cause and remedy**
One channel does not work when volume at maximum with a test signal applied to the center terminal of volume control VR2 of the dead channel.	A) Blown speaker fuse 1. Replace the fuse.
	B) Defective in transistor Q516/L/R, Q517L/R of the main amp board. Locate and correct the defect.
	C) Break in copper foil of printed circuit board. Repair or replace.
	D) Short in speaker output terminal. Repair the short.
	E) Defective resistor R56, R58, or R564L/R. Replace the defective resistor(s).
	F) Defective capacitor C44, C46, or C525L/R. Replace the defective capacitor(s).
One channel does not work when VOLUME at maximum with a test signal applied to the center terminal of VOLUME control VR2 of the dead channel	A) Defective IC109. Replace.
	B) Defective resistor R227, R230, R224, or R237. Replace the defective resistor(s).
	C) Defective capacitor C113, C114, C118, or C120. Replace the defective capacitor(s).
	D) Defective balance volume (VR2). Replace.
Speaker works normally but headphones inoperative	A) Headphone plug does not mate with jack. Replace the plug.
	B) Defective resistor R140, or R141. Replace.
All the inputs work normally except CD/AUX(TV/VCR) input	A) Poor contact in CD/AUX(TV/VCR) input jack. Repair or replace the jack.
	B) Defective CD/AUX switch(es), TV/VCR switch(es), or IC110. Replace.
Phono input inoperative	A) Poor contact in phono input jack. Repair or replace the jack.
	B) Defective phono switch or IC109. Replace.
Tape out 1,2 has no effect	A) Defective contact in tape out jack. Repair or replace the jack.
	B) Defective contact in tape monitor switch(es). (tape 1,2) Repair or replace.
FM inoperative	A) Defective front-end. (Q118, Q119, Q120, or Q138) Replace the switch.

Table 11-1. Continued.

Symptom	Cause and remedy
	B) Defective FM switch. Replace the switch.
	C) Defective transistor Q121, Q125, Q125 IC122, or IC123. Replace the defective transistor(s) or IC(s).
	D) Defective coil T103, T104. Replace the coil(s).
	E) Defective lead-in. Repair or replace the lead-in.
	F) Ceramic filter CF1, CF2, CF3 defective. Replace the defective ceramic filter(s).
	G) Defective controller circuit component. Replace.
Poor multiplex separation	A) Improper adjustment. Readjust VR101, and VR103. (Refer to MPX Alignment).
	B) IC102 defective. Replace.
	C) Variable resistor VR101, or VR103 defective. Replace the variable resistor(s).
STEREO indicator does not light	A) Defective indicator in LCD. Replace.
	B) Improper adjustment of VR101 of tuner board. Make adjustment. (Refer to MPX Alignment)
	C) Defective IC123, resistor R444, or Q135. Replace the defective component(s).
FM volume not sufficient	A) If volume of both L and R channels are not enough: Front-end section defective, or faulty IC122, coil T103, T104, or defective capacitor C275 of tuner board. Locate and replace the defective component(s).
	B) If sound of one channel is not enough. Defective T106, or T107. Replace.
FM MUTE has no effect	A) Defective FM MONO/FM Mute off switch. Replace the switch.
	B) Defective IC122, Q124, Q125, Q128, Q131, Q132, or Q137. Replace the defective component(s).
AM inoperative	A) Damaged IC120 of tuner board. Replace.

Table 11-1. Continued.

Symptom		Cause and remedy
	B)	Defective T108, T109, Q115, or CF4 of tuner board. Replace the defective component(s).
	C)	Resistor R164, R369, R372, R377, or R380 defective. Replace the defective component(s).
	D)	Capacitor C227, C228, C237, C240, or C295 defective. Replace the defective capacitor(s).
	E)	Defective AM switch. Replace.
	F)	Defective varicap diode D122. Replace varicap diode(s).
	G)	Damaged AM loop antenna. Repair or replace.
	H)	Defective controller circuit component. Replace.
LOUDNESS has no effect	A)	Defective loudness switch. Replace.
	B)	Defective resistor R56, R57, R58, R59, R69, R70, C43, C44, C45, or C46. Replace the defective component(s).
MONO not effective	A)	Defective mono switch. Replace.
TAPE 1/2 not effective	A)	Defective tape monitor switch. (tape 1/2) Replace.
	B)	Poor contact in tape in jack. Repair or replace jack(s).
	C)	Defective resistor R243, R244, R247, or R248. Replace the defective resistor(s).
BASS control has no effect	A)	Variable resistor VR6 defective. Replace.
	B)	Defective R1, R4, R5, R71, C1, C2, C6 or C7. Replace the defective component(s).
TREBLE control has no effect	A)	Variable resistor VR4 defective. Replace.
	B)	Defective R21, R22, R23, R25, R26, R27, C3, C19, C20, or C25. Replace the defective component(s).
OUTPUT POWER METER inoperative	A)	Defective R136, R137, C73, or C74. Replace.
	B)	Defective output power meter driver IC 106, or IC107. Replace.

Table 11-1. Continued.

Symptom	**Cause and remedy**
Excessive noise using PHONO input	A) Defective IC109. Replace. B) Defective C113, C114, C118, C120, R224, R227, R230, or R237. Replace the defective component(s).
MIDRANGE control has no effect.	A) Variable resistor VR5 defective. Replace. B) Defective R11, R12, R13, R14, R18, R20, C5, C16, C17, or C18. Replace the defective component(s).
AUTO tune inoperative (UP/DOWN) (AM or FM)	A) Poor contact in Up/Down key. Repair or replace. B) Defective IC202. Replace. C) Defective Q127, R128, or R139. Replace the defective component(s). D) Defective LCD101. Replace. E) Defective tuner circuit component. Replace. F) In case of RM only, improper adjustment of FM front-end. Replace.
MANUAL tune inoperative (UP/DOWN) (AM or FM)	A) Poor contact in Up/Down key. Replace. B) Defective IC202, IC701. Replace. C) Defective Q701, Q702, Q703, or Q704. Replace the defective component(s).
MEMORY settings (keys 1-10) inoperative	A) Poor contact memory keys 1 – 10. Replace. B) Poor contact in memory set key. Replace. C) Defective IC202. Replace the defective component.
LCD inoperative.	A) LCD defective. Replace. B) Defective IC101, or IC202. Replace. C) Defective X-TAL2. Replace.
Noise VOLUME control	A) Defective IC901. Replace. B) Defective capacitor C901, or C903.

Replace the defective capacitor(s).

Table 11-1. Continued.

Symptom	**Cause and remedy**
SIGNAL strength indicator LCD not functioning	A) Defective LCD101. Replace.
	B) FM reception, VR105, D130, R416, or IC122. Replace the defective component(s).
	C) AM reception, VR102, D125, D139, or R391. Replace the defective component(s).
High-Blend has no effect	A) Defective switch SW27. Replace.
	B) Defective IC202, Q126, or Q136. Replace the defective component(s).
LCD inoperative	A) LCD defective. Replace.
	B) Defective IC202, or IC101. Replace the defective component(s).
	C) Defective resistor R93 or R96. Replace.
REMOTE CONTROL UNIT inoperative	A) Weak battery. Replace.
	B) Defective. Replace.
	C) Defective IC202 (CPU board), or IC108. Replace.

12
CHAPTER

VCR mechanical and electronic problems

Although VCR repair can be quite difficult without a schematic, there are a lot of mechanical and electronic problems that can be solved without one. Mechanical problems can be seen, felt, and heard, while electronic symptoms might require added service time. Take the electronic problem, and isolate and locate the defective component. Look for defective ICs and transistors that produce many electronic troubles. IC and transistor parts cause most service problems within the VCR.

Carefully observe the entire chassis, and try to isolate the various circuits. Power circuits are easily located by power transformers, electrolytic capacitors, and diodes. The tuner stands out on the PC board (Fig. 12-1). The location of the tape heads, capstan, and motors are easily found inside the VCR.

Locate the corresponding ICs and transistors tied to the various parts. Leaky and overheated ICs can be located by touch and with critical voltage measurements. Case histories might help solve other VCR problems. Record each case history for future reference. Remember, several VCR brands might be the same inside.

Symptoms

Connect the VCR to the TV chassis, and determine what problems occur. Several symptoms can be caused by a common defective component within both the play and record modes. Compare these symptoms with other VCR models. Use a simple block diagram or another schematic to tie all the sections together (Fig. 12-2). VCR modules can be interchanged to isolate various problems. You might find that the chassis that is on the bench is the same as the diagram in your cabinet.

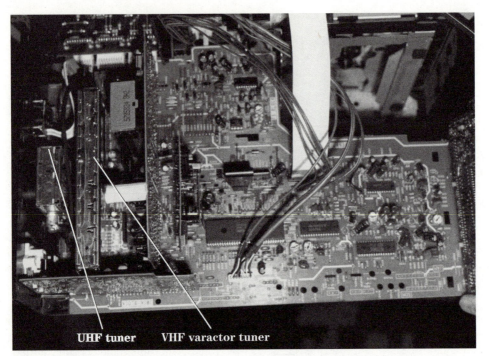

UHF tuner   VHF varactor tuner

12-1 The varactor VHF and UHF tuners are mounted side by side in RCA VR470.

Head cleaning

Dirty video tape heads can cause a loss or deterioration of the video signal. Video dropout and audio distortion can result from a clogged tape head. Some manufacturers have included automatic head cleaning systems in their machines. Clean the tape heads, guides, rubber pinch rollers, and threading mechanism of oxide. These surfaces can be cleaned with wet, dry, or magnetic systems.

The wet fabric and dry abrasive systems can cause some physical damage to tape heads and rubber pinch rollers. The best cleaning method is to use a chamois cleaning stick with solvent. Hold the top of the head with your fingers and rotate the head with the cleaning stick held horizontally (Fig. 12-3). Don't move the stick vertically or you might damage the head. Periodic cleaning of tape heads and moving parts can help prevent future repairs or damage.

Cleaning the audio head

Clean up the audio head found in VHS VCRs with a cotton swab and alcohol. Remove the top cabinet or cover, dip the cotton swab or stick in 90% isopropyl alcohol, and clean the audio control head (Fig. 12-4). Be very careful not to damage the upper drum and other tape running parts. Clean the head horizontally, not vertically. To prevent tape damage, wait until the head area is dry before operating the unit. A dirty audio tape head can cause distorted, intermittent, or no sound.

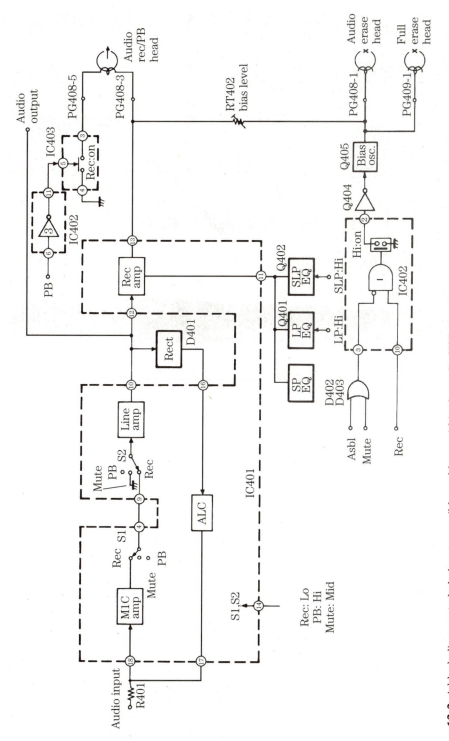

12-2 A block diagram to help locate possible problems within the audio VCR circuits.

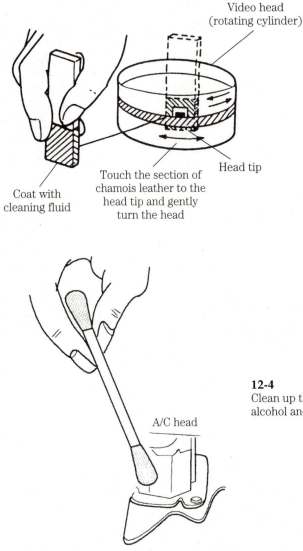

Video head
(rotating cylinder)

Coat with
cleaning fluid

Touch the section of
chamois leather to the
head tip and gently
turn the head

Head tip

12-3
Clean the tape head with
chamois leather moving
horizontally with the video
head.

A/C head

12-4
Clean up the audio tape head with
alcohol and a cleaning stick.

Normal load and speed problems

Loading problems can be caused by a defective mechanical loading bracket, a broken or loose rubber belt, the loading gear and cam assembly, basket gears, or the loading motor. Check the voltage at the motor terminals. Suspect a regulated zener diode or transistor when there is improper motor voltage. If voltage is present but there is no motor movement, check the continuity of the motor winding. Usually the loading motor is located near the front of the VCR (Fig. 12-5).

Electronic loading problems can be caused by defective ICs, transistors, resistors, or loading sensors. Don't overlook a defective loading or mode switch. Suspect

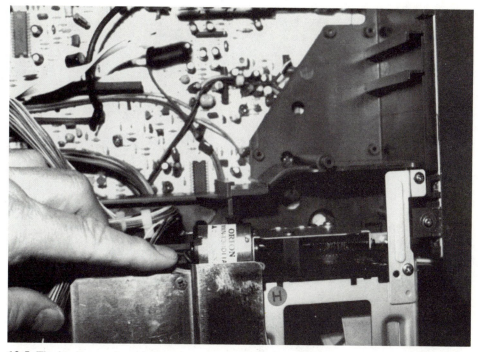

12-5 The loading motor rotates a cam gear assembly to load and unload tape cartridges.

a defective IC when the tape loads and then ejects. If the unit goes into fast forward after normal loading, check the alignment of the load switch. When the tape loads with the reel rotating and then ejects, suspect a defective end sensor.

In a Fisher FVH805 VCR, the cassette would not load. After removing the top cover, gum wrappers were found wrapped around the loading motor pulley. Foreign material pushed into the cassette opening caused the rubber belt to bind and break (Fig. 12-6).

Dead: no operation

Check for a blown fuse when nothing operates in the VCR. Remove the top and bottom covers to uncover the fuse holder (Fig. 12-7). Check for dc voltage across the main filter capacitor. With no dc voltage measurement, check the bridge rectifiers and the power transformer in the low-voltage power supply. Check for possible open or leaky voltage regulators.

Loading up

When the cassette will not seat or load, suspect foreign material within the opening. Sometimes small hands poke small components inside the door opening. In some VCRs, when the cassette is halfway inserted, the loading motor takes hold and loads the cassette. Check for a broken motor belt or binding gear assembly.

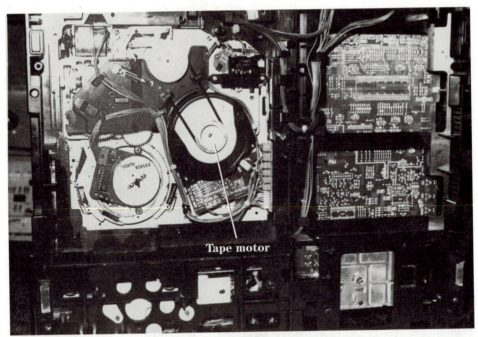

Tape motor

12-6 A loading motor drive belt was found broken in a Fisher FVH805 VCR.

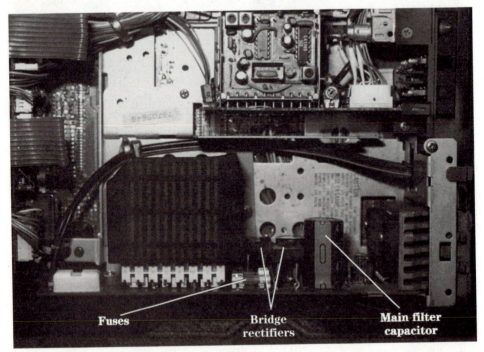

Fuses **Bridge rectifiers** **Main filter capacitor**

12-7 Check the fuses, bridge rectifiers and check the voltage across the main filter capacitor for a dead VCR.

Locate the loading motor that drives the front mechanism, which is located near the front or center of the VCR. Usually, the loading motor lays horizontally with a drive belt and gear train (Fig. 12-8). Check the voltage across the motor. Make sure that the loading switch is not open or defective. Replace the motor if voltage is found upon the motor terminals and there is no motor rotation. Remove the motor, and check again. Check the jammed gear assembly if the motor runs without the belt or gear train attached. Check the loading motor driver IC for load and unload. Measure the supply voltage upon the motor driver IC.

12-8 The loading motor is found in the center of the chassis of an MGA VCR.

VCR in TV chassis

Today, some VCR decks are located at the bottom of a small TV chassis. Most built-in VCR decks were installed in 13" screens or smaller. Today, the VCR can be found in the 19" portable TV. The VCR might have its own power supply or, in some models, operate from the TV chassis. The TV picture tube sits above the VCR deck (Fig. 12-9). Dismount the VCR deck for maintenance and standard adjustments.

Intermittent problems

Intermittent problems within the VCR are difficult to locate and require a lot of service time. Poorly soldered connections of ICs or transistors or cracked PC board

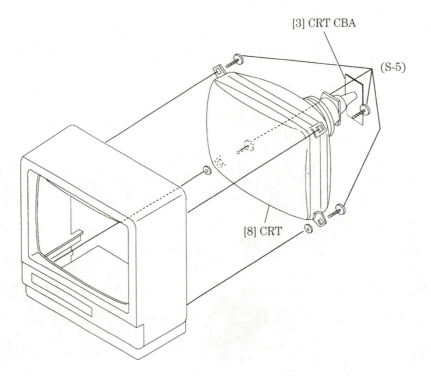

[3] CRT CBA

(S-5)

[8] CRT

12-9 The VCR is mounted in the bottom slot of a Realistic 13" portable TV.

wiring can cause intermittent operation. Check for a dirty or worn mode switch when all functions are erratic. Intermittent or erratic tape speed can be caused by a defective connecting IC, capacitors, or burned resistors. Intermittent record and playback functions can be caused by cracked or broken PC board wiring. Inspect and probe where boards are fastened with standoffs (Fig. 12-10).

Intermittent cylinder movement can be caused by poor or bad connections. Resolder the corresponding IC or processor pin connections. Check for a defective mode recording assembly when intermittent recording is a problem. Intermittent transistors, ICs, and cracked or burned resistors can result in intermittent audio. Check for poorly soldered IC pins or display ICs when you have an erratic or intermittent display assembly.

Belt replacement

Loose, worn, or broken drive belts can cause slow or improper speeds in VCR functions. Check all belts for shiny surfaces, cracks, or broken areas. A shiny belt indicates slippage around a capstan motor pulley. Check the drive belts and wheels for erratic or intermittent movement. Clean the belts when you clean the tape heads.

Replace the belts with original part numbers, when available. Universal belts and drive wheels can be ordered from local or mail-order firms. Sometimes exact drive wheels, idler pulleys, pinch rollers, and VCR gear assemblies can be obtained from

PC board

Plug
wires

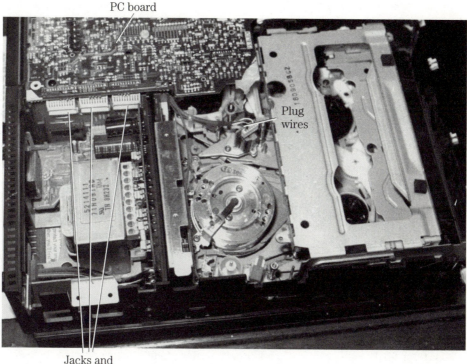

Jacks and
plug connectors

12-10 Inspect the terminals of plugs, jacks, cracked PC boards at chassis standoffs, and wire leads for intermittent VCR operation.

large mail-order firms. Several VCR belt kits for different manufacturers offer convenience, economy, and a wide range of replacement parts. Pick up a belt gauge for a quick, accurate method of measuring the thickness, width, and inside circumference of broken or worn belts.

Eats tape

Check for a dirty tape head or sticky rollers or spindles when the tape is being pulled. Eating of tape can be caused by a defective mechanical center bracket assembly, clutch assembly, friction gear, or function gear assemblies. Pulling of tape can be caused by a defective base assembly, clutch assembly, or transmitting arm unit. Chewing or eating of tape can be caused by a defective control IC. Clean all the tape path parts when you clean tape heads, to prevent eating or pulling of tape (Fig. 12-11).

Cylinder problems

The cylinder motor drives the video heads at the required speed, and servo problems affect the quality of the picture. Check voltages on the cylinder drive motor IC and

12-11 Clean up the tape heads, audio head, and tape movement components of the tape path.

voltages applied to the motor terminals. Locate the motor driving the video heads or drum assembly at the bottom of the unit. Take continuity test measurements (Fig. 12-12). Check for B+ 5 V on pin 10 and 12 V applied to the motor terminals (pin 5).

No cylinder movement can be caused by a defective drive IC or improper voltage and connections on the CBA unit. Suspect a defective drum CBA or lower cylinder when the heads will not spin. Intermittent or erratic speed can be caused by a drive IC or poor CBA connections.

Suspect a defective upper drum or cylinder with a poor or snowy picture in playback mode. Check the upper cylinder for noise in the picture during playback and white lines during recording mode. Don't overlook a defective cylinder motor when diagonal lines appear in the picture.

Capstan speed problems

The capstan motor provides tape motion in play, record, and fast-forward modes. The capstan motor is driven with a drive speed and system control IC. No tape motion can be caused by a defective capstan motor, drive belt, driver, or servo IC (Fig. 12-13). Check for supply voltage at the driver and system control IC. Measure the voltage applied to the motor terminals during play mode.

Suspect a defective or worn belt if the capstan motor is rotating at the proper play speed. Erratic or improper speed can be caused by a defective drive or control IC. Check for poorly soldered connections on CBA.

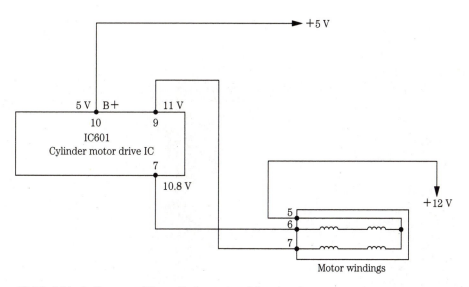

12-12 A block diagram of the cylinder motor drive circuits.

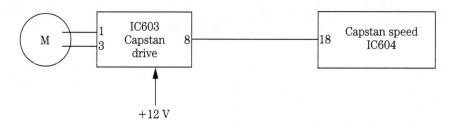

12-13 A typical block diagram of the capstan motor driver and speed IC.

Starts to play then stops

Check the capstan driver or control IC when the VCR starts to play and then shuts down. Inspect and clean the mechanism position switch in some models. This problem can be caused by a defective gear drive assembly or bracket assembly. Suspect a defective capstan motor if the VCR shuts down instantly.

Tape loads then shuts down

Suspect a defective drum motor or driver and control IC when the drum loads and then shuts down. Check for mechanical failure of a bracket assembly or a defective gear drive assembly or mechanism position switch when the VCR begins to play and then shuts down. Check the capstan driver IC and control system when the unit starts to play and shuts down. Trace the capstan motor leads back to the drum or drive IC for voltage tests and location.

Tape will not eject

Look for defective basket gears when the tape will not eject or the VCR keeps ejecting the tape after loading. A defective lift arm assembly or worn loading belt can keep the tape from ejecting. Check the loading motor and control IC when the tape keeps ejecting. A defective reel motor might not unload the tape. Suspect a right end sensor when the tape loads, the reel rotates, and then it ejects the tape.

Improper rewind

Notice if the VCR operates in the fast-forward mode, then load a tape and press the rewind button. No mechanical rewind can be caused by a worn, loose, or slipping capstan motor drive belt in the rewind position. Look for a defective idler pulley, AY gear assembly, or mechanism switch position. Don't overlook a defective capstan motor assembly.

Suspect a defective switching regulator, AY gear assembly, or motor drive IC when the unit will not rewind or fast forward. Check the signal at the capstan high and low pin terminals of the system control IC and those applied to the capstan motor assembly. Look for the capstan motor assembly that drives the capstan/flywheel with a large belt (Fig. 12-14). Trace the wires to the motor drive IC. A loud grinding noise can be caused by a worn clutch wheel in fast-forward operations.

12-14 The capstan motor drives several components with the drive belt in a Fisher FVH805.

Erratic tape speed

With erratic speed, check for an intermittent capstan driver or control system IC. Monitor the voltage applied to the capstan motor and on the driver IC (Fig. 12-15). A defective capstan motor can produce erratic tape speed. Inspect the drive belts for looseness and slippage. Measure the supply voltage on the capstan driver IC.

12-15 Suspect a broken drive belt or capstan motor when the capstan will not rotate the tape.

Dead VCR

Go directly to the low-voltage power supply when the VCR will not function at all. Check the main ac fuses. Measure the dc voltage at the main regulator IC. In some models, a low-voltage regulator transistor will be found outside the main power supply. Locate regulator ICs and transistors (Fig. 12-16). Usually the main regulator IC provides several voltages to the various VCR circuits.

Most power-supply problems are caused by regulators, diodes, burned resistors, or filter capacitors. Intermittent dc voltages can be caused by defective voltage regulators, poor terminal connections, resistors, and poor pin connections of IC regulators. Resolder all pin terminals before replacing the suspected IC. Monitor the various dc voltage sources to determine if the intermittent symptom is produced in the low-voltage power supply.

Locate the largest power transformer within the VCR (Fig. 12-17). Check around the transformer, or trace the wiring to silicon diodes or bridge rectifier cir-

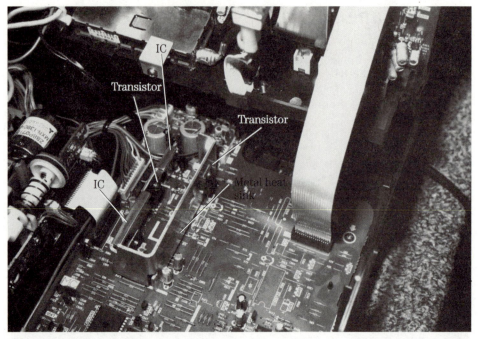

12-16 Check for open or leaky regulator transistors if the cassette will not load or the VCR shuts down.

12-17 Locate the power transformer and electrolytic filter capacitors to locate the bridge diodes in the low-voltage power supply.

cuits. Trace out the positive voltage from the rectifiers to the large IC voltage regulator. Check the voltages on all terminals. Sometimes each transformer winding has a small ac fuse. If there is no or low voltage at the regulator, suspect leaky diodes and open fuses.

In an RCA VLT250 VCR, no voltage was found at the 12-V source. After replacing the main ac fuse, no voltage was found at pin 5 of IC851. Another 2-amp fuse was found blown (FU852) in the input of the voltage regulator. Both fuses opened after replacement. IC851 was suspected of leakage and was replaced with part number 163823 (Fig. 12-18).

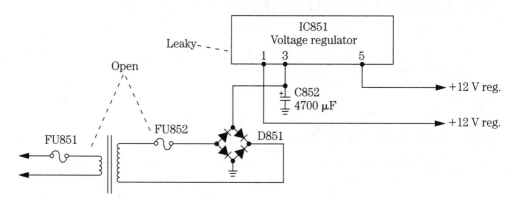

12-18 A leaky IC851 opened fuses FU852 and FU851 in the RCA VLT-250.

Goldstar GHV1265-M shutdown

In this Goldstar VCR, the unit might shut down at once or play a few minutes. At other times, the speed of the tape would change in play mode. Monitor the voltage applied to the capstan motor. When the voltage remains constant, suspect the capstan motor. Sometimes tapping the end bell of the motor will cause it to change speeds. This defective capstan motor caused speed problems and shutdown.

Voltage regulators

Voltage regulators within the VCR chassis produce many problems. The main voltage regulator is tied to the silicon diodes or bridge rectifiers. Low regulator voltages will occur with ICs or power-type transistors. In the RCA VLT250/260, the main voltage regulator provides 12 V and 16 V. IC605 provides a 5-V source from the 12-V line (Fig. 12-19). The A/D converter (IC971) supplies a 10-V source, while IC901 furnishes a 5-V source from the 10-V line.

A defective voltage regulator can produce intermittent or improper voltages to the various circuits. An open voltage regulator IC or transistor has zero output voltage. A leaky regulator can have low or high dc output voltage. Regulator transistors and ICs that are operating red hot should be replaced. Resolder each terminal pin if it is an intermittent problem.

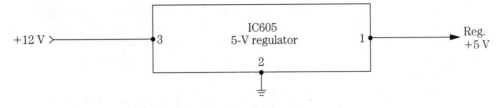

12-19 IC605 provides regulated +5 V in the RCA VLT250/260.

Emerson VCR910 distorted picture

Picture distortion with tracking lines was caused by a sluggish back tension lever. Mechanically, the back tension lever should move freely. Check the lever, and see if it is difficult to move. If it is not, remove the back tension lever. Clean off the old grease, and wash out the lever pivot. Lubricate the lever with light grease, and reinstall it.

No audio in playback

Clean the audio play/record and erase heads when you clean the video heads. A dirty erase or play head can produce noisy sound.

In playback mode, check the main audio IC that is connected to the play/record head assembly. Locate the audio head on the tape path. Trace the wires to the audio playback and record IC. Check the supply voltage pin on the IC. With an external audio amp, check the audio output pin that goes to the line amp or input pin of the test point. Suspect the audio output IC amp when there is no signal out and the voltages are normal. Don't overlook the play/record mute switch IC or head switch IC (Fig. 12-20).

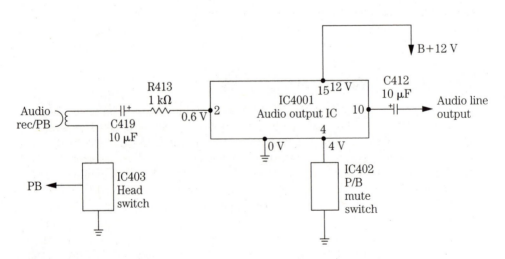

12-20 A block diagram of the audio playback circuits.

Check for open coils, capacitors, and transistors when the audio drops out in audio circuits. For intermittent audio, check the audio tie points, terminal blocks, and poor terminal IC and transistor connections. Resolder all IC terminals. Poor audio and snow in the picture can be caused by a defective head amp PCB.

In a Goldstar, the drum will not rotate

The drum would not rotate in a Goldstar 6HV1240M VCR. Check the drum CTL signal if voltage is applied to the drum motor. Replace the lower drum assembly. If the lower drum is okay, check the servo IC and circuits. Check the supply voltage to the servo IC.

No picture in playback

Clean the video heads with a chamois and alcohol or cleaning solvent. Trace the video head wires back to the prerecord amp. Check the universal replacement manual for the right replacement and also what stage the IC functions in. If the IC replacement is not listed, trace the signal with a scope. Check the IC number in a similar VCR chassis or schematic.

Check the supply voltage applied to all IC and transistor components within the playback mode. A poor picture in playback can be caused by a prerecord amp IC or luma process IC (Fig. 12-21). Check the voltages on all IC and transistors to determine if a component is leaky. Inspect the PC board for burned or open resistors.

Snow and noise in the picture or a faint picture can be caused by a defective upper cylinder and head amp IC. Check the tuner-demodulator circuits if there is only half of a picture. If the picture is fuzzy and noisy, resolder the pins on the head amp board and check the VHF block assembly. Check for a missing control pulse when there is noise in the playback mode. Replacing IC components within the video circuits can solve most playback audio and picture problems.

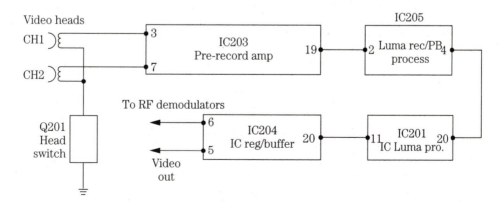

12-21 A typical block diagram of the video play circuits.

The capstan motor in an Emerson VCR951

No capstan motor rotation was found in this Emerson VCR. A quick continuity measurement of motor winding was normal. No voltage was found across the motor terminals in play or rewind modes. The servo motor IC2001 was replaced, resuming capstan motor operation.

Poor recording

When the cassette will not record, check the safety tab at the rear of the cassette. If the tab is broken, the record button cannot be pressed down. This safety tab protects valuable recordings. Also check the record sensor switch.

Clean the video, audio, and erase heads when there is poor or fuzzy recording. Keep video heads clean. Does the VCR playback normal audio and video? Usually the same IC circuits are used in both record and playback modes. If the video does not record, suspect additional circuits such as the luma record/playback process IC, transistor record switches, video record amp, or supply voltages (Fig. 12-22). Also check the safety switch, record subassembly, and lower drum assembly when there is no video recording.

Check the audio play and record circuits with poor or no audio recording. If the audio playback is poor and distorted, suspect the head switching IC or transistor or the bias oscillator circuits. Test for critical voltage measurements on these components. Check all bias oscillator circuits when previous audio will not erase.

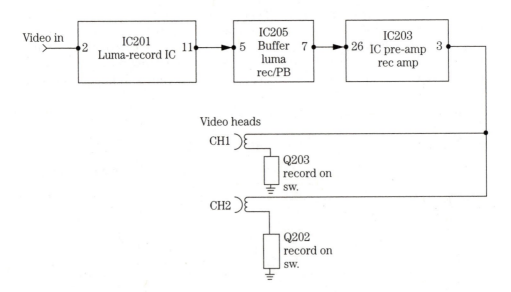

12-22 A block diagram of the video recording circuits.

No sound or no video

No sound or video was noted in a Fisher FVH940 VCR player in playback mode. A burned R83 (68 Ω) resistor was found. Upon checking the same circuit, a leaky electrolytic capacitor C18 (470 µF) was located. Replacing C18 and R83 restored the voltage, solving the no sound or video symptom.

No audio erase

The audio bias oscillator circuits are similar to those found in the cassette player. A bias signal is applied to the full erase, audio erase, and audio record/playback heads. A waveform test at either head will indicate if the bias oscillator is functioning.

Check the voltages on the bias oscillator transistor. Look for a small transistor near the bias transformer. Trace the bias lead from the full erase or audio erase head to locate the small bias transformer. Resolder all transformer board connections when there is intermittent erasing. Check for leaky capacitors across the bias transformer when there is no waveform at the full erase head (Fig. 12-23).

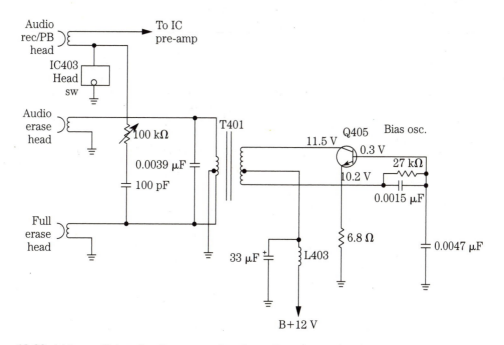

12-23 A bias oscillator circuit connected to the audio and erase heads.

Shutdown problems

Shutdown problems in the VCR can be caused by a defective take-up reel sensor, mode switch, interuptor sensor, loading motor IC, regulator IC, electrolytic capacitor, and parts in the low-voltage power supply. When the VCR will play for 2 to 10 seconds and then shuts down, check the regulator power IC, take-up reel sensor, regulator IC, and reel sensor assembly located under the take-up reel assembly (Fig. 12-24). If the chassis shuts down after 1 to 5 seconds, check the sycon and servo ICs.

12-24 Shutdown after 1 to 5 seconds can result from a defective take-up reel sensor, which is located under the take-up reel.

Replace the servo IC when the VCR loads and then shuts down immediately. If shutdown occurs with no function operation, check the sycon IC. When the VCR shuts off immediately, suspect a regulator IC. If the VCR shuts off after playing awhile, check for a bad end sensor.

When the VCR shuts down during play or fast forward, replace the defective take-up reel sensor. Check for a hot loading motor IC when the unit plays and will not reverse. Sometimes the loading motor IC, belt, and motor must be replaced. If the unit operates for a few seconds and then shuts off, replace the motor driver IC. Replace the capstan motor when a change in speed is noted and then the unit shuts down. Replace the take-up reel sensor if the unit shuts down during play and fast-forward modes. Check the center position of the mode switch when fast forward and rewind slow down. When the VCR shuts down and will not accept a tape, replace the loading motor driver IC and loading motor. Intermittent shutdown can result from poor terminal connections around servo and loading ICs.

No display features

No timer or intermittent timer/counter display can result from a defective timer IC. Check the supply voltages on each IC terminal. Replace the timer IC if there is erratic on-screen display. Scope the oscillator pin on the timer IC.

Service schedule of components

Clean all parts for the tape transport (upper drum with video head/pinch roller/audio control head/full erase head) using 90% isopropyl alcohol. After cleaning the parts, make all deck adjustments. Check the following components within the VCR for standard maintenance (Table 12-1).

Table 12-1. The service schedule of components that need serviced or replaced. (Radio Shack.)

Service Schedule of Components

H: Hours ○: Check ●: Change

	Deck	Periodic Service Schedule			
Ref. No	Parts Name	1,000 H	2,000 H	3,000 H	4,000 H
B2	Cylinder assembly	○	●	○	●
B3	Loading motor			●	
B6	Pinch roller arm assembly		●		●
B8	Pulley assembly		●		●
B21	Belt LDG		●		●
B26	Clutch block assembly		●		●
B27	Band break assembly		●		●
B28	Main brake S assembly		●		●
B29	Main brake T assembly		●		●
B30	T break arm assembly		●		●
B31	AC head assembly			●	
B32	Reel assembly			●	
B37	Capstan motor		●		●
B52	Belt FWD		●		●
B54	Ground brush assembly			●	
✱ B73	Full erase head			●	
☆ B86	F break assemly		●		●

VCR symptoms

The following VCR symptoms might help you service the models given and other units made by the same manufacturer as well as other brands of VCRs.

Improper loading in a Goldstar 6HV1280 VCR

Isolation Notice if the cassette carriage goes into load by itself with no tape loaded.

Repair Replace the defective mode select switch.

No cylinder movement in a Samsung VR2000

Isolation Take waveforms on IC601 (Fig. 12-25).

Repair No cylinder movement was caused by a missing 30-Hz pulse on pin 1 of IC601.

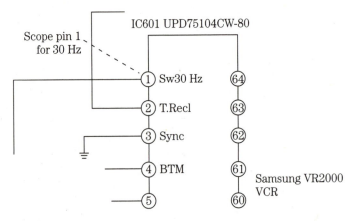

12-25 When there is no cylinder movement, scope pin 1 for a 30-Hz pulse in a Samsung VR2000.

Erratic capstan speed in a Magnavox VR3223AT01

Isolation Take voltage measurements on the capstan motor drive (IC405) (K405BA6219B).

Repair Found erratic voltage at pin 2 of IC405. Replaced IC405 (Fig. 12-26).

Tape loads then shuts down in a Sharp VC6847U

Isolation Locate the change mode switch (mechanical position switch), and take resistance measurements (Fig. 12-27).

Repair Used an ohmmeter to check for linear resistance between 01, 02, and 03 when rotated. Replaced switch.

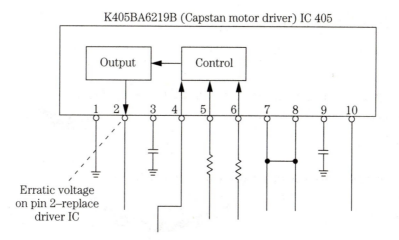

K405BA6219B (Capstan motor driver) IC 405

Erratic voltage
on pin 2–replace
driver IC

12-26 A defective capstan motor IC caused erratic speed in a Magnavox
VR3223AT01.

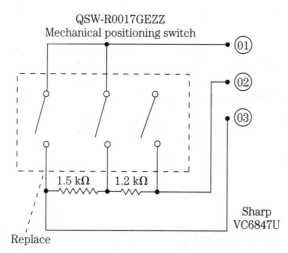

QSW-R0017GEZZ
Mechanical positioning switch

1.5 kΩ   1.2 kΩ

Sharp
VC6847U

Replace

12-27
Replace the mechanical mode
switch when the tape loads and
the VCR shuts down in a Sharp
VC6846VB.

The tape will not eject in a Sharp VC-6846UB

Isolation Check the operation of the reel drive unit.
Repair Changed the reel drive unit (NPLYV01076GEZZ) (Fig. 12-28).

Noise in playback in a Goldstar GHV1240M

Isolation Check for correct tape contact on the spindle posts.
Repair Check P3 and P4 posts for correct tape contact. Adjust height as necessary to detent position (Fig. 12-29). This symptom can occur in many models.

NPLYV01076GEZZ

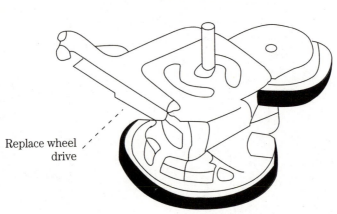

Replace wheel drive

12-28 When the tape will not eject in a Sharp VC6846UB, replace the wheel drive assembly.

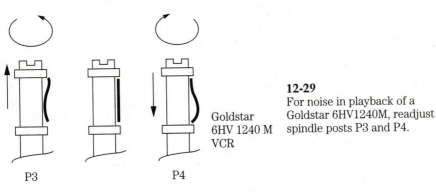

Goldstar 6HV 1240 M VCR

P3

P4

Readjust P3 and P4

12-29
For noise in playback of a Goldstar 6HV1240M, readjust spindle posts P3 and P4.

No auto-rewind in an RCA VMT390A

Isolation Locate the supply end sensor (Fig. 12-30).

Repair Check to see if pin 6 (P64) goes low (0 V) when the end of the tape passes the end sensor (Q141). Replace the supply end sensor.

Will drift off channel and will not tune in most channels on a Fisher FVH720 VCR

Isolation Check the tuner and IF module.

Repair Replace (C09) 0.47-µF capacitors in the tuner IF module.

No picture in playback in a Sylvania VC4213AT01

Isolation Locate the regulator transistor (Q701).

Repair Replace Q701 and C713 for low +5 V at the emitter of the regulator transistor (Fig. 12-31).

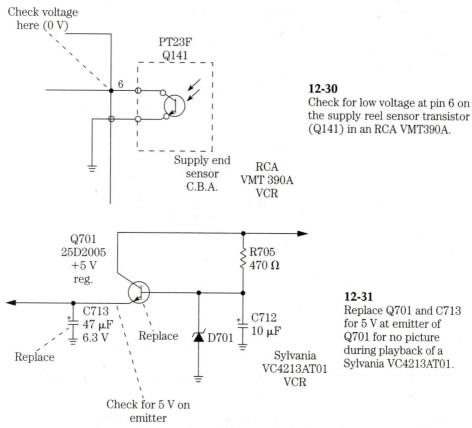

Check voltage here (0 V)

PT23F
Q141

6

Supply end sensor C.B.A.

RCA VMT 390A VCR

12-30
Check for low voltage at pin 6 on the supply reel sensor transistor (Q141) in an RCA VMT390A.

Q701
25D2005
+5 V
reg.

R705
470 Ω

C713
47 μF
6.3 V

Replace

C712
10 μF

D701

Sylvania VC4213AT01 VCR

Replace

Replace

Check for 5 V on emitter

12-31
Replace Q701 and C713 for 5 V at emitter of Q701 for no picture during playback of a Sylvania VC4213AT01.

No audio in playback in a Sanyo VHR 2250

Isolation Scope the audio playback circuits.

Repair Scope pin 6 of IC2001 (LA7095) for in audio and pin 11 for out audio. Replace IC2001 (Fig. 12-32).

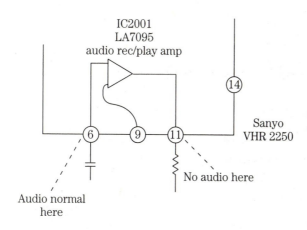

IC2001
LA7095
audio rec/play amp

14

6   9   11

Sanyo VHR 2250

No audio here

Audio normal here

12-32
Check for no audio at pins 6 and 11 of IC2001 in a Sanyo VHR2250.

Noise in video playback in a Quasar VH5041

Isolation Scope the video head amplifier.

Repair Check for video on pins 11 and 12 of IC3002 (AN3312), the head amplifier IC (Fig. 12-33). Replace AN3312.

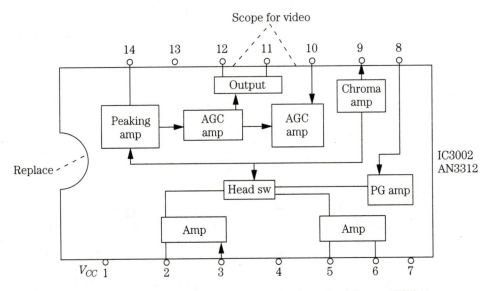

12-33 Replace IC3002 when there is noise in the playback mode of Quasar VH5041.

Color missing in the record mode of an RCA VR250

Isolation Locate chroma FL304 and FL303.

Repair Scope for color at pins 3 and 4 (FL304). The signal was normal on pin 3 but not on pin 4. Replace FL304 (Fig. 12-34).

Will not play back in color in a Quasar VHS041

Isolation Locate the chroma processor (IC8102).

Repair Scope for pulse loss at pin 18 of IC8102. Replace the chroma processor (IC8102) (Fig. 12-35).

Will not record audio in an Emerson VCR874

Isolation Locate the VIF/SIF audio IC (IC6801).

Repair Scope for audio on pin 3 of IC6801. Replace LA753BN IC (Fig. 12-36).

Tape counter in an Emerson VCR900 will not rotate

Isolation Check the tape counter assembly.

Repair Remove the take-up reel, clean off excess grease under the take-up reel, and lubricate with light grease.

Figure 12-37 shows a UL listing number on a consumer electronic product that identifies the manufacturer who made it.

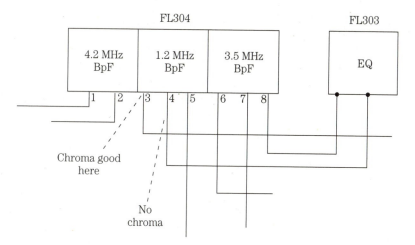

12-34 Check pins 3 and 4 of chroma FL304 in an RCA 250 with missing color in record mode.

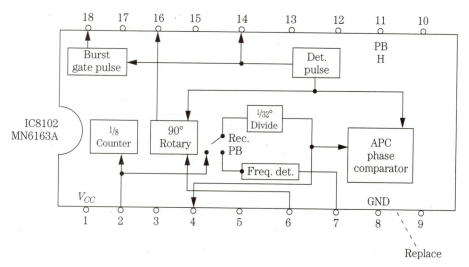

12-35 Replace IC8102 when there is no color during playback of a Quasar VHS041.

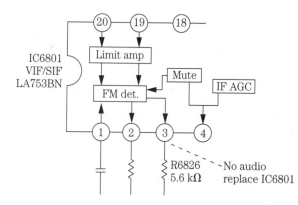

12-36

Replace IC6801 when there is no audio in record mode of an Emerson VCR874.

UL listing number to VCR manufacturer (Unofficial)

UL Number	Manufacturer	Brand Names
146C	Goldstar	
153L	NEC	
16M4	Samsung	Supra, Multitech, Unitech, Tote Vision, Cybrex, GE, RCA Sears
174Y	Toshiba	Sears
238Z	Hitachi	RCA, GE, Penney, Pentax
270C	Sony	
277C	JVC	
282B	Sharp	
289X	Emerson	
333Z	Symphonic	Teac, KTO, Realistic, Multitech, Funai, Porta Video, Dynatech, TMK
336H	RCA	
347H	NAP	
43K3	Kawasho	
403Y	Fisher/Sanyo	Realistic, Sears
436L	Quasar	
439F	JVC	Zenith, Kenwood, Sansui
444H	Zenith	
44L6	TMK	Emerson, Lloyds, Broksonic
504F	Sharp	Wards, KMC
51K8	Portavideo	
536Y	Mitsubishi	Emerson, Video Concepts, MGA
540B	GE	
570F	Sony	Zenith
623J	Sampo	
628E	Samsung	MTC, ToteVision
679F	Panasonic	RCA, GE, Magnavox, Quasar, Canon, Philco
723L	Sanyo	
727H	Hitachi	
74K6	Funai	
781Y	NEC	Dumont, Video Concepts, Vector, Sears
828B	Panasonic	Olympus
843T	Magnavox	
86B0	Goldstar	Realistic, JC Penney, Tote Vision, Shinton, Sears, Memorex
873G	Mitsubishi	
41K4	Portland	

12-37 UL listing to VCR manufacturers. (Courtesy of *Electronic Servicing & Technology* magazine)

13
CHAPTER

Testing the remote control circuits

Today just about every electronic product operates with a remote control. At first, the TV chassis was operated with a mechanically transmitted sound that turned the power on and off and the volume and channels up and down. These components within the TV set were operated by different motors. Later the mechanical transmitter was replaced with a supersonic unit. Today the infrared transmitter does everything, including scanning stations up and down, selecting individual channels, turning the power on and off, volume selection, muting of sound, and menu selection (Fig. 13-1).

Too many hands in the pot

The remote control transmitter can be damaged by rough treatment. Although children might be able to operate the remote control better than some adults, little ones can cause a lot of damage. They carry the remote around and drop it on the floor, pour soda pop into it, stick sharp objects into the buttons, and sometimes use it as a hammer. Besides hiding under newspapers, blankets, and rugs, the remote control can easily be stepped upon. Actually, the remote should be left in a safe place so that little hands cannot play with it.

A mistreated or dropped remote cannot operate with dislodged batteries, a broken case and PC boards, or sprung battery contacts. This can cause intermittent or no operation. Most remote control transmitters operate from 2 or 3 AAA or AA batteries.

Remote functions

When a button is pressed on the remote control, a signal is sent out and picked up by a sensor attached to a receiver that controls the many functions. You might find the remote control in TV sets, large AM/FM/MPX receivers, VCRs, DSS systems, CD players, and the car combination tape, radio, and CD trunk players (Fig. 13-2). The

13-1 The RCA Dimensia digital control remote operates the color TV set and VCR.

13-2 The single remote operates a five-disc CD Magnavox changer.

remote control can operate two or more electronic products. Besides the combination remote, a universal remote can be adjusted to any remote receiver function.

Although the tone-type remote is no longer in service, you might encounter a supersonic RF remote transmitter. The RF remote generated an electronic signal to control certain frequencies to be radiated through a speaker or transmitter. The frequency chosen was between 44.75 kH and 47 kH. The frequency avoids most erroneous signals that might trigger the remote receiver. Of course, the garage door opener, door and telephone bells, and other supersonic signals can trigger some TVs. Erroneous signals from various electrical and RF generating devices can trigger the remote control receiver.

Infrared remotes

The infrared remote operates in the line of sight (Fig. 13-3). The infrared signal does not bend or go through subjects. If a chair, foot, or person is found between the TV set and remote, the remote will not trigger the remote sensor. When a person walks between the infrared remote and TV receiver, while selecting a different channel, nothing happens. The infrared remote will not trigger another remote product in a different room.

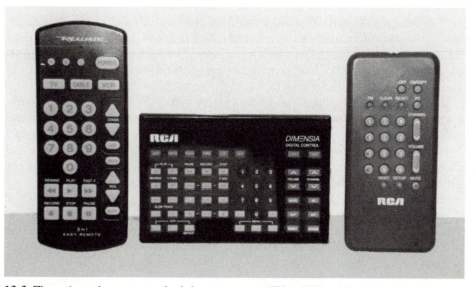

13-3 The universal remote on the left operates any TV or VCR, while the one in the center controls the TV set and VCR and the small one on the right controls only the operation of a 13" portable RCA TV.

The infrared remote sends out an infrared signal from one or two transmitting LEDs. This infrared signal is picked up by an infrared photo transistor sensor, amplified by transistors or one IC component, and fed to a system control IC. The system control might consist of a microprocessor, CPU, and system control IC in the

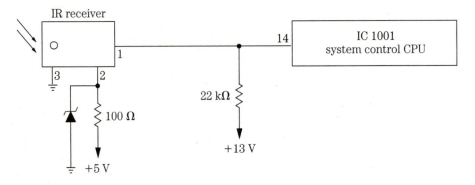

13-4 The infrared receiver picks up the remote-control signal and sends it to the IC1001 system control CPU.

latest TV circuits. The micon or processor chip sends the remote control messages to the various circuits to be operated (Fig. 13-4). In the sound circuit of some remote-controlled receivers, a small motor rotates the volume control up and down.

Nothing operates

The dead remote can be caused by weak or dead batteries. Most remotes are operated from a series of batteries with a total of 3, 4.5, 6, and 9 V. The infrared remote might operate from two AAA or two AA batteries. The early remotes operated from 4.5- and 9-V batteries. The high-powered remotes, as the RCA DSS disk system, operate two transmitting LEDs with four AA (6 V) batteries.

Preferably, check the batteries with a battery tester (Fig. 13-5). If not available, check each battery under load with the low-voltage range of the DMM. Hold down one button, and test the voltage across each battery. Replace all batteries that test 1 V or less. Weak batteries can cause intermittent or weak operation. It is best to replace all batteries when one or more is found weak or dead. Always, replace all the batteries at the same time. Replace remote batteries with high-energy batteries.

Heavy-duty power cells or ultra-alkaline batteries have a relative service life of from 2 to 10 times longer than ordinary carbon batteries. Wipe the new battery contacts by rubbing against a cloth, towel, or pant leg before installing. Make sure that the positive (+) terminal is inserted properly. Most battery holders have marked positive and negative terminals. Double-check the battery polarity.

No see—Aim high

Check the output of the infrared transmitter to determine if the transmitter or the receiver is defective. You cannot see the infrared signal from the remote transmitter, and the remote does not make any noise. Aim the remote high above any object between the operator and the TV set or unit to be operated. The infrared remote can be checked against a broadcast receiver (Fig. 13-6). Hold the remote close to the receiver, and press down on each button. Each time a button is pressed, you will

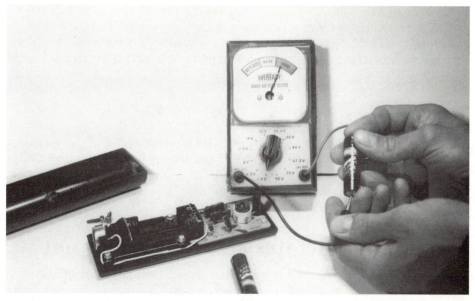

13-5 Check the remote batteries with a battery tester or under load with a voltmeter.

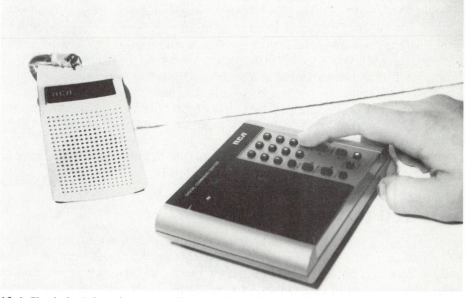

13-6 Check the infrared remote with a portable radio to see if it is sending out signals.

hear a gurgling sound in the portable radio. However, this does not indicate how strong the infrared signal is.

Another method is to use an infrared indicator card in front of the remote transmitter. The infrared card will change to a different color if the transmitter is working.

These infrared indicator cards can be obtained through TV and parts distributors. The RCA infrared card for checking the output of remote control transmitters is stock number 153093. These infrared cards will only indicate if infrared signal is present, but not how strong or weak the signal. It only indicates that the remote control is operating.

Try another one

Simply try another remote to see if it will turn on the TV set, VCR, or CD player. Usually, there are several different remotes in the house. Most remotes have a function that will operate another receiver in the TV or VCR. If the subbed remote operates one or two functions, you can assume that the sensor and receiver are operating. Then check the defective remote transmitter.

Sometimes operates and sometimes not

An intermittent remote can irritate the operator. Sometimes, during the frustrations of trying to get it to operate, the remote is slammed down or dropped. Dropping the remote can cause intermittent operation. A weak battery can cause intermittent operations. Poor battery connection can cause intermittent operation and, when tapped, the remote might come to life.

First, check and test all the batteries. Replace all the batteries if one is found weak or dead. If the remote is still intermittent, check the battery terminal connections. Bend or pull out the small flat spring battery terminal or wire springs. Clean each terminal with sandpaper. If batteries are left in the remote control and the remote is not used for several months, the batteries might leak or corrode the battery terminals.

Loose components or cracked PC boards can cause the remote to work intermittently. Inspect the PC wiring under a strong light. Check all the coils, transistors, and capacitors for broken or loose leads. Resolder connections on the PC board. Make sure the wires to the battery terminals are intact and soldered. When large defective components are found inside the remote transmitter, send it in for repair or exchange at the manufacturers and tuner repair depots.

Infrared power meter

The infrared power meter that is used to check the laser optical assembly in a CD player can test out those infrared remote transmitters. Start with the lowest scale of the power meter (0 to 0.3 mW). If the results are poor, move up to another scale. The laser power meter might have a 0.3-mW, 1-mW, and 3-mW range with switchable wave lengths settings of 633 nM and 750 to 820 nM. The laser power meter used in the infrared test is a Tenma meter number 72-670 (Fig. 13-7). Also a leader laser power meter 70-510 can be ordered by mail from:

MCM Electronics
650 Congress Park Dr.
Centerville, OH 45459-4072

13-7 Check the remote control operation with a laser power meter.

Place the remote 3" away from the pick-up probe, and register the measurement on the power meter. Likewise, check a powerful remote control unit, and write down the measurement. By making comparison tests with new remotes, the defective or weak remote measurements can be compared to the new infrared measurements. Move the remote back and forth to acquire the best reading. A weak or dead remote will have a low or no measurement compared to a normal remote transmitter. Then, check batteries and contact terminals to obtain a higher measurement. Remote comparison tests with the laser power meter can indicate if the remote or receiver is defective.

Infrared remote circuits

An early remote transmitter might consist of one transistor, a coil, small capacitors, batteries, and a sonic transducer. Three transistors were added with two infrared LEDs in later models. Today the single infrared remote might contain one IC, a transistor, and an LED sensor. Later two infrared transmitting LEDs were used in the power remote controls.

The infrared remote transmitter found in a Radio Shack VCR unit consisted of an IC microprocessor, xtal, TR1, LED1, and two batteries in series (Fig. 13-8). This small remote provides 64 different operations. Besides operating the speed, play, and stop modes of a VCR, the remote control unit operates the TV channels up and down, turns the TV monitor volume up and down, and changes TV monitor modes. The rest of the operations are applied to operating the VCR (Fig. 13-9).

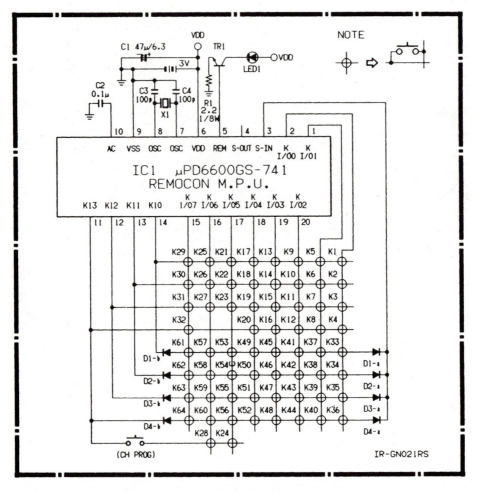

13-8 A simple infrared remote circuit. (Radio Shack)

In the deluxe receiver remote transmitter, IC1 controls all the various functions with IC2 operating 11 different infrared transmitting LEDs. IC3 provides operating voltage to the LEDs and micon IC1. Q1 through Q4 provide power to the infrared LEDs. There are 73 different push button switches tied into IC1 for the many different functions (Fig. 13-10). There are 11 fixed diodes found in the many operations.

Universal remotes

The universal remote control transmitter can operate most TVs, VCRs, audio and video components, CD players, and deluxe AM/FM/MPX receivers (Fig. 13-11). The General Electric RRC500 does the work of three remotes, while the RRC600 model does the work of four remotes. The RRC600 controls up to four infrared audio/video products, with over 200 key combinations, program sequencing, LCD display, and a low-battery indicator.

AVAILABLE KEY FUNCTION OF REMOTE CONTROL UNIT

	Key No.	Function		Key No.	Function		Key No.	Function		Key No.	Function	
D	K1	PAUSE/STILL	USE	K17	9	USE	K33	INPUT		K49	TV CHANNEL DOWN	
	K2	SLOW		K18	8	USE	K34	AUTO TRACKING	USE	K50	TV MONITOR	
	K3	SPEED	USE	K19	7	USE	K35	AUTO TRACKING		K51	TV MONITOR	
	K4	AUDIO OUT		K20	1--	USE	K36	SLOW		K52	AUDIO OUT	
C	K5	FF	USE	K21	6	USE	K37	CAPTION		K53	TV CHANNEL UP	
	K6	CHANNEL SCAN		K22	5	USE	K38	SLOW DOWN	USE	K54	QTR ON TIME	USE
	K7	STOP	USE	K23	4	USE	K39	SLOW DOWN		K55	TV/VCR	
	K8	REC	USE	K24	ERASE	USE	K40	TRACKING DOWN		K56	QTR ON TIME	
	K9	INDEX		K25	3	USE	K41	CAPTION MODE		K57	TV VOLUME DOWN	
B	K10	MEMORY	USE	K26	2	USE	K42	BLANK SEARCH		K58	QTR OFF TIME	USE
	K11	PLAY	USE	K27	1	USE	K43	BLANK SEARCH		K59	DISPLAY	
	K12	REW	USE	K28	AUTO	USE	K44	INPUT		K60	QTR OFF TIME	
	K13	0	USE	K29	CLEAR	USE	K45	TV POWER		K61	TV VOLUME UP	
	K14	CHANNEL UP	USE	K30	PROGRAM	USE	K46	SLOW UP	USE	K62	TV/VCR	USE
	K15	CHANNEL DOWN	USE	K31	POWER	USE	K47	SLOW UP		K63	VCR PLUS	
	K16	RESET	USE	K32	ADD		K48	TRACKING UP		K64	TV/VCR	
A	D1	—	USE	D2	—	USE	D3	—		D4	—	

13-9 The many different key functions of the remote control unit. (Radio Shack)

This remote, like all other universal control units, can be adjusted to work on all remote-control products. These remote-control units can be purchased at electronic TV dealers, hardware stores, mall stores, and Radio Shack (Fig. 13-12). The universal remote operates on batteries and can be checked with the remote infrared tester, like other remotes. The RCA DSS system remote cannot be operated with a universal remote control transmitter.

Servicing the remote control unit

The biggest problem with the remote is to simply test and change the batteries. Clean or repair the battery holders for dead or intermittent operation. Check each transistor with the transistor tester or diode test of DMM. Test the infrared diode like any silicon diode. Check it out on the diode test of DMM. Check the various voltages on the control IC (Fig. 13-13). The push buttons can be tested with a low-ohm resistance test, with one lead to the common terminal and other across the switch at the control IC.

Remote or receiver

When the remote has been tested or subbed with another unit and the electronic product does not respond, the problem can be in the infrared receiver or power standby circuits. In today's TV receiver, the standby voltage is on all the time, even when the set is turned off. The remote receiver must be alive to receive the commands of the remote control to turn the electronic product on.

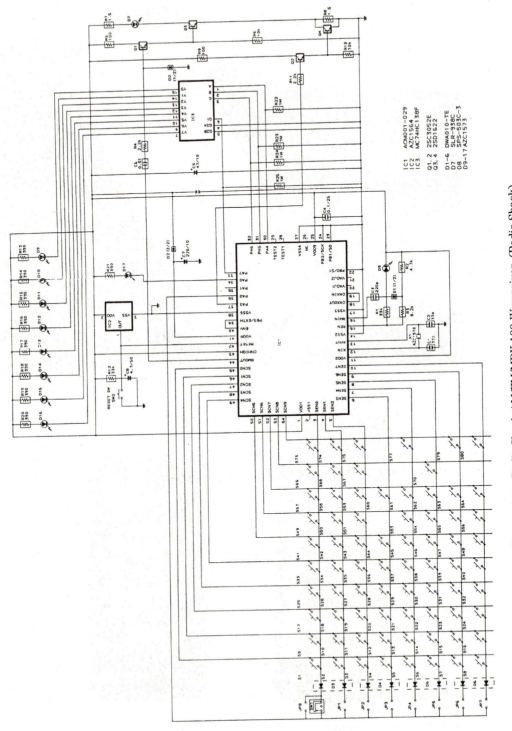

13-10 The deluxe remote control operating a Radio Shack AM/FM/MPX 100-W receiver. (Radio Shack)

13-11 The General Electric universal remote transmitter.

13-12 The Realistic universal remote control to the left side of an RCA TV remote.

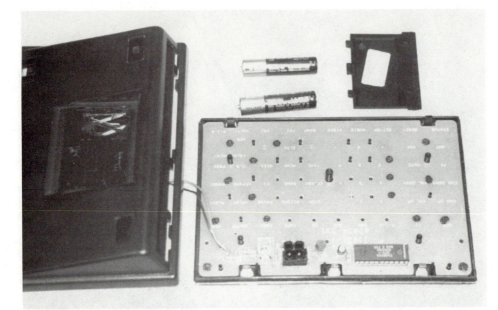

13-13 The various components found on the second PC board in a RCA remote transmitter.

The remote-control receiver sensor is found in the front panel of any electronic product. This infrared sensor is actually an infrared phototransistor that picks up the infrared signal and connects to the remote receiver in the TV set, VCR, or CD player. The infrared sensor is usually placed in a shielded container to prevent outside noise and unwanted signals from striking the sensor.

In earlier infrared receivers, small transistors were used to amplify the weak signal to the system control IC. You might now find an IC component between infrared sensor and decoder. One large IC might be found with a transistor amp before feeding the remote signals to the system control IC (Fig. 13-14). In turn, the system control IC operates the various functions of the remote transmitter.

Troubleshooting the infrared receiver

Go directly to the infrared remote receiver when the remote transmitter is found good but there is no operation of the TV, VCR or CD player. Check the dc voltage applied to the sensor and receiver IC. Usually, the voltage applied to the IR pre-amp IC is from 5 to 12 V, depending upon the standby voltage of the TV receiver, VCR, or CD player. If no voltage is found at the infrared receiver or pre-amp IC, suspect a defective standby circuit (Fig. 13-15).

Locate the IR receiver circuit by locating the IR sensor and tracing the wire cable to the PC board and receiver IC. Check the infrared photo transistor with the diode scale of a DMM. You can use a transistor tester to check the photo IR transistor. Make sure the cable wires are correctly connected to the receiver board. After taking voltage measurements, check the infrared transmitted signal at the receiver.

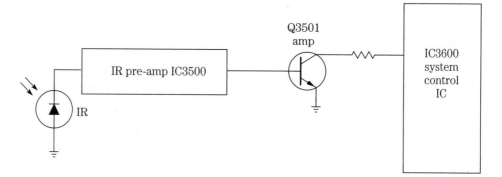

13-14 Here the IR sensor picks up the remote control signal, feeds it to the IR pre-amp IC3500, and to a transistor amp before the system control IC.

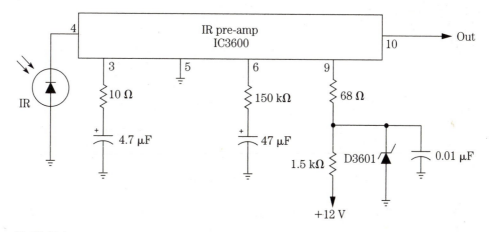

13-15 If there is no voltage at pin 9 of IC3600, suspect the standby power-supply circuits.

The signal picked up from the infrared sensor can be checked, indicating that the infrared sensor is operating, with a scope signal waveform. Usually, a test point is found after the infrared sensor, or the signal can be scoped after the pre-amp and detector circuits of the receiver. The remote-control transmitter Play button must be pressed down at all times to receive the transmitted signal. Suspect a defective infrared sensor if no scope waveform is found. Again check all dc voltages on the suspected IC. Test each part that is connected to the IC before installing a new one.

Sometimes, broken or loose cable connections produce intermittent remote-control functions in the receiver circuits. Scope the remote signal, and take accurate voltage and resistance measurements on the IC and transistor components to locate the defective component.

Denon DCM560 remote sensor

In many of the remote control CD players, a remote infrared sensor unit is mounted on the front panel PC board. Infrared sensor (RM601) has a +5 V applied

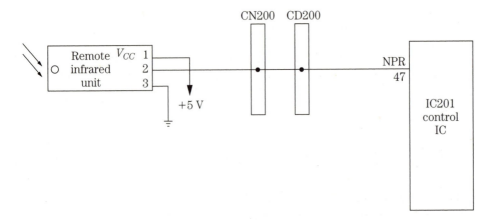

13-16 The infrared remote signal is applied to pin 47 of IC201 in a Denon DCM560 receiver.

to terminal 1 with pin 3 grounded. The infrared output signal is found at pin 2 to connector CN200 (pin 11). The NPR signal is applied to pin 47 of the control IC201 (Fig. 13-16).

Check the remote signal at pin 47 of IC201 when the remote transmitter is normal and the remote receiver will not function. If no signal at this point, check the voltage source (+5 V) at pin 1 of RM601 unit. Then check the signal at the remote output pin. Most infrared sensor units are replaced instead of trying to repair the infrared unit.

RCA CTC145 TV chassis remote receiver

In this TV chassis, the remote sensor is fed into a transistor IR amp, second IR amp, third, and fourth IR amp in series transistors. All of the IR amp transistors and sensors are enclosed in a shielded area. The top shield cover must be in place for the remote to function (Fig. 13-17). A +12-V source is fed to the IR receiver and analog interface microprocessor (U3300).

When the IR receiver does not function, check the dc voltage (+12 V) to the IR amp stages. Test the IR detector (CR3404) with a diode test of the DMM. Take an in-circuit transistor test of each transistor. A scope test at the fourth IR amp output or at pin 36 of U3300 should indicate whether the remote receiver is functioning. The IR test point is found at FB3304 and FB3305. Suspect IC U3300 if IR receiver is functioning but there is no remote control. Check for standby +5 V on the microprocessor.

Standby power supplies

Check the standby power supply when low or no voltage is found at the IR receiver and system control IC. The standby power supply is on all the time, so the remote control functions can operate when the TV, VCR, or CD player is turned off.

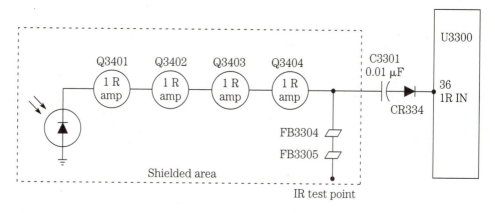

13-17 All working components of the infrared receiver are enclosed in a shielded area of an RCA CTC145 chassis.

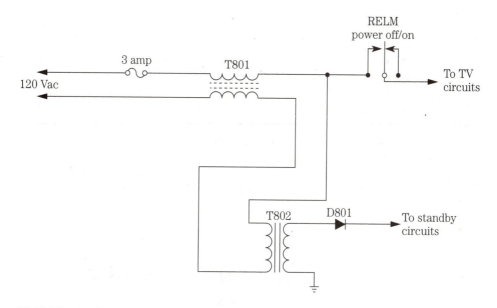

13-18 The simple standby power transformer is tapped off the ac power line after the fuse in a TV chassis.

Usually, the standby power transformer is connected within the ac circuit after the fuse and before the ac TV power switch (Fig. 13-18).

The simple standby power-supply circuit consists of a small step-down power transformer with half-wave diode rectification and filter capacitor. You might find transistor or zener diode regulation or a combination of both. The standby voltage might appear as +5, +10, or +12 V (Fig. 13-19). In some TV chassis, a run voltage is also taken from the low-voltage standby circuits.

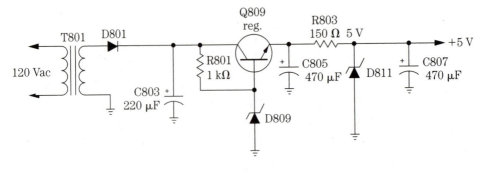

13-19 A typical 5-V standby power supply.

Servicing standby circuits

When low or no voltage is found at the remote receiver or system control IC, suspect the low-voltage supply in the standby circuits. No voltage is usually caused by an open regulator transistor or silicon diode. Really low standby voltage can result from a leaky regulator transistor, a leaky zener diode, or both.

Locate the small stepdown transformer within the TV, VCR, or CD player chassis. Check the ac voltage to the silicon rectifier or across the secondary winding. This voltage should be from 10 to 20 Vac. Test the silicon diode with the diode test of DMM.

If dc voltage is measured up to the regulator transistor and low voltage out at emitter terminal, suspect a leaky transistor. Check the regulator transistor in-circuit. Notice if the zener diode appears quite warm. Test all the diodes in the standby circuits with a diode tester or DMM. When voltage is normal at the collector terminal and low at the emitter terminal, check for a defective electrolytic capacitor. Clip a known electrolytic capacitor across each one in the power supply until the voltage returns to normal. Pull the power cord, and clip a capacitor across the suspected one with the unit turned off.

Intermittent remote: Sanyo 91C90 TV

Sometimes the remote control operated perfectly and, at other times, would start and stop. A new battery was installed in the transmitter, but the results were the same. The remote was checked against another model and was normal. The intermittent remote was checked by a laser power meter, and the remote operated even when tapping the remote case.

Simply moving the receiver cables caused the remote to act up. The transmitter was placed near the sensor of receiver in the TV chassis with the volume turned up. When the remote went into the intermittent mode, no tuner or volume up and down were noticed. The remote seemed to act up most of the time when the pre-amp remote board was tapped or moved.

The small pre-amp board was removed and inspected for possible poorly soldered connections. Each component was moved and prodded. After several min-

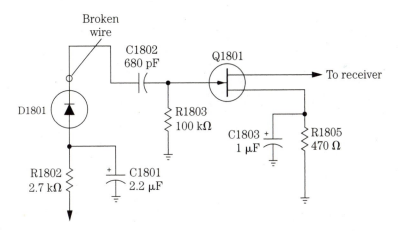

13-20 A broken wire on a remote sensor in a Sanyo 91C90 TV resulted in intermittent remote operation.

utes, a broken base wire from the sensor diode (D1801) was located (Fig. 13-20). The wire had snapped off.

No remote receiver operation

The remote-control transmitter tested normal with no remote action in a Toshiba CT317C TV set. Either the sensor or IC signal amp was defective. The receiver worked perfectly with the TV push button manual operation. After locating the infrared sensor, the wires were traced to ICR01. The sensor tested normal with the DMM diode test.

Critical voltage measurements were made upon each terminal of ICR01, and all of them seemed to be off and low in voltage. The supply voltage was only 1.7 V and should be around 5 V (Fig. 13-21). Pin terminal 9 of ICR01 was unsoldered with soldering iron and solder wick, leaving the terminal open. When the TV was plugged in this supply voltage rose to 5.7 V, indicating ICR01 amp or components tied to it were leaky. All components were tested at each terminal and seemed to be normal. ICR01 part number 23119588 (UPC1474HA) was ordered out from the Toshiba distributor.

Dead remote control receiver

The remote control would not turn on the Sharp 19SB60R portable TV. The remote was tested and appeared good. Very little voltage was found upon the receiver terminal pin 2. A voltage test was made upon the +5-V standby voltage and was fairly normal. R1016 (100 Ω) resistor appeared quite warm (Fig. 13-22). A low resistance measurement was made at pin 2 and receiver shielded area with only 27 Ω to ground.

Pin 2 was removed from the remote receiver, and the voltage was still low. Upon checking the wiring, a zener diode appeared burned and quite warm. Replacing leaky D1052 solved the no remote operation.

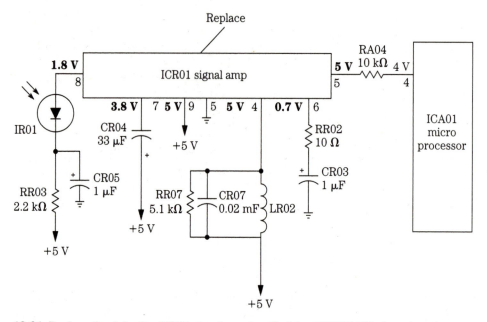

13-21 Replace the defective ICR01 signal amp in a Toshiba CT317C TV when there is no remote action.

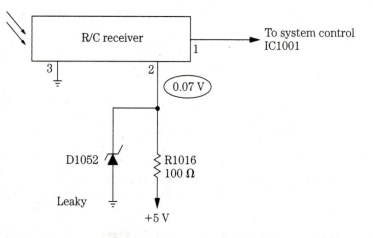

13-22 A leaky zener diode D1052 in a Sharp 19SB60R TV caused the remote receiver to not operate.

No standby voltage: Goldstar CMT2612

The remote control would not operate in a Goldstar CMT2612 TV. The remote control unit was okay. When voltage was checked on the receiver IC, none could be measured. The standby voltage source was traced to a small power transformer. Voltage measurements were made upon a silicon rectifier and were quite normal.

When the dc voltage was measured upon the transistor regulator (Q803), the collector voltage was good without any voltage at the emitter terminal (Fig. 13-23). Q803 was tested in the circuit and appeared open. After removing the regulator transistor from the circuit, a regular transistor test proved Q803 was open. The D816 zener diode checked normal. Q803 was replaced with a ECG152 universal transistor.

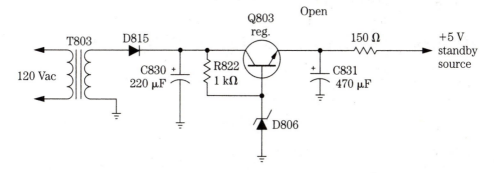

13-23 An open Q803 in the standby power supply caused no remote operation in a Goldstar CMT2612.

No remote: Dead TV

No remote operation and a blank raster was found in an RCA CTC157 chassis. Because the picture tube had a blank raster and no remote action, the symptom was traced to the low-voltage standby circuits. In this model, the 12-V standby and run voltage is found in the standby power-supply circuits.

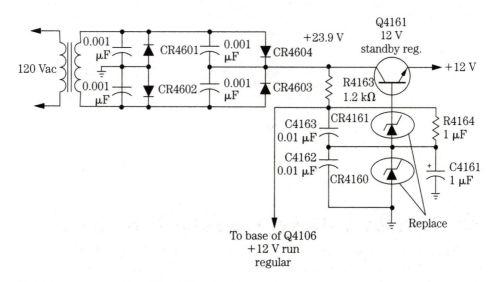

13-24 Replace both CR4160 and CR4161 if the remote won't work and there is blank raster in the standby power supply of an RCA CTC157 chassis.

The standby transformer was located upon the chassis and ac voltage was measured to the bridge rectifier circuits. 24.7 V was found upon the collector terminal of the 12-V standby regulator (Q4161). Really low voltage was found at the emitter terminal (Fig. 13-24). Q4161 was tested good with in-circuit transistor test. CR4161 and CR4160 were found leaky and replaced. The 12-V standby voltage was now measured upon the emitter terminal of Q4161 regulator. CR4161 and CR4160 were replaced with a universal ECG5071A zener diode.

Remote control troubleshooting chart

Check Table 13-1 for troubleshooting the remote control, receiver, and standby circuits.

Table 13-1.

Symptom	Repair
Dead remote control	Check and test the batteries. Check the remote against the indicator card. Test the remote with a laser power meter. Sub another remote.
Intermittent remote operation	Clean the battery termianls. Check for loose batteries. Remove the batteries, and clean the end terminals with sandpaper. Bend the terminals out for a tight fit. Suspect the remote has been dropped several times. Inspect the PC board for cracks or loose parts on the PCB.
Weak reception	Check batteries—Replace all the batteries when one is found bad. Test the remote on an indicator or power meter.
Remote normal—No action	Check the voltage source to the IC or transistor amps. Take waveform tests. Test each transistor in the circuit. Take voltage tests on the IC amp or receiver. Test the sensor unit with a diode test of the DMM. Suspect the standby voltage supply.
Standby power supply	Check the output voltage supplied to the IR receiver. Test the output voltage on the standby supply. Low or no voltage at the regulators. Test the transistors and zener diode regulators with a diode test of the DMM. Remember the standby voltage must be present or the remote won't work.

14
CHAPTER

Servicing the boom-box cassette/CD players

Today, the boom-box cassette player contains high-powered amps, larger speakers, AM/FM/MPX receiver, and cassette and CD players. The boom-box player might have detachable speakers and a dual cassette tape player. One cassette player is used for record and play, while the other is just for playing the cassettes. You might find a top-loading CD player in the larger boom-box machine (Fig. 14-1).

The average boom-box portable cassette recorder has AM/FM/MPX stereo digital tuning, high-speed dubbing, an LCD digital display, and a three-band equalizer. The audio can come through a pair of stereo headphones or three or more speakers. The portable player can be operated from the power line or six C or D batteries.

A deluxe boom-box cassette player might have some of the following features: digital tuning, 11 AM and 10 FM presets, high-speed dubbing, auto reverse, stereo wide system, a built-in mike, and a CD player. The top-loading CD player can have a 16-track programmable memory, repeat, and an LCD track display. Synchro-start makes CD-to-tape dubbing and tape duplication easy. You can copy tapes at normal or high speed, auto-play two tapes in sequence, and record from the CD, AM-FM radio, or live with a built-in electronic microphone. Often, the CD player is top-loaded (Fig. 14-2).

The dual tape decks are mounted side by side or one at the top and the other at the bottom. Tape deck 1 is used for recording and playback. Tape deck 1 can record from tape deck 2 or duplicate deck 2. Continuous playback of both decks can be provided with tape deck 2 first and deck 1 last. Each cassette deck can be controlled by its own push buttons. Most deluxe players have soft-cassette eject features.

Required test instruments

Most test instruments needed to service the cassette boom-box player are found on the electronic service bench. The DMM and capacitor tester can test transistors, capacitors, and diodes and make accurate voltage and resistance tests (Fig. 14-3). A

14-1 The table-top CD player has front-loading operations, and most boom-box portables contain top loading.

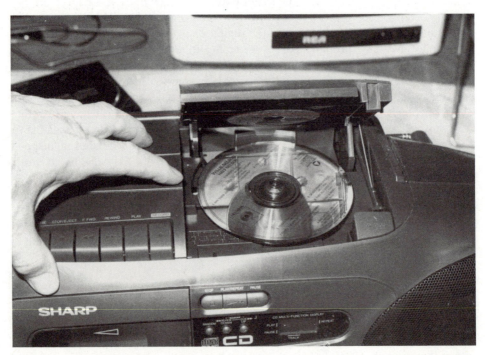

14-2 This Sharp CD boom-box player loads the disc from the top.

14-3 The digital multimeter (DMM) and capacitor tester can solve many defective circuits and components in the boom-box player.

battery-operated soldering iron is ideal when soldering up SMDs, transistors, ICs, and microprocessor PC circuits.

The oscilloscope, external audio amp, audio signal or function generator, and frequency counter can be used for signal tracing and critical adjustments. The scope can be used to signal trace audio, locate distorted circuits, adjust the tape head, and check the waveform of the bias oscillator in record mode. A frequency counter can locate speed problems and amplifier adjustments and check frequency response of the amplifier stages. The function or audio signal generator can eject audio signals, sine and square waveforms, to locate distortion, low-level signals, and weak and distorted stages.

Here is a list of required test instruments:
- Digital multimeter (DMM)
- VOM or Fet-VOM
- Capacitor tester
- Oscilloscope
- Audio, function generator or noise generator
- Frequency counter
- External audio amplifier
- Noise signal injector
- External dc power supply
- Solder station and battery-operated iron (Fig. 14-4)
- Test speakers
- Test tapes
- Transistor tester

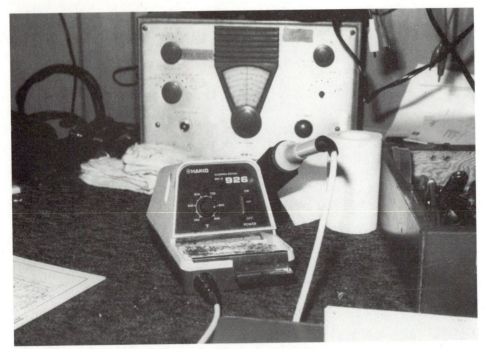

14-4 A soldering iron station is ideal when servicing the boom-box player.

For those who specialize and service only sound equipment, additional test equipment is needed:

- ac millivolt meter
- Wow and flutter meter
- Distortion analyzer
- Hi-fi stereo amplifier
- Laser power meter
- Variac or variable isolation transformer
- Speaker dummy loads
- CD test disc
- Stereo generator

Although the bulk of boom-box cassette players can be serviced with only a DMM, VOM, transistor tester, audio signal generator, and variable line transformer, a few added test instruments can make quick and accurate audio tests. Today, a single test-bench-type DMM with voltage, resistance, current, transistor, and diode tests might also include a capacitor and frequency tester and can be used to repair a lot of CD players. The distortion, wow, and flutter meters are quite expensive and are used in larger audio-specialized service centers.

Without a schematic

Servicing the digital tuning of the AM/FM/MPX circuits is more difficult without a schematic. Often a schematic diagram is not required for the audio stages. Simply compare each audio circuit with the normal one. The same components, voltage, and resistance measurements can be compared to one another in the stereo audio circuits. In fact, you can easily locate a weak or distorted stage by comparing the injected audio signal with a dual-trace scope.

After removing the back cover, look the chassis over for the components that possibly cause the audio symptoms. Survey the entire chassis or board for burned or broken components. Inspect the PCB for overheated components. Check for hot spots on the PC wiring side. An overheated resistor will have a dark area around the PC donut connection. Often, the output components, such as power output transistors and ICs, are located on a heat sink.

Check for the location of a small power transformer, large filter capacitor, and silicon diodes for a power-supply symptom (Fig. 14-5). Notice if a transistor and zener diode regulators are mounted nearby. A quick voltage test across the largest filter capacitor can indicate if the low-voltage power circuits are functioning.

14-5 Most power transformers are found mounted off to one side of the power-supply circuits.

If there is no physical evidence, start at the volume control and signal trace the audio signal in either direction. When audio is found at both stereo circuits at the volume controls, suspect a defective AF or output component for weak, distorted, or no audio. Use both channels to test the different audio components at different points in the stereo amp circuits.

Locate the stereo audio output transistors or ICs on separate heat sinks (Fig. 14-6). Take critical voltage tests on each output transistor. Check the input and output signal of a dual-IC amp with the scope or external amp. You might find all of the AF and power output stages in one large IC. Today, in high-powered output amplifiers, a large dual-IC might contain the entire audio system of both channels.

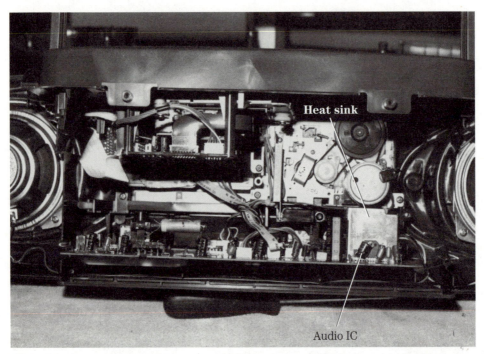

14-6 Locate the audio output IC on the boom-box heat sink, and take critical voltage measurements.

You can locate the weak, distorted, or dead channel at the speaker. Simply trace back the speaker wiring to the PC board or electrolytic coupling capacitor or directly coupled to two audio output transistors or an IC. Take critical voltage and resistance measurements on each transistor and IC. Test each transistor in-circuit for leakage or open condition. Remove the suspected transistor, and test it out of the circuit.

Solder up

Cleanly soldered joints and connections are a must in audio repair. Be very careful when soldering up transistors, ICs, microprocessors, and SMD components within the audio circuits. Poorly soldered connections can cause intermittent or dead audio symptoms. Sloppily soldered connections can bond two different circuits together and cause damage to the solid-state components and can burn out small resistors.

Use a small pointed soldering iron when working around small components that are tucked tightly beside each other in small areas. You can insert a small tip in the battery-powered soldering iron and get down to the soldered connection without melting other components. Be especially careful when removing and installing miniature SMD parts. Use a strong light and magnifying glass to check each soldered connection.

Accidental erase

To save a valuable recording, break out the rear tab of the cassette. When a cassette with the tab broken out is inserted into the player, accidental erasure is prevented by a lever mechanism that keeps the record button from being pressed down (Fig. 14-7). If a certain cassette will not record, first check the tab at the back. If you later desire to use this recorded cassette, simply place scotch tape over the opening.

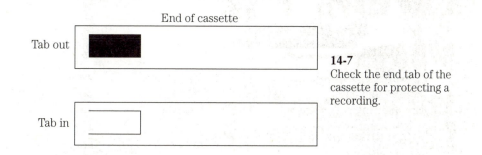

14-7
Check the end tab of the cassette for protecting a recording.

Continuous tape playback

A boom-box cassette player with two different tape decks might have continuous playback features. For extended playback of both cassettes, load both cassette holders. Place the tape that you want to play first in tape deck 2. Then, place the second cassette in deck 1. Rotate the function switch to the tape position. Depress the play button of deck 2, then depress the pause button. Now depress the play button for tape deck 1. Make sure that the dubbing switch is off. When deck 2 has completely played, the automatic stop button in deck 2 engages. The pause button will release, and tape deck 1 will begin to play. When deck 1 is completed, the automatic stop shuts off the player. Always follow the manufacturer's instructions by the numbers with continuous-play operations.

ac power to you

Most of the boom-box players can operate from the ac power line or batteries. The batteries are switched into the B+ circuit when the ac cord is pulled out from

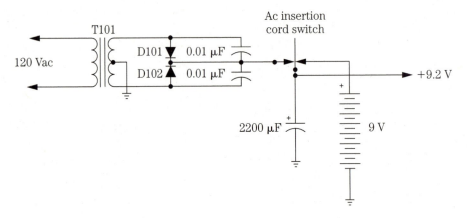

14-8 Check for a defective insertion switch when the cassette player will not operate on batteries.

the player (Fig. 14-8). The step-down ac voltage might contain full-wave rectification with two silicon diodes or a bridge circuit. The bridge circuit might have four separate diodes or one single bridge component. The dc ripple voltage is smoothed out with a large filter capacitor.

A boom-box CD player might operate from the same dc circuits as the cassette player, except better voltage regulation is found here. The bridge circuit might provide 12 to 18 Vdc to a diode-transistor regulator source. The 12-Vdc source can be fed to Q203 with a +10-V diode-transistor regulation circuit (Fig. 14-9). Often, single electrolytic filter capacitors are found at each voltage output. C203, C207, C208, and C209 provide filtering action in each voltage source. Dual-voltage regulation of Q203 and D203, Q207, and D207 provide a +5-V regulation.

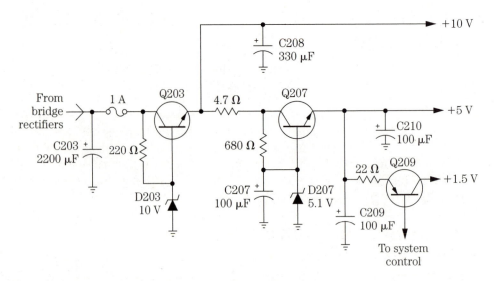

14-9 You might find transistor and diode regulation in the low-voltage circuits.

Boom-box circuits

The AM/FM/MPX tuner system might be controlled with a digital tuning system and varactor tuning. A variable-tuned dc voltage from the control system is applied to the varactor tuning diodes in the AM-RF/FM-RF, AM-OSC/FM-OSC, and AM/FM mixer circuits. In digitally controlled circuits, the digital tuning controller tunes in both the AM and FM circuits (Fig. 14-10).

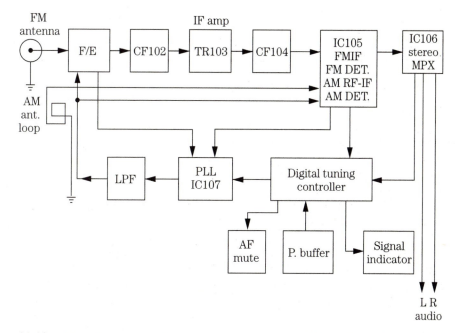

14-10 A block diagram of a digital tuning control in the AM/FM/MPX radio circuit of a boom-box player.

The cassette player consists of two different stereo tape heads and one erase head circuit. The erase head is excited by a dc voltage or a bias oscillator stage. Because the tape head signal is rather weak, two separate AF transistors or one dual IC pre-amp circuit amplifies the weak tape head signal. Usually, in cassette record and playback circuits, the switching circuits are found in the input of each stage and the output circuits (Fig. 14-11). In fact, the pre-amp IC circuits might be used as the complete recording circuits. Usually, Dolby circuits are found between the pre-amp and AF amp audio circuits in the expensive players.

Boom-box amp circuits

The amplifier section of a stereo boom-box player might consist of a pre-amp stage for the cassette and CD player. In the early boom-box players, the audio output circuits consisted of power ICs with transistor AF and driver transistors. Today,

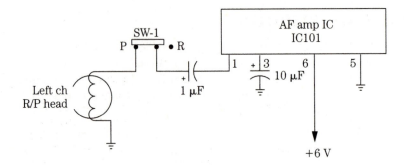

14-11 Check for a dirty or worn record/play switch when the player will not record.

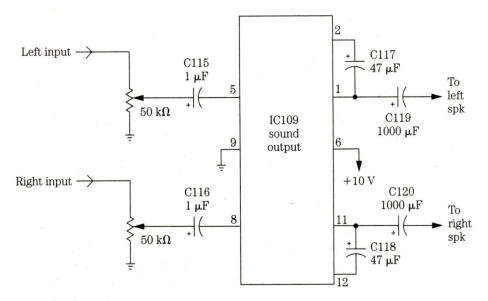

14-12 The inexpensive boom-box player might contain one dual sound output amp IC for both stereo channels.

one large IC might include the AF and output amp circuits of both channels, while in some inexpensive players, separate output ICs are found in low-wattage amplifier circuits (Fig. 14-12).

The defective amplifier might include a weak, distorted, or intermittent audio circuit. Sometimes in one large IC, you might notice a weak and distorted audio channel or an intermittent and distorted IC. The dual-output IC can cause defective symptoms in both channels or only in one channel. When both channels contain sound problems, suspect a defective dual-IC or improper voltage source.

Without a schematic, check the boom-box chassis for a large output or check both ICs on a heat sink of the PC chassis. Another method is to actually trace the sound circuits back to the correct output IC from left or right speakers. Look for a 100- to 1000-μF electrolytic capacitor in the speaker circuits (Fig. 14-13). Trace

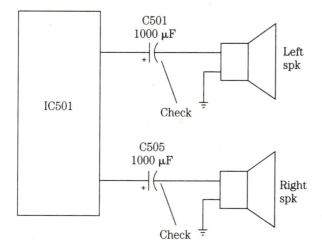

14-13
Trace the weak, dead, or distorted audio signal from speaker terminals to the large electrolytic coupling capacitor to the correct IC amp terminals.

from the speaker wire to the common ground and the corresponding series electrolytic capacitor. Use the low-ohm scale of the DMM, and trace from the other terminal (usually positive) to an output terminal upon the audio IC. Both audio output IC terminals can be checked in the same manner.

A quick method to locate the defective audio output IC is with the radio or cassette playing and checking for sound at each output terminal. Notice if one terminal has loud audio and the other channel is weak or dead. Likewise, check for a weak input terminal at each stereo channel. If a weak audio signal can be heard on more than one terminal output, suspect one large IC amp. The audio signal can be traced from each volume control center-tap with an external audio amp or scope (Fig. 14-14).

When a dead stage is located by signal tracing, take critical voltage measurements. In a dual-output IC amp, locate the voltage supply source terminal. Critical voltage measurements upon each terminal can locate a defective IC. Then take critical resistance measurements from each terminal to common ground. When a low resistance measurement is noted, check for possible leakage or a ground connection. Leaky bypass capacitors to ground from the various IC terminals can cause distorted and weak sound. A change in the bias resistance can produce distorted sound.

Audio comparison tests

Compare the audio signal of one channel to the other at a given point in the sound circuits to check for distorted, weak, and intermittent symptoms. Start at the volume control center terminal, and notice if both channels are normal. If one channel is weak or distorted, proceed toward the pre-amp stages. Signal trace the audio output circuits from the volume control to the input terminal of each audio circuit. Suspect an open electrolytic capacitor or grounded volume wiper control for distorted or weak audio signal.

If both audio channels are normal up to the input terminals of the dual IC or separate IC output, check the signal at the output speaker terminals. When one channel is distorted, weak, or dead, suspect a defective IC, supply voltage, or components be-

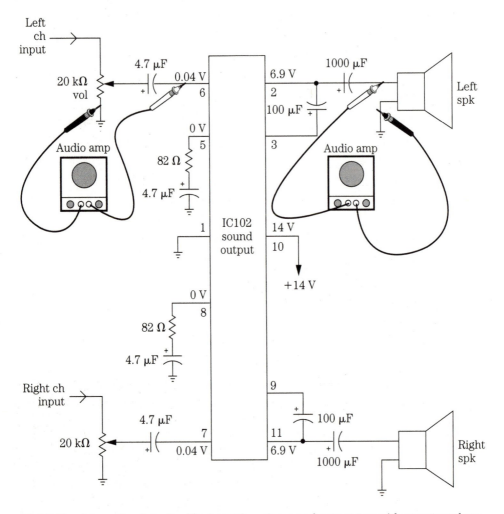

14-14 Signal trace the defective IC amp with an input and output test with an external amplifier.

tween the IC output terminal and the electrolytic coupling capacitor (Fig. 14-15). These coupling capacitors have been noted to cause intermittent, weak, and distorted sound. By comparing the audio signal with the external amp on each audio component, the defective component or stage can be located.

Separate transformer

Often, the small ac power transformer is located off of the regular PC board chassis. The transformer is bolted to the main chassis (Fig. 14-16). All transformer wires are then connected to a separate PC board or main chassis board. A small fuse, silicon diodes, resistors, and electrolytic capacitors are mounted upon the separate PC board.

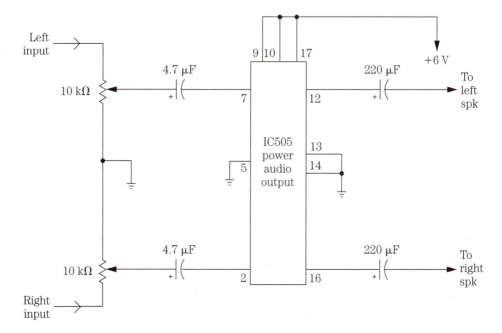

14-15 Check a defective IC, supply voltage, or components between the IC output terminal and the electrolytic coupling capacitor for weak or distorted audio in the boom-box player.

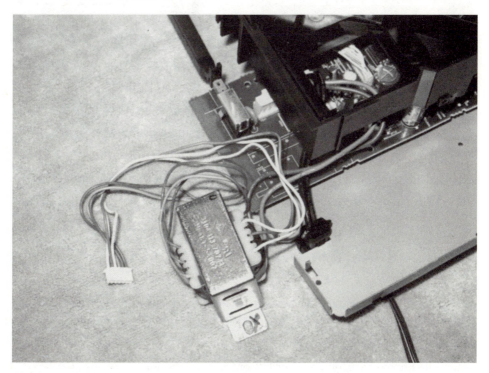

14-16 Locate the ac power transformer mounted off of the main PC chassis.

The warbler

A howling or warbling sound might occur when the function switch is rotated, indicating worn or dirty switching contacts. Because most sliding switches have silver contacts, the silver will tarnish after several years and cause a poor contact. Clean up each switch contact with cleaning fluid. Inject the spray tube right on each switching contact. Spray each contact with cleaning fluid. Rotate the switch back and forth to help clean the switching contacts.

Speaker problems

Several speakers are found in the boom-box player. You might find two 5" speakers in a small boom-box player. The larger boom-box contains three or more speakers on each side or in each stereo channel (Fig. 14-17). Often woofer, midrange, and small tweeter speakers are found in each channel.

14-17 Check for a small tweeter speaker mounted on the top area of the plastic cabinet.

Speaker damage can result from dropping the player on sharp or solid objects, sticking sharp instruments through the speaker grille, and having the player volume turned wide open. The woofer speaker can be damaged with extreme volume. Either the voice coil is frozen against the magnet piece or torn away from the speaker cone. When the boom-box is operated or left out in the rain and inclement weather, the cone becomes warped and the voice coil can drop down against the center magnet.

Notice if the speaker coil moves with the music. A frozen cone will not move and will sound muffled or tinny. A loose cone around the outside ring of the speaker will produce a blatting sound. Small vibration noises can result from holes poked into the cone or loose particles found in the cone area. Small holes and tears can be repaired with speaker cement. Glue around the top rim for vibrating noises. Replace the speaker with a warped cone or loose spider.

Carefully inspect the speaker cone area for an open connection. Usually, the flexible speaker wire will break at the outside cone connection. Place a hot iron tip on the connection, and melt off any paint over the connections. Be careful not to damage the speaker cone. Resolder the speaker cone wire and the flexible speaker wire connection. Use solder flux for a good connection. Apply speaker glue over the connection. Sometimes the flexible wire will break at the outside speaker terminal connection. Replace the speaker with an open or frozen voice coil and a warped cone.

PM speakers should be replaced with the same physical size and voice coil impedance. Do not replace an 8-Ω speaker with a 16-Ω voice coil as the volume will be somewhat lower. Check for correct speaker impedance with the low-ohm scale of a DMM. For instance, an 8-Ω voice coil will measure around 7.5 Ω. The actual coil resistance will be a little lower than the required speaker impedance (Fig. 14-18).

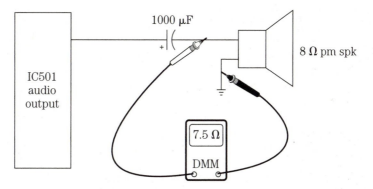

14-18 Check the speaker voice coil for open winding with the DMM. An 8-Ω speaker might register 7.5 Ω.

A 5" PM speaker should be replaced with a 5" speaker. Sometimes the speaker area will designate what size can be replaced. Choose a PM speaker with the same size magnet. The defective speaker can be replaced with a universal replacement, providing the size, impedance, and magnet size are the same as the original. The odd-shaped and tweeter speakers should be ordered with the same part number.

Headphone amplifier

Both stereo channels are amplified by a headphone amplifier IC and are fed to an earphone jack. There might be a mute circuit in the stereo output headphone circuits. In some models, a volume control might control the volume of the earphones

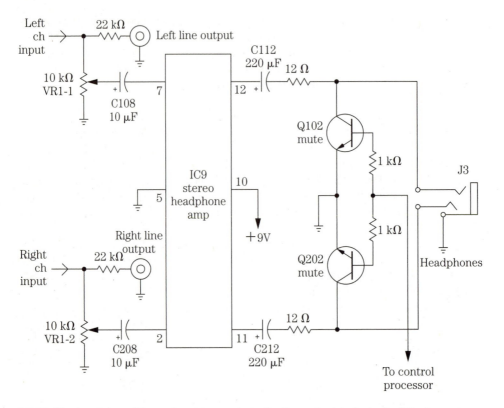

14-19 The headphone IC amp input connects to the line output jacks in the CD player.

connected before the input terminals. The stereo input terminals connect to the line output circuits (Fig. 14-19).

When stereo line output signal is found and there is no headphone reception, trace the earphone jack wires back to the PC board that leads to the headphone IC amp. Take voltage measurements on all IC terminals. Suspect a defective dc-dc converter or power supply source if there is no voltage at the CD player supply source terminal. You can signal trace the audio signal from the line output to the input terminal of the headphone amp IC with an external amplifier. Likewise, check the stereo output terminals with the external audio amp.

Erratic and intermittent sound can result from a defective audio amp, headphone amp IC, coupling capacitors, and poor headphone jack connections. A dirty or worn headphone jack can cause erratic or no reception in one audio channel. Spray cleaning fluid down into the jack area, and work the plug back and forth. A poor contact upon a grounding terminal can cause erratic earphone reception. Do not overlook a defective mute transistor circuit ahead of the earphone jack.

Boom-box motor circuits

In boom-box players with two different cassette players, you might find one or two different motors. Often, one cassette deck is used for playback and the other for both playback and recording. When one motor is found to operate both decks, check for a large drive belt with several pulleys (Fig. 14-20). A stretched motor drive belt can cause poor cassette speeds in both players. Replace the motor belt. The cassette motor is usually located on the same cassette deck. A defective drive motor might be intermittent, change speeds, or result in slow or no rotation. Check the suspected motor by taking voltage measurements across the motor terminals. No supply voltage might indicate a defective switch, voltage regulator, or isolation and voltage dropping resistor. Suspect an open motor winding with correct supply voltage at the motor terminals (Fig. 14-21).

14-20 A large drive belt rotates both cassette players with one motor in a GE boom-box player.

Take a quick resistance measurement across the motor terminal. Most dc motors have a resistance below 10 Ω. The open motor might have an open armature winding, an unsoldered terminal, or a brush hang-up condition. Replace the defective motor instead of trying to repair it. Order the original part number.

When improper or no voltage is found at the motor terminals, trace the wiring to an isolation resistor or coil, right up to the motor switch. Check for voltage across the motor switch with the switch in the off position. Go directly to a zener diode or transistor regulator in the motor circuit. Trace the motor wires to a voltage supply source. An open transistor regulator will have no voltage measurement, while a leaky regulator can have either a low- or high-voltage measurement. The leaky zener diode output voltage is always low.

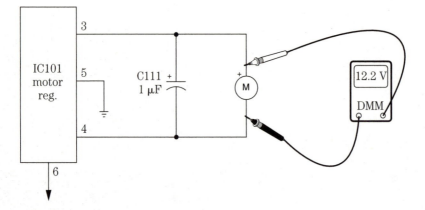

14-21 Suspect an open motor winding when there is normal dc voltage at the motor terminals.

Intermittent motor operation might result from bad brushes, a poor armature, and loose motor connections. Tap the end of the motor with the end of a screwdriver, and notice if the motor changes speed. Replace the intermittent motor. Notice if the motor belt is riding upon the flange of the motor pulley, producing a faster speed. Inspect the motor drive belt for oil or grease spots for slow speeds. Clean up with alcohol and a cleaning stick.

CD player operation

Besides digital tuning with 30 presets and super-bass horn or electronic bass, the boom-box player might consist of a CD player with top or front loading (Fig. 14-22). Some of the larger deluxe CD players have a remote control to operate the boom-box functions. The CD functions might include program, repeat, intro-scan, and synchro-starts of the cassette deck. The real deluxe boom-box player might have a triple play CD changer, equalizer presets, dual cassette deck, and detachable speakers.

Most boom-box CD players contain AM/FM/MPX circuits, dual cassette players, and a top-loading CD player. You might find a JVC boom-box player with a three-play CD changer and find a five-disc turntable in a jumbo Zenith boom-box. The CD discs are loaded at the top in most players, while front loading is found with changer or turntables.

Take a look

Servicing the CD player is more difficult than the cassette player. Carefully look the chassis over before attempting to make repairs without a schematic. A momentary shutdown of CD circuits can result from a defective optical assembly or RF amplifier component. No disc movement can be caused by a defective disc motor, motor drive IC, and no signal from a central processor circuit. Try to obtain another similar schematic for another boom-box CD player to understand how the various circuits are tied together (Fig. 14-23).

14-22 Manual top loading is found in the portable boom-box player.

Poor or improper tracking and focusing can result from an open coil, improper voltage, a defective focus and tracking drive IC, or the digital signal processor. The defective SLED or disc motor can be caused by a common servo driver and servo microprocessor. Weak or distorted audio can result from a defective analog circuit and improper voltage applied to the D/A IC and audio circuits. Intermittent or erratic headphone reception can be caused by a dirty or worn headphone jack, and on it goes.

Location of motors

A portable CD player might have only two motors: a disc motor and a SLED motor. Top loading is found in most CD boom-box players, eliminating the loading motor. The disc motor rotates the CD being played, while the SLED motor slides the optical assembly towards the outside rim of the compact disc. The front-loaded boom-box CD players might have a motor that operates disc loading. Both the disc and SLED motors can be found on the optical assembly.

The disc motor is directly coupled to the disc platform and is located underneath the rotating holder. There is no holder or flapper in the portable CD player. The SLED, carriage, or sliding motor moves the optical assembly along on one or two sliding round rods (Fig. 14-24). The loading motor moves the sliding tray in or out of a front-loading CD player. A magazine or turntable motor rotates a three- or five-disc changer to load and play different discs. The JVC CD boom-box player has a three-disc changer, while a large Zenith boom-box CD player contains a five- and six-disc turntable changer.

BLOCK DIAGRAM

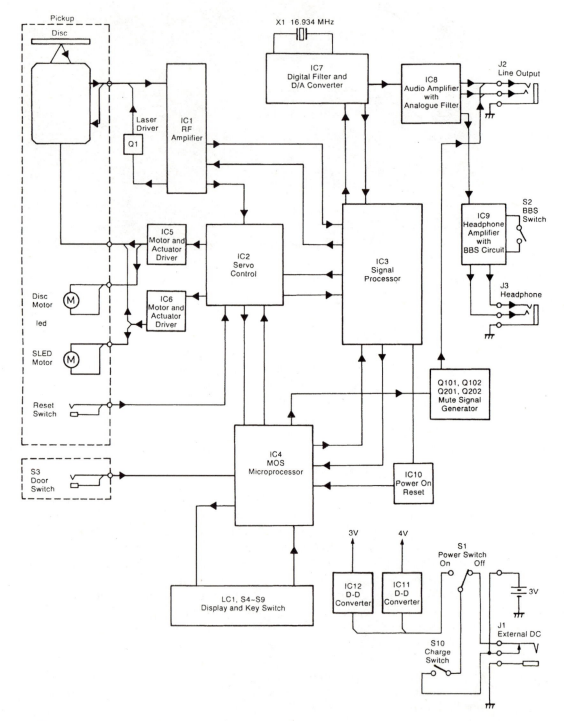

14-23 A block diagram of the boom-box player showing how the various circuits are tied together.

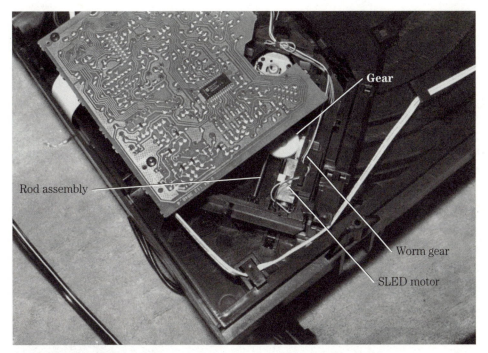

14-24 The SLED motor rotates a worn gear to move the optical assembly upon one rod assembly in this CD player.

The defective CD motor

The defective CD motor can become intermittent, rotates and stops, hums with no rotation, and will not rotate at all. A dead motor can be caused by an open armature winding, defective brush assembly, and no applied voltage. The intermittent motor operation results from improper brush seating, poor connections, a poorly soldered armature connection, and improper operating voltage. The slow- and fast-speed operation can result from a defective motor and intermittent applied voltage. The no motor rotation can be caused by a dead motor, poor belt movement, and a jammed motor gear assembly (Fig. 14-25).

Check the dead or intermittent motor operation with a low-ohm continuity test across the motor terminals. Remove one motor lead for a correct test. Rotate the motor pulley by hand, and notice if the reading is intermittent or open. The best intermittent measurement is done with a VOM-type meter. Critical resistance and voltage tests are measured with a DMM. Suspect an open motor winding when normal voltage is found across the motor terminals (Fig. 14-26). Remember that these CD motors are dc and are operated from batteries, dc-dc converter, and ac-dc power supply circuits.

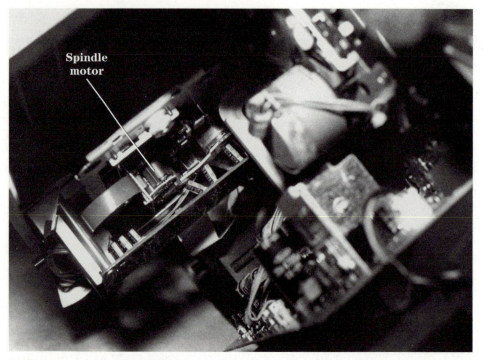

14-25 Drop the CD player mechanism down, and check the dead spindle motor with an open winding.

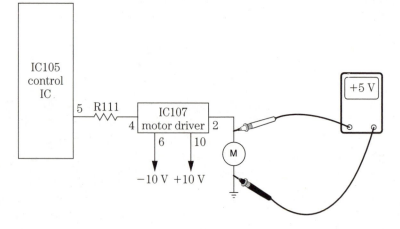

14-26 Check the voltage across the motor terminals for correct voltage with the CD motor in operational mode.

Motor driver circuits

Most CD motors are driven by a driver transistor or IC circuit. In the early CD motor circuits, two or more transistors operated the dc motor; however, today an IC driver or one or two motors might be operated from a single IC. The servo IC might operate both disc and SLED motors in a small boom-box CD player (Fig. 14-27).

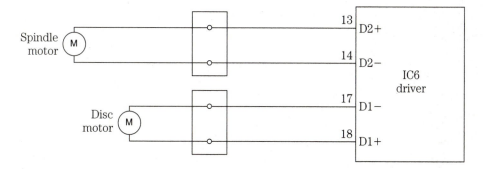

14-27 Here one servo control IC controls both SLED and disc operations.

A signal voltage is applied to the driver transistors or IC to supply voltage across the motor terminals. Check the input and output voltage of the motor driver IC. In a no-operation symptom, zero voltage is found at the motor terminals. If the motor operates in both forward and reverse direction, a different polarity of voltage is applied to the motor terminals by the driver IC.

Suspect a defective IC or transistors, low or no supply voltage, or an improper input signal when the motor will not rotate. Check the supply voltage applied to the driver IC. Measure a positive and minus voltage applied to the transistor driver circuits. Test each driver transistor for open or leaky conditions. The overheated IC or transistor might have gray, dark body or terminal areas with a leaky symptom. Inspect PC terminals for burned areas on the PC board. Replace the defective IC with the exact part number.

CD voltage sources

The portable CD boom-box player can be powered by batteries, a dc-dc converter, and a step-down ac power supply. Most voltage sources are +5, -5, +9, -9, +12, and −12 V. You might find a dc-to-dc IC converter circuit supplying a ±5-V source. In larger players, the dc voltage source can be supplied from an ac power line.

Locate the ac power line components upon the PC board. Take critical voltage measurement at the largest electrolytic capacitor (Fig. 14-28). Often the dc circuits are mounted close by. You might find that the small step-down power transformer is mounted off of the main PC board. Trace the secondary leads to the PC board.

Test each diode for leaky or open conditions when there is no voltage at the large electrolytic capacitor. Most CD boom-box power supplies contain a bridge rectifier circuit. Remove one end of the diode from the circuit if one shows signs of leakage.

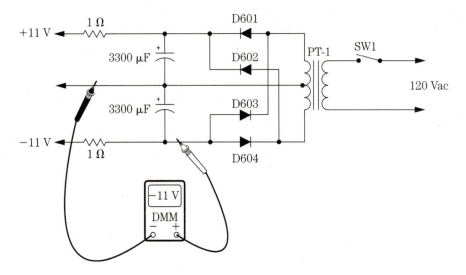

14-28 Check the low-voltage source with a voltage test across the largest filter capacitor.

Double-check the silicon diode out of the circuit. Make sure that each power voltage source contains a positive and negative voltage. The main dc power source might contain a transistor IC or zener regulator or both (Fig. 14-29).

dc-regulated circuits

The dc source might contain a transistor voltage regulator in the optical, microprocessor, and servo IC circuits. Check the limiting resistor at the transistor regulator for an open condition. No output voltage from the transistor regulator might result in an open transistor. A leaky or open zener diode within the base circuit of the transistor regulator can provide a lower dc voltage output. You might find IC regulators in larger boom-box power voltage sources. Check for IC regulators mounted on metal heat sinks.

Test the transistor in the circuit with a diode transistor test using a DMM or transistor tester. If in doubt, remove the transistor regulator from the circuit and then test it for open or leaky conditions. A voltage test at the input and no regulated voltage indicates an open transistor. Check the zener diode as any other silicon diode on the DMM. Do not overlook a low dc voltage source caused by an open electrolytic capacitor (Fig. 14-30). These small microfarad capacitors have a tendency to lose capacity or dry up after a few years. Sometimes the terminal lead inside the electrolytic capacitor breaks off at the foil, indicating an open capacitor.

Take a voltage measurement on each side of an IC regulator for input and output voltage. The center terminal should be at ground potential. The output terminal voltage is always lower than the input. No regulated output voltage might indicate a leaky or open IC regulator.

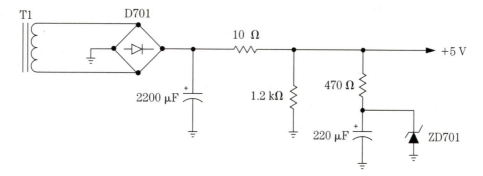

14-29 Check the main power supply source for a transistor, IC, or zener diode regulation circuit.

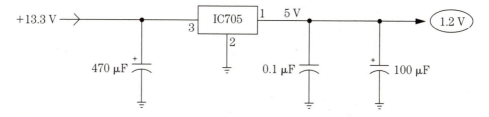

14-30 Defective electrolytic capacitors in the low-voltage sources result in lower supply voltages.

The CD pickup

The optical pickup assembly contains the pin photo diodes, monitor diode, laser diode, focus and tracking cards, and disc and SLED motors (Fig. 14-31). Within the boom-box CD section, most components are mounted on the optical assembly and one PC board. The laser diode shines on the pits and landings on the bottom side of the (CD) disc and is picked up by the photo-detector diodes. The digital signal is then fed to the RF amplifier IC or transistors.

The laser beam can be harmful to your eyes. Never look directly at the laser beam. Keep your head at level 30 cm (centimeters) away from the optical pickup. Remember, you cannot see the laser beam.

When the laser diode becomes weak or does not emit a beam, no music can be heard. The laser emitting power can be measured with a light power meter or can be indicated by an infrared card. The infrared card was intended for use to indicate if a remote transmitter is operating. Although the infrared card will only indicate the laser diode is functioning, the infrared power light meter will show if the laser diode is weak or normal (Fig. 14-32). The laser voltage and current can be checked by measuring the voltage across a resistor, then determining the correct current of the laser diode.

Check the laser diode applied voltage when no signal is found at the RF amplifier. This voltage ranges from +5 to +10 V and is supplied from a dc source of the low-

Compact Disc Mechanism

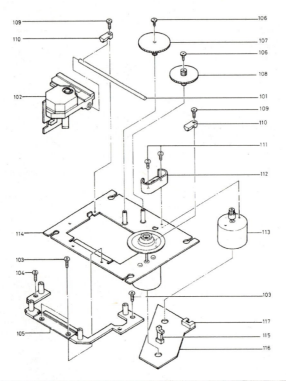

Ref. No.	Description	RS Part No.	Mfr's Part No.
551	Assembly, Mechanism, CD		1471-105-0-00
101	Shaft, Slide		4-910-431-02
102	Pick-up, KSS-150A(H) (RP)		8-848-046-51
103	Screw, +PTT, M2×6		7-685-783-01
104	Screw, +PTT, M2×4		7-685-781-01
105	Holder (J), Chassis		2-641-444-03
106	Screw, Special, M1.7×3		3-303-809-31
107	Gear, A		2-641-404-02
108	Gear, B		2-641-403-06
109	Screw, Special, +STP, M2.6×8		2-641-447-01
110	Clamp, Shaft		2-641-448-02
111	Screw, +P, M2×4		7-621-255-25
112	Cover, Gear		2-641-434-01
113	Motor Assembly, Sled		X-2604-770-1
114	Motor Assembly, Spindle		X-2641-337-1
115	Switch, Leaf		1-570-822-21
116	PCB, Motor		1-622-824-14
117	Pin, Connector, 4pin		1-564-720-11

14-31 The optical assembly might contain a laser diode, photo diodes, and focus and tracking coils.

14-32 Check the laser diode output with an infrared light power meter.

voltage power supply or dc-dc converter. When taking voltages and waveforms from the optical assembly, place a piece of tin foil or a CD across the lens assembly to prevent eye damage. Remove the tin foil when taking waveforms on the RF amp IC. Remember, a normal eye pattern waveform indicates that both RF amp and optical assembly are functioning.

The RF amplifier

The transistor or IC RF amplifier circuit amplifies the weak RF signal from the optical photo diodes to the signal processor and servo circuits. Transistors were found in the RF amplifier in the early CD chassis, while today, IC components are used as the RF amplifier. In fact, you might find the RF amp and signal-processing circuit in one component (Fig. 14-33).

When no RF signal is fed to the servo circuits, the whole CD chassis shuts down. The CD player might come on and both focus and tracking cards move, but then it shuts down. Usually, if the RF section is working and the chassis shuts down, a temporary RF waveform can be taken at the RF amp output terminal. Although the RF signal might disappear at shutdown, a quick waveform might indicate that the RF amp is functioning and that the shutdown was caused by other components.

Most RF amplifier circuits contain a test point for both voltage and RF or eye-pattern waveform (Fig. 14-34). Start at test point TP1, because most test points

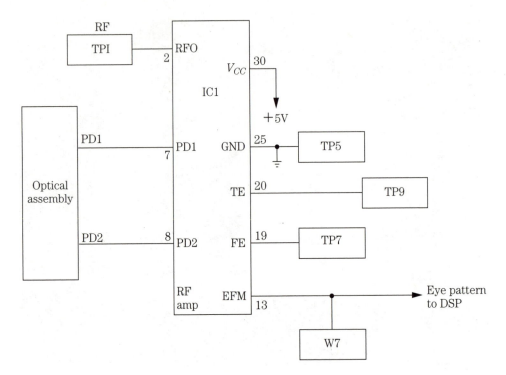

14-33 The weak photo diode signal is applied to the RF amplifier with an eye pattern output.

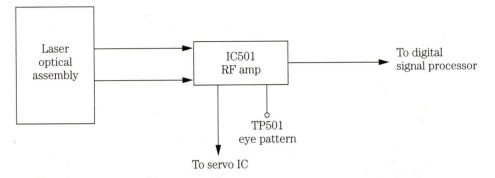

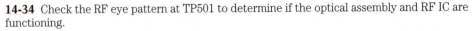

14-34 Check the RF eye pattern at TP501 to determine if the optical assembly and RF IC are functioning.

start at the front-end circuits. A quick scope test on each test point can identify the eye pattern or RF waveform. If no test points are found on the PC board, check each terminal of the RF amplifier. Identify the RF IC by tracing the two input signal wirings from the optical assembly. Remember, the CD player will shut down with no RF or eye pattern.

Take voltage tests at each test point. Check the voltage at each pin terminal of the RF amp when a schematic is not available. Often, the dc voltage found powering the RF amp is 5 to 10 V. When there is improper or no supply voltage at the RF amplifier IC, the CD chassis will shut down. Check the dc voltage source from the dc-dc converter circuits or low-voltage power supply.

Digital signal processing

In small boom-box CD players, the digital signal processor (DSP) IC might include an input signal from the PLL circuit and RF eye pattern. The digital output signal is applied to the D/A analog IC, focus and tracking servo, RAM, disc and SLED servo circuits, reemphasis circuits, and a bit microprocessor (Fig. 14-35). The large CD player might have a separate digital signal processor IC, servo processor, and control IC. The large digital signal processor has over 80 pin terminals.

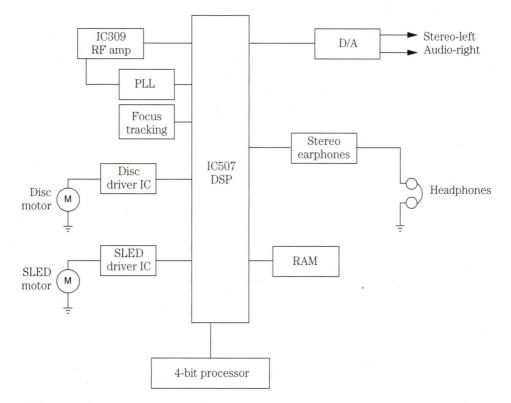

14-35 The digital signal processing IC507 provides signal to several circuits in the boom-box player.

The focus servo circuit provides a signal to control the focus and tracking coils. Usually, the focus and tracking servo circuits are contained in one IC. The servo IC can have a supply voltage of 5 to 10 V. You can tell if the tracking and servo coils are moving when the CD player is turned on, even in quick chassis shutdown. When checking the coils for continuity, notice the coil movement when the DMM leads are touched to the coil terminals. If you reverse the DMM leads, the coils will move in the opposite direction. Take voltage measurements on the servo or driver IC with normal coil continuity.

The digital/analog signal is sent to a digital/analog IC, which provides audio reception. The right and left stereo channels are developed within the digital/analog circuits. Suspect a defective DSP microprocessor when more than one channel does not function. Take voltage tests on each pin terminal. Most signal pin terminals will have a zero voltage measurement until the signal is activated. The DSP microprocessor supply voltage can be from 5 to 10 V. Make sure that the large microprocessor is defective before attempting to remove it from the surface board area.

D/A circuits

The digital/analog (D/A) IC converter changes a digital signal to audio. The D/A circuits might be contained in a separate IC or combined with other circuits. Besides converting the signal to audio, there are two separate stereo sound channels at the D/A output circuits. This weak audio signal from the D/A IC is amplified, passed through a low-power filter (LPF) network, and passed on to a left and right channel stereo channel (Fig. 14-36). The right and left stereo channels can be applied to a headphone amplifier and earphone jack or to separate line output jacks.

The audio signal can be signal traced with a scope or external amplifier from output jacks up to the stereo output of the D/A converter IC. The two audio amplifier circuits might be contained in one IC component. A dead or weak amplifier IC can be signal traced from the earphone or line jacks. Notice that the audio signal will be weaker as the test probe moves towards the D/A IC. When weak or no signal is found at the output of D/A IC, check the D/A converter IC for a correct supply voltage.

The D/A converter and IC amplifier might receive the dc source from a dc-dc converter power supply. Check the dc voltage from the batteries or ac supply that provides voltage to all circuits. If you hear weak or no CD audio from either channel, suspect the B+ voltage source. Clean the switch contacts if music is erratic or intermittent. If only one channel of audio is missing or weak, signal trace the audio in the normal channel and compare. Check for correct voltages on the weak audio IC with the normal channel.

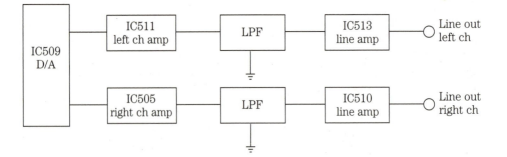

14-36 The D/A IC509 provides analog or audio signal at the audio amplifiers and line output jacks.

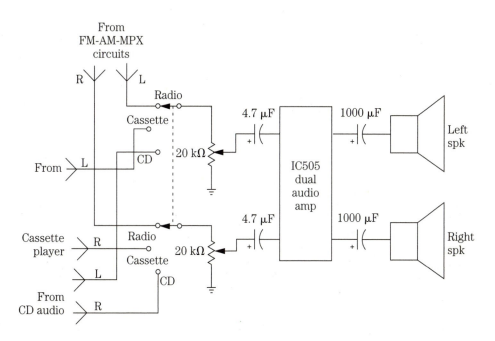

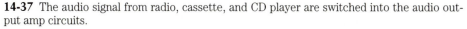

14-37 The audio signal from radio, cassette, and CD player are switched into the audio output amp circuits.

In the boom-box CD player, a channel switch might be found in the audio output of each channel. The selector switch might switch in the CD, tape, and radio audio signals to a stereo output audio circuit (Fig. 14-37). Often, the same audio output amplifier circuits are used for all three functions, right up to the PM speakers. Spray cleaning fluid down inside the switch area when either channel appears erratic or intermittent.

15
CHAPTER

Twenty actual tough dogs

The tough dog consumes a lot of service time and is difficult to locate and isolate. Remember, a tough dog for one technician might be easy for the next. Tough-dog problems should be tackled early in the morning when the mind is fresh and clear. The following tough dogs actually occurred. These problems are more difficult without a schematic (Fig. 15-1).

Cassette players

No recording

A Soundesign 5928 cassette player would not record on the left channel. The right recording level meter worked, but the left recording meter was dead.

At first, the bias oscillator and circuits were suspected, but the right side was normal. Clean the tape heads when recording or playback is dead in one channel. A scope test on the left channel tape head indicated a bias signal. An external amp was applied to the tape head and pre-amp (Q202). A weak signal was found at the collector terminal. Critical voltage tests found R113 (22 kΩ) was open (Fig. 15-2).

Intermittent recording

Both channels were normal in the playback mode of a Soundesign 4285. At first, the erase head was suspected. A scope test at the erase head indicated that the bias circuits were good. The scope was left at the tape head to determine if the bias oscillator was intermittent.

After two hours, the bias oscillator waveform collapsed, indicating trouble within the bias oscillator circuits. The terminal connections for T203 were resoldered, and Q211 was tested in the circuit. The results were the same. Q211 was suspected of being intermittent and was replaced. The contacts of SW201 were cleaned, with the same result. All resistance and voltage measurements seemed normal (Fig. 15-3).

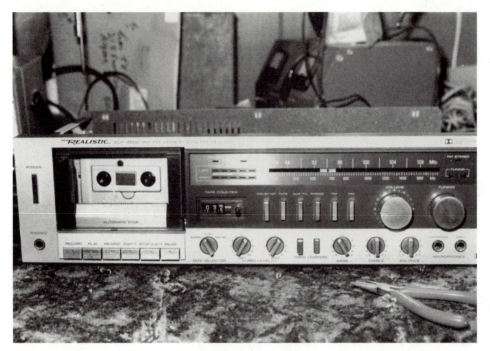

15-1 The tough dog can occur in a TV chassis, cassette player, or the AM/FM/MPX receiver.

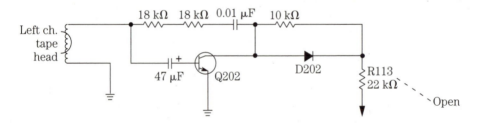

15-2 An open R113 prevented recording in the left channel of a Soundesign cassette player.

C243 was checked with a capacitor tester and measured 230 μF. When C242 was tested, the reading at first was normal and then zero. Moving C242 around with an insulated tool indicated the capacitor was intermittent. Replacing the 0.01-μF capacitor solved the intermittent problem.

Weak volume

A Sanyo RD-7 cassette deck had very weak volume on both channels. There was no level meter movement on either channel. Both channels were very weak, so the problem had to be something in common with the low-voltage power supply or IC. The signal traced from IC551 to R821 was normal. The other side of R821 and the line output jack were dead.

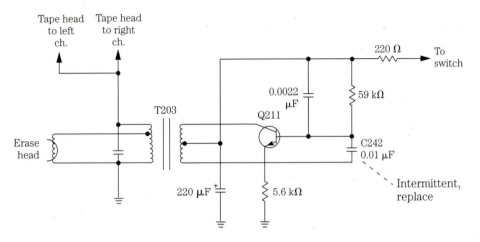

15-3 C242 (.01 µF) capacitor caused intermittent recording in the bias oscillator circuit.

R821 was traced to the emitter terminal of Q803. Because a negative voltage was applied to the base terminal and the collector was at ground potential, Q803 was assumed to be some type of electronic muting switch (Fig. 15-4). Q803 shut off when a negative voltage was applied to the base terminal, and to get Q803 to open, a more positive voltage must be applied.

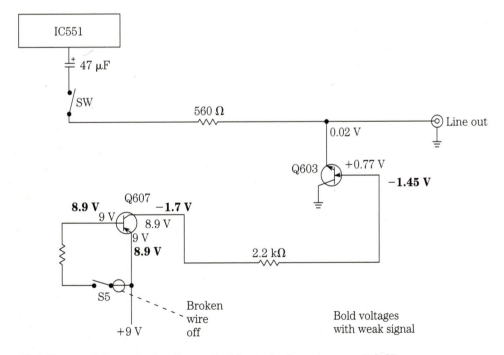

15-4 Very weak line output audio resulted from a broken wire on switch S5.

The negative voltage was traced through a 2.2-kΩ resistor to the collector terminal of Q607. Voltage tests on the collector terminal were −1.7 V and, on the base and emitter, 8.9 V. Q607 was tested for leakage and appeared normal. Checking the base and emitter circuits found that both wires ran to the play switch. Here a broken wire was found at the emitter side of S5. Although a dirty switch contact can cause the same problem, attachment and resoldering of the terminal wire solved the muting switch problem.

Intermittent

The symptom in a MacDonald AM/FM/MPX dual cassette player was intermittent recording. The tape heads were cleaned. All tape head connections were resoldered. A scope probe was placed on the erase head and found no oscillator waveform in recording mode. No doubt the bias oscillator for the tape heads was not operating.

The hookup leads to the erase head were traced back to the bias oscillator transformer. The oscillator bias transistor (2SC2001) was located. A supply voltage of 23.5 V was fed to the bias transformer (Fig. 15-5).

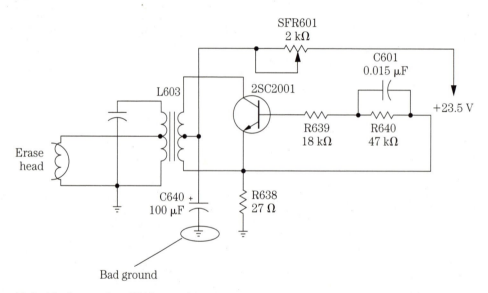

15-5 A bad ground on C640 caused intermittent recording in a MacDonald AM/FM/MPX dual cassette player.

Although voltage measurements on the bias oscillator were fairly normal, no oscillator waveform was found in the primary winding. Both windings were checked and had normal continuity. The oscillator transistor was replaced with an NTE85 universal replacement with no results. All resistors were checked by removing one lead from the circuit. C640 and C601 were subbed to no avail.

After spending several hours with the cassette player in recording mode, the player was shut down to take different resistance measurements. When a resistance test was made across C640 (100 μF) capacitor, no ground return was noted. A poor negative connection was found on C640. After resoldering the ground terminal of C640, the bias oscillator had a waveform and now operated in the recording mode.

Car radio

Weak FM reception

Only the FM stations were weak in this car radio with normal AM and cassette operation. The RF FM stage was suspected, but it was good. The signal was traced to the output diodes of T102. Where the signal separated at MC1310, both channels were very weak. Perhaps MC1310 was leaky or there was no supply voltage.

Because the signal was not coming out of either pin 4 or 5, was it at the input of the IC? An external amp was applied to each terminal, and only a weak signal was found on pins 2, 4, and 5 (Fig. 15-6). No doubt pin 2 was the input terminal. A weak signal was found on one side of C114 and strong audio on the other. C114 was found to be open. These small coupling capacitors have a tendency to dry out and cause a loss of signal.

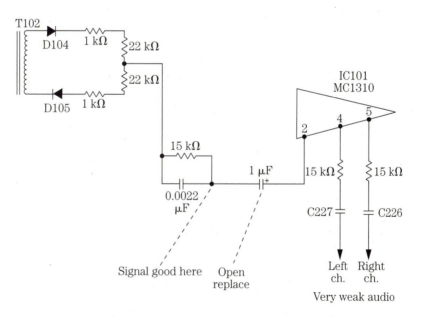

15-6 An open 1-μF coupling capacitor produced weak FM signal to IC101.

Only received a local station

In this Delco 90BPBT1, you could hear most stations, but they were very weak, while one local station was fairly good. The AM signal was traced from the antenna to IC DM-32. Voltage tests on the IC were fairly normal with a 7-V power supply. DM-32 was replaced, but with the same results.

Continuity measurements were taken of all the input coils, and they seemed normal. Resistance measurements were taken on each IC pin to ground. Pin 13 had a 42-Ω leakage to the chassis ground. After trying to make a sketch of the RF circuit, with terminal 13 tied to the RF tuning coil to the 7-V supply (Fig. 15-7), another voltage test was taken. Only 0.3 V was found at pin 13, indicating the C5 was grounded. When the top of C5 was disconnected from the RF coil, the voltage increased. C5 was ordered out and replaced. Very seldom do you find a defective trimmer capacitor.

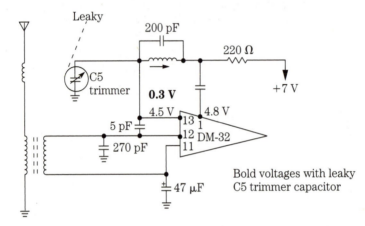

15-7 A defective trimmer capacitor (C5) caused weak FM reception.

AM-FM-MPX receiver

FM garbled

The FM was garbled on all stations in a Sharp SG174U AM-FM-MPX receiver. Even when a local FM station was tuned in, it was weak and not clear even with added volume. When the receiver was moved, something appeared microphonic. Tapping the chassis would make the noise appear and quit. When tapping around on the MPX section, the unit would really act up.

The FM signal was traced to the base terminal of FM-MPX transistor (Q201) (Fig. 15-8). Q201 tested good with in-circuit tests. Voltage and resistance measurements were assumed normal because a schematic was not available. While prodding around on the MPX circuits and when VR201 was touched, the signal popped in and out with a microphonic-type noise. VR201 was sprayed with cleaning fluid, and readjustment solved the garbled FM problem.

Weak right channel

There was very little volume in the right channel of the audio amplifier in a MCS3230 J.C. Penney receiver. Because weak audio can be caused by dried-up small coupling capacitors, an external audio amp was used for signal tracing. To determine if the audio amps were defective or input channels, a 1-kHz audio signal was injected

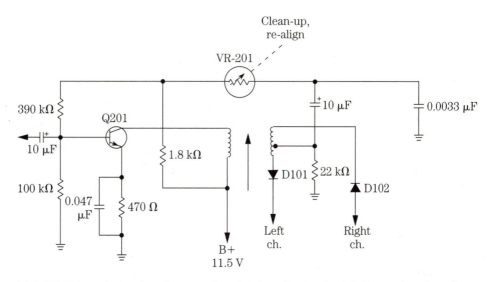

15-8 VR-201 caused microphone and garbled audio in the MPX circuits of a Sharp AM/FM/MPX receiver.

at each volume control. Both channels seemed normal from the volume control to the speaker terminals.

No doubt the weak channel was in the audio input circuits. Coupling capacitors C204 (1 μF) and C206 (4.7 μF) were shunted with a 4.7-μF electrolytic capacitor with no increase in volume. Voltage measurements upon both directly coupled transistors seemed to be way off. The supply voltage was 19.5 V with the collector terminal tied to it. The base terminal measured 17.7 V, and the emitted on transistor Q204 measured 17.5 V (Fig. 15-9).

The collector terminal of Q202 was high at 17.7 V. A continuity measurement indicated that the collector of Q202 was directly coupled to the base of Q204. The base and emitter terminals of Q202 were fairly normal with a bias voltage of 0.58 V. The bias voltage on the emitter and base of Q204 was only 0.2 V.

Q202 and Q204 were tested in the circuit, with Q204 indicating leakage. The resistance measurement between base and emitter was 820 Ω. The base terminal of Q204 was removed from the circuit and tested again. A low leakage was found between the base and emitter of Q204. Q204 was removed from the circuit and tested for leakage between the base and emitter.

While Q204 was out of the circuit R220 (2.2 kΩ), R218 (820 kΩ), R216 (4.7 kΩ), and R210 (220 kΩ) were tested for correct resistance. Q204 was replaced with a GE-20 universal replacement and solved the weak right channel.

Cassette deck

Defective automatic stop

The automatic stop would not function in recording in a Lloyds BB8692. When recording, an auto-stop switch is activated, the motor will stop, and a light will come

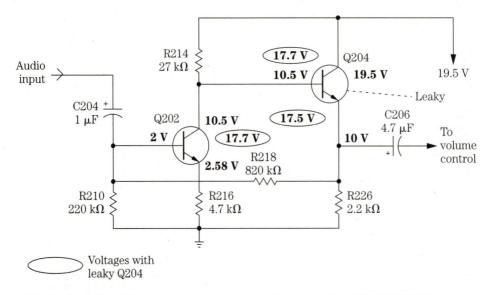

Voltages with leaky Q204

15-9 A leaky Q204 AF transistor caused a weak right channel in a MCS3230 J.C. Penney receiver.

on after each recording. To start the recording on channel 2, remove the cartridge, press the record button, and insert the cartridge. The automatic stop will shut off the motor after each recording. In this case, the motor would run all the time (Fig. 15-10).

Because the tape player was quite old and there was no schematic, the motor and shut-off circuits were traced to locate the possible defective part. All switches were cleaned at the same time as the tape heads. Q19 was the dc electronic switch,

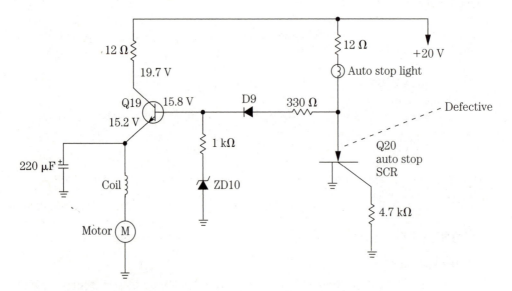

15-10 A defective SCR (Q20) caused the automatic stop to malfunction.

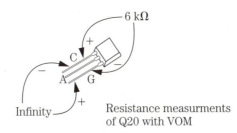

15-11
The normal resistance
measurements of SCR (Q20).

6 kΩ

Infinity

Resistance measurments
of Q20 with VOM

and Q20 was the automatic SCR stop component. Q19 was checked in-circuit with a
resistance test on Q20 (Fig. 15-11). After several hours of frustration, Q19 was replaced, with the same results. When Q20 was replaced, the automatic stop worked
perfectly. Because semiconductors cause the most troubles, replacement might
solve the problem.

Stops after a few minutes

After a few seconds of operation, the Sharp RT1165 tape deck would shut down.
Each time the unit would start up, but would shut down. On other models with automatic shutdown, a magnetic switch or rotary magnet would stop and shut the unit
down. A binding or broken belt can cause the problem. After several minutes, a
round magnet was found behind the counter assembly. Here the belt rotated the indicator for only a few seconds.

The rotating magnet was above a magnetic switch or electronic circuit. Right below the magnet was a solid-state component that looked like a transistor or IC. The
VH1D838A part was ordered from the Sharp service center and replaced. This
solved the shut-off problem.

Intermittent shutoff

The automatic shutoff was not working correctly in a J.C. Penney 683-3338D
cassette deck. The player cut off during playing. The next time, it worked perfectly.
After several hours of checking parts and transistors, there was no doubt some component was intermittent. All the switches were cleaned, and Q903 and Q904 were replaced with no results.

Because the J.C. Penney repair center was just across the street in the mall, a
telephone call to the service technician resulted in critical information for this tape
deck. A 33-Ω resistor was to be added in series with switch S905 (Fig. 15-12). Sometimes working with your fellow technicians can help in solving tough-dog repairs.

Unusual distortion on both channels

This unusual distortion problem was found in a Channel Master HP-6863. Distortion was found in both audio channels. When the volume was increased ,the distortion was greater. Often, when distortion is found in both channels, you will find
poor power-supply voltage or a bad dual output IC or speaker coupling capacitors.
Voltage measurements on separate power ICs were fairly normal.

The input signal on pin 7 of IC101 was normal on both channels. Output signal
on pin 12 was fairly normal until the volume was turned up. Could both ICs break

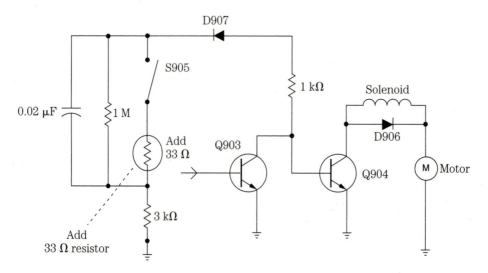

15-12 Add a 33-Ω resistor in series with switch S905 to prevent intermittent shut-off.

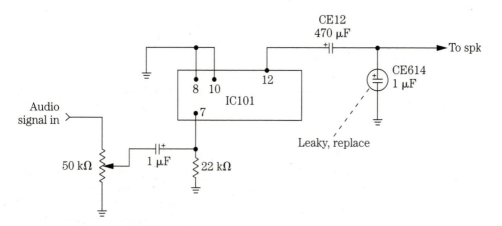

15-13 A leaky CE614 caused distortion in the audio channels.

down under a stronger signal? The ICs were sprayed with coolant, and nothing happened. When CE614 was sprayed, the distortion became greater (Fig. 15-13). CE614 was replaced with a 1-μF, 16-V capacitor. It was unusual to find both CE614 and CE613 defective in the output of each channel.

Amplifiers

SX-950 amp shutdown

After a few seconds, the protection circuits would shut down the chassis. Two low-ohm resistors in the output circuits were burned. Both were replaced with 0.5-

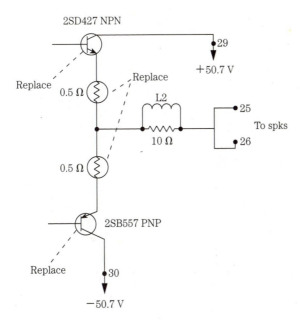

15-14
Leaky output transistors and bias resistors caused a Pioneer SX-950 amplifier to shut down.

Ω, 2-W resistors. Usually, when the bias resistors open, the output transistors are leaky or shorted. The 2SD427 transistor was replaced with an ECG280 and the 2SB557 (PNP) with an ECG284 universal replacement (Fig. 15-14).

When the chassis was powered up, both resistors were red hot. The power cord was pulled at once. The AF and driver transistors were checked ahead of the output transistors and tested good. Some dc voltage was found at the speaker output terminals, indicating an unbalanced dc coupling circuit.

Sometimes taking resistance measurements on output transistors and speaker terminals to chassis ground can help locate a leaky circuit (Fig. 15-15). The resistance measurements from the collector terminals of the output transistors to ground were checked with the normal channel and were good. Without a speaker or resistor load connected, the resistance from speaker terminals 25 and 26 showed a high leakage of 14.7 kΩ, and the normal channel showed an infinity reading.

All base, collector, and emitter terminals in the defective channel were normal with the good channel. When resistance tests were made at the two AF transistors, the base terminal of Q4 measured 91 Ω across a 1-kΩ resistor. The 1-kΩ resistor had overheating marks and was replaced (Fig. 15-16). A 330-μF electrolytic capacitor next to it was replaced. A leaky Q4 (2SA726S) was replaced with an SK3450 universal replacement. Q2 was found to be leaky, with 1.6 kΩ from the collector to base terminals. Q2 was replaced with another SK3450 transistor.

The Pioneer SX-950 right channel was repaired by replacing overheated transistors and bias resistors (Q2 and Q4), a burned 1-kΩ resistor, and a leaky capacitor (330 μF). No doubt the leaky capacitor destroyed Q2 and Q4, placing –64.5 V on the transistors. In turn, the two output transistors were destroyed, shutting down the protection circuits.

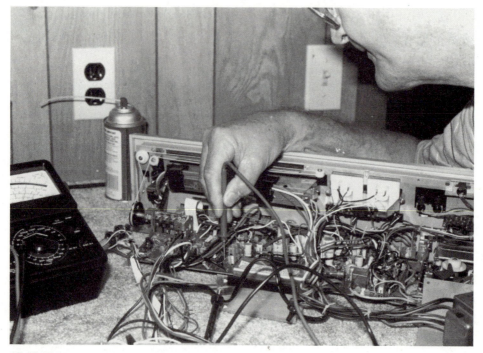

15-15 Take resistance measurements at the ground of the output transistors and ICs to locate a defective component.

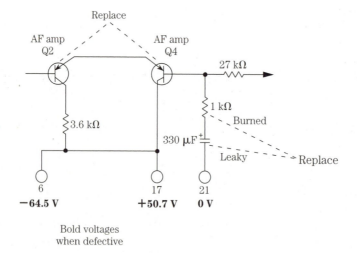

15-16 Replace leaky Q2, Q4, 330-µF capacitor, and 1-kΩ resistor to repair a high-wattage amp.

Noisy Sansui 2000 amp

Sometimes the left channel became noisy when the on/off switch was turned on. Other times, the chassis operated perfectly. Often the noise could be shocked in by rapidly turning the on/off switch. The amp might operate for one or two hours before acting up. When the unit shorted at the volume control, the noise was still there, indicating that the noisy component was in the AF, driver, and output stages. Remember, testing the suspected transistor will not show a noisy transistor.

To check the bank of audio output transistors and to see which channel is noisy, remove the mounting screws on the output transistors. When the noise stops, you have found the noisy channel. Tracing the signal back from the speaker to the input stages might take a little longer.

Because transistors and IC components cause most frying noises, especially after operating for a long time, short the base and emitter terminals together. Make sure that you have the correct terminals, or you can damage the transistor. Power output transistors are easy to locate. The metal body of the transistor is the collector terminal. The collector terminal will have the highest voltage.

The noisy transistor turned out to be an AF TR801 (Fig. 15-17). When the base and emitter terminals were shorted together, the noise was still present. The noise should quit when the two terminals are shorted together if the noise is generated ahead of the tested transistor. Troubleshooting noise in high-power amps takes time, especially without a schematic.

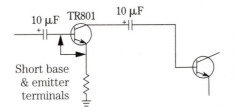

15-17
Noisy AF and driver transistors can be located by shorting the emitter and base terminals together.

Damaged speakers

The chassis of a Pioneer SX80 amp had been serviced before with a new output, and both driver transistors were replaced. A voltage check on the speaker terminals measured –24.7 Vdc, which burns, opens, or destroys the voice coil in the speaker. The balanced output IC or transistor dc circuit should be at zero potential at the speaker terminals.

Critical voltage measurements on the defective left channel output IC had a –24.7 V on pin 8, –26.9 V on pin 0, and –24.1 V on pin 1 (Fig. 15-18). Comparing these same voltages with the normal right channel, pin 0 should be 1.4 V, and pin 1, –1.4 V. Pins 3 and 8 are at 0 V.

When a speaker load resistor (10 Ω, 10 W) was placed across both right and left speaker terminals, the 10-Ω resistor on the left channel became red hot. The 10-Ω resistor was removed from the left channel, and the volume control was turned down to zero. Because the right channel was normal, the resistor remained to protect the output transistors with a speaker load.

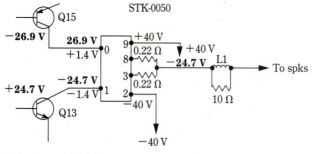

15-18

A high –24.7 V was placed on the speaker voice coil with a leaky output IC.

Bold voltages with defective transistors

To help in locating parts and voltage measurements, a rough sketch was made with accurate voltage measurements recorded. A magnifying glass helped to locate the front-end transistors and parts because they were so close together. Some of these components are not marked on the wiring board. The transistor terminals were drawn and marked on a large piece of paper.

Transistor terminals were located by taking transistor tests with the DMM to find the common base terminal. Voltage and resistance measurements can be checked against the normal right channel. Voltage measurements on the driver transistors were way off, but the voltage measurements on the Darlington transistors were fairly normal. These transistors are flat and have five terminals.

When the voltage of Q5 was taken, the collector measured 28.5 V, the base measured 0 V, and the emitter measured 28.5 V. The emitter voltage was way off compared to the normal channel (Fig. 15-19). Q5 tested good. The wiring was traced through R239, VR5, and R237. Going a bit farther, a loose tie bar or wire was found with a poor soldered connection. By pushing around on the tie wire, voltages were normal and there was zero voltage at the left speaker terminals.

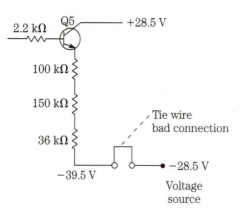

15-19

A poorly soldered connection of tie wire placed a –24.7 V on the speaker terminals.

TV

Dark bars to the left

Sometimes experience tells us that horizontal black bars coming in with the signal can result from pickup noise from the TV antenna. Horizontal white scanning lines at the top or center of the picture are caused by defects in the vertical circuits. Vertical jail bars on the left side of the picture can be open electrolytic bypass capacitors in the AGC or horizontal circuits. Heavy horizontal dark bars in the raster can occur with defective filter capacitors. Unusual lines found in the raster or picture are difficult to find without a schematic (Fig. 15-20).

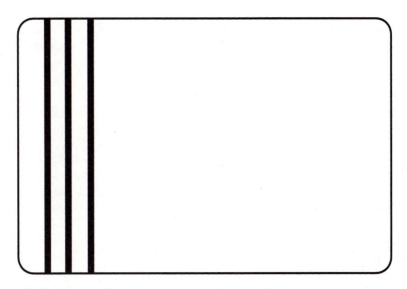

15-20 Defective filter capacitors in the horizontal output transistor circuit can cause vertical bars in the raster.

First, determine if the bars are in the raster, picture, or both. Tune the station off channel, and turn the color down to determine if the noise is picked up by the antenna or within the set. If the dark bars are still on the left side, go directly to the horizontal circuits.

Usually, vertical lines that vary with the signal originate in the video or AGC circuits. Dark vertical bars on the left side are caused by dried-up or open electrolytic capacitors in the voltage supply of the horizontal output transistor circuits.

Because an exact schematic was not handy for a Sony KV-2644R chassis, voltage tests were made on the horizontal flyback circuits (Fig. 15-21). The voltage was above 125 V and fairly normal. The B+ voltage was traced to a small isolation coil,

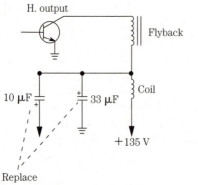

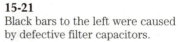

15-21
Black bars to the left were caused
by defective filter capacitors.

and two electrolytic capacitors were found behind the flyback. When shunted, the
vertical bars disappeared. The capacitors were replaced.

Destroys fuse and output transistor

In this RCA CTC140 chassis, the main fuse was blown, and after it was replaced,
it opened when the high voltage tried to come up. Either the horizontal output transistor or low-voltage diodes were causing the fuse to blow. Because it took several
seconds for the fuse to open, the horizontal output should be checked.

Looking at the rear of the RCA chassis, the horizontal output transistor usually
sticks out at the back so that you can see it. However, it doesn't in this chassis. A leak
test was made across the large filter capacitor, but there was no leakage. Finally, the
horizontal output transistor was traced from the flyback to a spot covered with a
heavy aluminum shield (Fig. 15-22). In fact, this flat transistor does not look like a
regular output transistor.

After removing the leaky transistor, it was replaced with an original RCA part
number (179743). Again the chassis was powered up, but nothing happened. The
chassis was quickly shut down, and leak tests were made. With the horizontal output
transistor out of the circuit, leakage was found from the flyback to the chassis. After
a couple of hours tracing the wiring from horizontal output transistor to yoke assembly, a 0.56-μF, 250-V capacitor was found to be leaky. This special capacitor was
replaced with the exact replacement (Fig. 15-23).

Intermittent width then shutdown

The customer complained that the picture would pull in on the sides, return to
normal, and play for several days. Then the picture would narrow, and the TV set
would shut down. This time, the chassis quit entirely. When the TV was checked the
low-voltage fuse was found open and, when replaced, would blow at once.

To determine if the yoke was at fault, the yoke plug was pulled. Then the fuse
would not blow. The high voltage measured only 12.5 kV with the yoke disconnected.
A fine vertical line indicated that there was no horizontal sweep in the RCA CTC149
chassis (Fig. 15-24). With the horizontal circuits functioning and 125 V at the horizontal output transistor, the trouble must be in the pincushion and yoke return circuit.

15-22 Locate the horizontal output transistor under a shielded cover in an RCA CTC140.

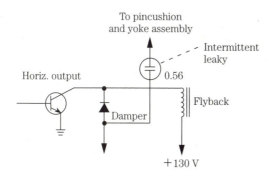

15-23
Intermittent capacitor (0.56 μF) destroyed the fuse and output transistor in a CTC140 chassis.

The TV was taken to the shop because tracing out the circuit would take a lot of time. After pulling the chassis and tracing the wiring from the flyback to the yoke return circuits, low resistance was measured across a 0.43-μF capacitor (Fig. 15- 25). The capacitor was removed and replaced.

Vertical foldover

A 0.5" inch of vertical foldover was found in a GE 19PC-F chassis. Adjustments of the vertical hold and height controls had no effect on the raster or picture. The foldover could not be adjusted out of the picture.

Both the vertical driver and output transistor bias voltage were normal at 0.6 volts, indicating that both transistors were normal. Voltage measurements on both

15-24 When the yoke lead was removed in the RCA CTC149 chassis, only 12.5 kV was found on the CRT.

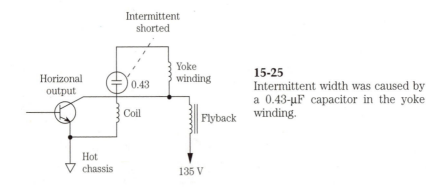

15-25
Intermittent width was caused by a 0.43-μF capacitor in the yoke winding.

collector terminals were suspected of being quite high, until the voltage supply source was measured at +115 volts. The input drive waveform from the sweep IC501 was quite normal at 0.75 volts p-p (Fig. 15-26).

Because a change in the capacity of the electrolytic capacitors in the sweep output or vertical feedback circuits can cause poor linearity and foldover problems, each one was shunted. When C613 and C614 were shunted with a 100-μF electrolytic capacitor, the foldover remained. Shunting C623 and C603 with a 10-μF electrolytic capacitor did not help at all. However, when C621 was subbed with a 100-μF capacitor, the foldover was removed, restoring the TV back in service.

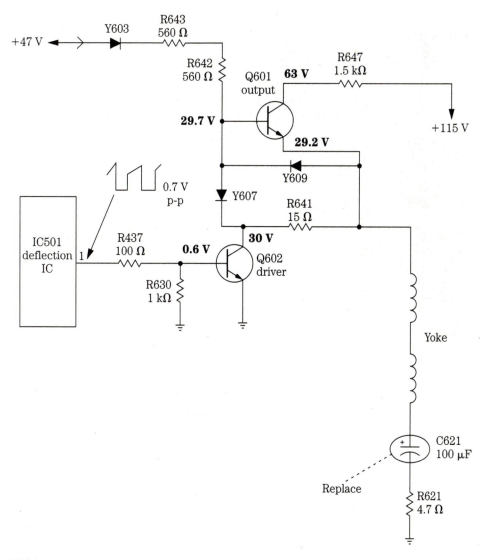

15-26 An open C621 caused vertical foldover in a GE 19PC-F chassis.

Intermittent bowing

The picture in an RCA CTC140 had a small bow in the center and would come and go. The intermittent bowing was going to be difficult to find because it was in and out (Fig. 15-27).

Sooner or later, the intermittent condition of any TV set will get worse, and now it acted up all the time. Although the bowing of the picture was hardly noticeable, it was not right. Because picture pull in or poor horizontal linearity is caused by components in the pincushion circuits, locating the defective part can be very difficult.

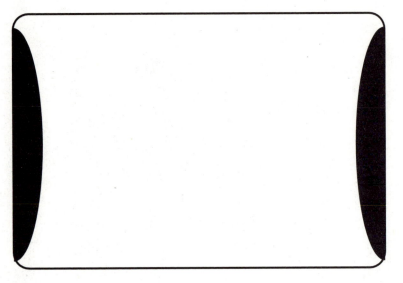

15-27 An intermittent pincushion output transistor produced an intermittent bowing picture.

Usually pincushion circuits consist of a transformer, a coil, and transistors. After locating the coils and transformer, the wiring can be traced to the pincushion transistor circuits (Fig. 15-28). The larger pincushion circuits consist of two or more transistors.

It took one hour after turning the chassis over several times to trace the pincushion circuits from flyback and yoke to pincushion transformer. A pincushion transistor collector terminal tied to a coil connection between the pincushion primary winding was found and tested. All transistors tested normal with in-circuit tests.

No doubt one of the transistors was intermittent, or there was a poorly soldered terminal of the pincushion transformer. All connections between the flyback and transistors were resoldered without any results (Fig. 15-29).

Because transistors can break down under load, the transistor was moved and tested. At first, it showed leakage and then returned to normal. Q4804 was replaced with an ECG152 universal replacement, and this solved the intermittent bowing problem.

Conclusion

Servicing any defective chassis takes a lot of time. Try to obtain a schematic of the chassis or one that is made by the same manufacturer to help isolate and locate a defective part. Repairing the tough-dog chassis takes a lot of valuable service time. Remember, the most troublesome components are transistors, ICs, electrolytic capacitors, and resistors, in that order. Try to locate the defective part with your eyes, ears, nose, and common sense when a schematic is not available.

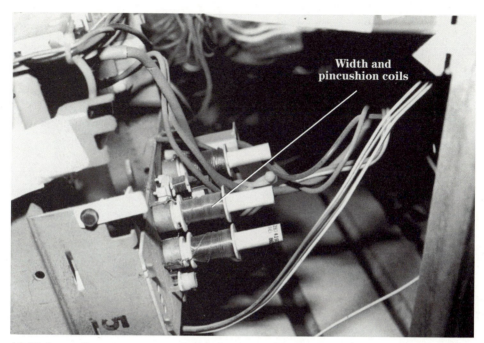

15-28 Locate the width and pincushion coils to find the pincushion circuits.

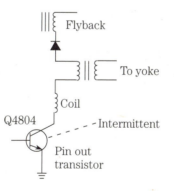

15-29
An intermittent pincushion output transistor (Q4804) causes bowing in the RCA CTC140 chassis.

Index

Illustrations are in **boldface.**

About the Author

Homer L. Davidson has written more than 35 books and more than 1,000 articles in the field of technician-level electronics troubleshooting and repair. His highly popular books include *Troubleshooting and Repairing Audio—Third Edition*, *Troubleshooting and Repairing Compact Disc Players—Third Edition*, *Troubleshooting and Repairing Camcorders—Second Edition*, and *Troubleshooting and Repairing Solid-State TVs—Third Edition*. He is currently the TV Servicing Consultant for *Electronic Servicing and Technology Magazine*.